A GUIDE TO ARCHIVES AND MANUSCRIPT COLLECTIONS IN THE HISTORY OF CHEMISTRY AND CHEMICAL TECHNOLOGY

COMPILED BY
GEORGE D. TSELOS
AND COLLEEN WICKEY

CENTER FOR HISTORY
OF CHEMISTRY
PUBLICATION NO. 7

Copyright ©1987 by the Center for History of Chemistry
All rights reserved
ISBN 0-941901-05-X

For additional copies or further information on CHOC's archive program, write

Center for History of Chemistry
215 South 34th Street
Philadelphia, PA 19104-6310

Telephone (215) 898-4896

CHOC's core programs are supported in part by grants from

The Air Products Foundation
American Cyanamid Company
ARCO Chemical Company
J. T. Baker Chemical Company
Celanese Corporation
Dexter Chemical Corporation
The Herbert and Junia Doan Foundation
The Dow Chemical Company
E. I. du Pont de Nemours & Company, Inc.
Eastman Kodak Company
Essex Chemical Corporation
Exxon Education Foundation
W. R. Grace & Company
Hercules Incorporated
Ernest Christian Klipstein Foundation
Merck and Company, Inc.
Monsanto Fund
Olin Corporation Charitable Trust
The Pfizer Foundation, Inc.
PQ Corporation
Rohm and Haas Company
Rorer Group Inc.
Sigma-Aldrich Corporation
SmithKline Beckman Corporation
Syntex Corporation
United States Borax & Chemical Corporation

TABLE OF CONTENTS

- iv   Preface
- 1   Archives and manuscript collections
- 184   Name index
- 186   Subject index
- 189   Geographical index

LIST OF ILLUSTRATIONS

- iv   Fred P. Nabenhauer of Smith, Kline & French
- v   Syntex announces cortisone synthesis
- vi   Carroll Hochwalt and Charles Thomas in the days before Monsanto
-      Dow Saran handbag
- vii   Du Pont Experimental Station
- viii   Charles A. Joy's chemistry class at Columbia
-      Priestley's pocket field balance
- 13   Arnold Beckman's pH meter patent drawing
- 27   Stephen M. Babcock's butterfat tester
- 54   Naphtha-fractionating tower at Exxon
- 63   The Rohm and Haas Bridesburg plant
- 91   Kodak Research Laboratories, Kodak Park
- 125   Ascorbic acid production at Pfizer
- 184   Max Tishler of Merck

Fred P. Nabenhauer of Smith, Kline & French Laboratories, a key figure in the development of Benzedrine. From John P. Marion, The Fine Old House *(Philadelphia, 1980). Courtesy SmithKline Beckman Corporation.*

Announcement of cortisone synthesis, Syntex, Mexico City, 1951. Left to right: A. L. Nussbaum (Harvard biochemist, one of Djerassi's first Ph.D. students at Wayne State); Mercedes Velasco; Gilbert Stork (Harvard organic chemist and Syntex consultant); J. Pataki; George Rosenkranz (holding test tube over Mexican yam); Enrique Natres; J. Berlin; Carl Djerassi; Rosa Yashin; Octavio Mancera; and Jesus Romo. Courtesy Carl Djerassi.

PREFACE

This *Guide* is an expanded and revised edition of the Center's 1984 publication *A Preliminary Guide to Archives and Manuscript Collections in the History of Chemistry*. The entries have been compiled from several sources: the National Union Catalog of Manuscript Collections (NUCMC) for the years 1959–1983, guides published by archives and manuscript repositories,[1] subject survey guides, and personal visits to repositories by Center staff. The entries represent material donated to or deposited in repositories across the country. Archives held by national, state, and local governments are not included, nor are archives held by organizations and companies on their own premises.[2] University archives are included in cases where they are combined with special collections or if the university has a number of well-known chemical collections.

SOURCES

For most published NUCMC entries, we simply photocopied the text from the catalog (after checking with the repository to determine the validity of the entry). Repositories also provided CHOC with information about collections not yet published in or reported to NUCMC, from which we constructed entries. In the absence of NUCMC information about collections, published guides or finding aids sent by repositories served as sources. In a few cases (e.g., the Hagley Museum and Library), we traveled to the repository to collect information about collections firsthand.

SCOPE

This *Guide* includes collections in the history of chemistry and chemical engineering, the chemical and pharmaceutical industries, and a number of related chemical process industries and businesses.

---

1. The following guides proved especially useful in our survey: Susan Sokol Blosser and Clyde Norman Wilson, Jr., *The Southern Historical Collection: A Guide to Manuscripts* (Chapel Hill: University of North Carolina, 1970); John B. Davenport, *Guide to the Orin G. Libby Manuscript Collection and Related Research Collections at the University of North Dakota, Grand Forks,* vols. 1 and 2 (Grand Forks: University of North Dakota, 1975, 1983); Richard C. Davis and Linda Angle Miller, eds., *Guide to the Catalogued Collections in the Manuscript Department of the William R. Perkins Library, Duke University* (Santa Barbara, Calif.: Clio Press, 1980); Clark A. Elliott, *A Descriptive Guide to the Harvard University Archives* (Cambridge, Mass.: Harvard University Library, 1974); Elliott, "Sources for the History of Science in the Harvard University Archives," *Harvard Library Bulletin* 22 (January 1974), 49–71; Andrea Hinding, ed., *Women's History Sources: A Guide to Archives and Manuscript Collections in the United States* (New York: Bowker, 1979); *Historical Records of Washington State: Records and Papers Held at Repositories* (Olympia: Washington State Historical Records Advisory Board, 1981); Kathleen Jacklin, ed., *Cornell University: Collection of Regional History and the University Archives, Report of the Curator and Archivist, 1962–1966* (Ithaca, N.Y.: Cornell University Libraries, 1974); Edwin T. Layton, *A Regional Union Catalog of Manuscripts Relating to the History of Science and Technology Located in Indiana, Michigan, and Ohio* (Cleveland: Case Western Reserve University, 1971); Robin E. Rider and Henry E. Lowood, *Guide to Sources in Northern California for History of Science and Technology* (Berkeley: Office for History of Science and Technology, University of California, Berkeley, 1985); John Beverly Riggs, *A Guide to the Manuscripts in the Eleutherian Mills Historical Library: Accessions Through the Year 1965* (Greenville, Del.: Eleutherian Mills Historical Library, 1970); Riggs, *Supplement Containing Accessions for the Years 1966–1975* (Greenville, Del.: Eleutherian Mills Library, 1978); Harry E. Whipkey, ed., *Guide to the Manuscript Groups in the Pennsylvania State Archives* (Harrisburg: Pennsylvania Historical and Museum Commission, 1976); and Helena E. Wright, *The Merrimack Valley Textile Museum: A Guide to the Manuscript Collections* (New York: Garland Publishing, 1983).

2. For a survey of relevant federal government records, see National Archives and Records Service, *Guide to the National Archives of the United States* (Washington, D.C.: U.S. Government Printing Office, 1974); Meyer H. Fishbein, "Archival Remains of Research and Development During the Second World War," in James F. O'Neill and Robert W. Krauskopf, eds., *World War II: An Account of Its Documents* (Washington, D.C.: Howard University Press, 1976), 163–179; and Sharon Gibbs Thibodeau, "Science in the Federal Government," *Osiris*, second series, 1 (1985), 81–96. For an initial guide to company records, see Linda Edgerly and Debbie Risteen, eds., *Directory of Business Archives in the United States and Canada* (Chicago: Society of American Archivists, 1976).

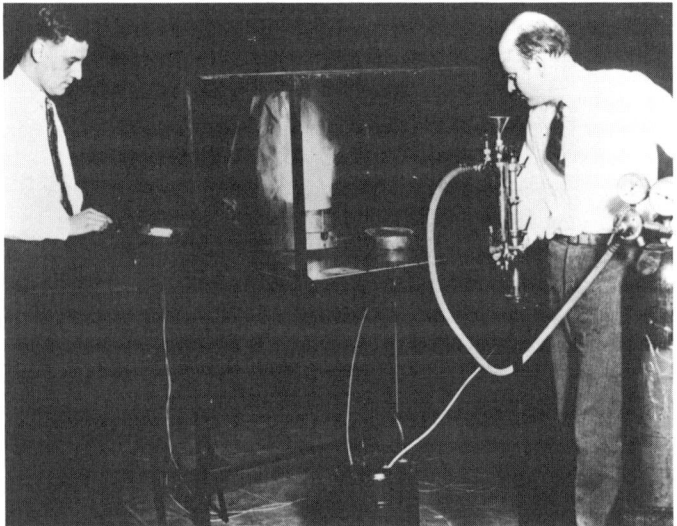

Carroll A. Hochwalt and Charles A. Thomas experiment with a new chemical fire extinguisher, Dayton, Ohio, 1928. By substituting potassium carbonate and chlorosulfonic acid for the soda-acid mixture typically used in fire extinguishers, Hochwalt and Thomas invented a fire extinguisher effective at extremely low temperatures. This project was one of the first successful consulting assignments for Thomas & Hochwalt Laboratories. In 1936, Hochwalt and Thomas sold their partnership to Monsanto, and their laboratories became Monsanto's central research department. Courtesy Carroll A. Hochwalt.

and metallurgy were included where information on specific collections came to light. However, though material on chemical research and education is also likely to be found in the papers of college and university presidents and deans, such collections were not included unless the individual was a prominent chemist (e.g., James Bryant Conant, entry C52). Aside from trustees' minutes, presidential archival files are the most likely of all university records to be preserved; we assumed that interested scholars would not overlook them.[6]

Business records were included not only for manufacturing and commercial chemical enterprises per se, but also for such chemical process industries as brewing and distilling, leather, ore processing, iron and steel making, paint and varnish, petroleum and sugar refining, and the extraction of nonmetallic chemical raw materials like potash, saltpeter, and phosphates. Entries relating to such businesses were selected depending upon how closely they related to chemistry. The greatest emphasis was placed upon material dealing directly with the chemical industry. Almost any collection of documents arising from manufacturing or commercial chemical businesses was included, even if it contained only financial records. Moving from core collections

We made a special effort to provide comprehensive information on the personal and professional papers of chemists, chemical engineers, and metallurgists. Also included are the papers of biochemists and, to a lesser extent, physiologists (when the latter had a biochemical emphasis, e.g., Jacques Loeb, entries L44 and L45).[3] The *Guide* also includes the papers of professors of pharmacy, pharmacology, and toxicology that came to light during our survey.[4] In addition, research areas on the borderline between between chemistry and physics are represented in some of the entries.[5]

Records of academic departments and schools of chemistry, chemical engineering, biochemistry, pharmacy, pharmacology,

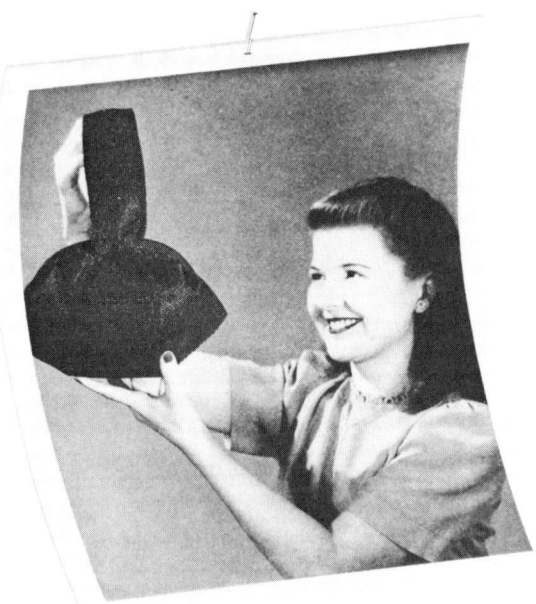

A 1946 fashion statement. This handbag made of Saran, a Dow polymer product, is just one of the panoply of consumer items made possible by the expansion of the polymer industry just after World War II. Courtesy Dow Chemical Company.

---

3. For a comprehensive guide to biochemical sources, see David Bearman and John T. Edsall, eds., *Archival Sources for the History of Biochemistry and Molecular Biology* (Philadelphia: American Philosophical Society, 1980); and Edsall and Bearman, "Historical Records of Scientific Activity: The Survey of Sources for the History of Biochemistry and Molecular Biology," *Proceedings of the American Philosophical Society* 123 (1979), 279–292.

4. Additional sources in these areas are discussed in John Parascandola and Elizabeth Keeney, *Sources in the History of American Pharmacology* (Madison, Wis.: American Institute of the History of Pharmacy, 1983).

5. There is much material relevant to the history of quantum chemistry in Thomas S. Kuhn, John Heilbron, Paul Forman, and Lini Allen, *Sources for History of Quantum Physics: An Inventory and Report* (Philadelphia: American Philosophical Society, 1967). In addition, the Center for History of Physics of the American Institute of Physics has produced a series entitled National Catalogue of Sources for History of Physics, which includes Joan Nelson Warnow, *A Selection of Manuscript Collections at American Repositories* (New York, 1969); Charles Weiner and Joan Nelson Warnow, *Source Material for the Recent History of Astronomy and Astrophysics: A Checklist of Manuscript Collections in the United States* (New York, 1971); Lawrence Badash, *Rutherford Correspondence Catalogue* (New York, 1974); Albert F. Gunns and Judith R. Goodstein, *Guide to the Robert Andrews Millikan Collection of the California Institute of Technology* (New York, 1975); and *Preliminary Finding Aid to the Archives of the Lick Observatory, from the Card Catalogue Maintained by the Lick Observatory Archival Staff*, University of California at Santa Cruz (New York, 1980). In preparation (under the new series title International Catalogue of Sources for History of Physics) are *A Guide to Archival Collections in the Niels Bohr Library*; *Catalogue of Sources for the History of Solid State Physics*; and *Catalogue of Sources for the History of Laser Physics*.

6. For more information about the value of college and university archives to scholars, see David B. Potts, "College Archives as Windows on American Society," *American Archivist* 40 (1977), 43–49.

into peripheral areas of the chemical process industries, commercial pharmacies, and so on, we exercised greater selectivity. Thus, collections of strictly financial data were not included for chemical process companies or commercial pharmacies. Collections relating to workers in the chemical industry were included, but not such collections relating to workers in the chemical process industries.

This guide, like most compilations of this sort, is not comprehensive. We could not contact every archival and manuscript repository, and not every repository we contacted responded to our inquiries and follow-up letters. The most frequent characteristic common to repositories represented here is that they reported chemistry-related collections published in NUCMC by 1983.

HOW TO USE THIS GUIDE

Entries are listed alphabetically and assigned sequential numbers within each section of the alphabet (e.g., A1, A2, A3). Collections are indexed by subject, name,[7] and geographical location, and are located by means of the guide entry number rather than by page number.

In all cases where we are aware of finding aids for listed collections, their availability is noted at the bottom of the entry. In the course of the survey we collected an extensive array of finding aids for many of the listed collections; these are available for use at CHOC. Researchers are urged to write directly to repositories to obtain current information regarding availability of finding aids, restrictions on use, and other matters affecting access to specific collections.

We hope that this guide will prove useful to historians of chemistry and chemical technology, to chemists and chemical engineers curious about the development of their disciplines, and to others interested in the role of chemical science and industry in the history of their communities and regions.

ACKNOWLEDGEMENTS

We thank Jeffrey L. Sturchio for his advice on sources and selection criteria; Anthony Giannakoulias for his assistance in the preparation of entries and indexes; Frances C. Kohler, Bruce V. Lewenstein, and the staff of the CHOC Publications Office; Jonathan Liebenau and John Parascandola for their advice on pharmaceutical entries; and all the archivists and others who responded to our survey and provided important information about their manuscript collections. We welcome information on relevant collections not listed here for inclusion in a future edition of this *Guide*.

GEORGE TSELOS
COLLEEN WICKEY
Labor Day 1986

---

[7]. In the name index we have listed only chemists, chemical engineers, biochemists, and well-known persons in related disciplines (e.g., physics and physiology, and Nobel laureates). We have indexed both names used as main entries and "hidden names," which did not appear as collection headings (i.e., persons about whom significant information had been collected by another person, or whose papers appeared as a subunit of a larger collection).

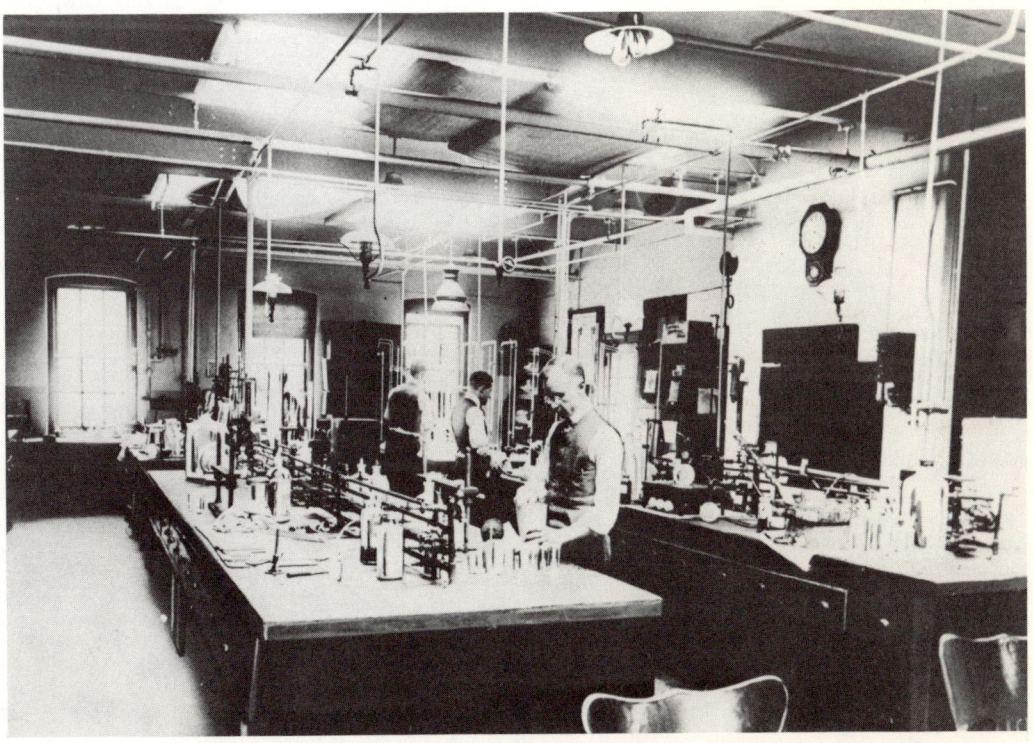

Chemists at work in the Du Pont Experimental Station circa World War I. Courtesy Hagley Museum and Library.

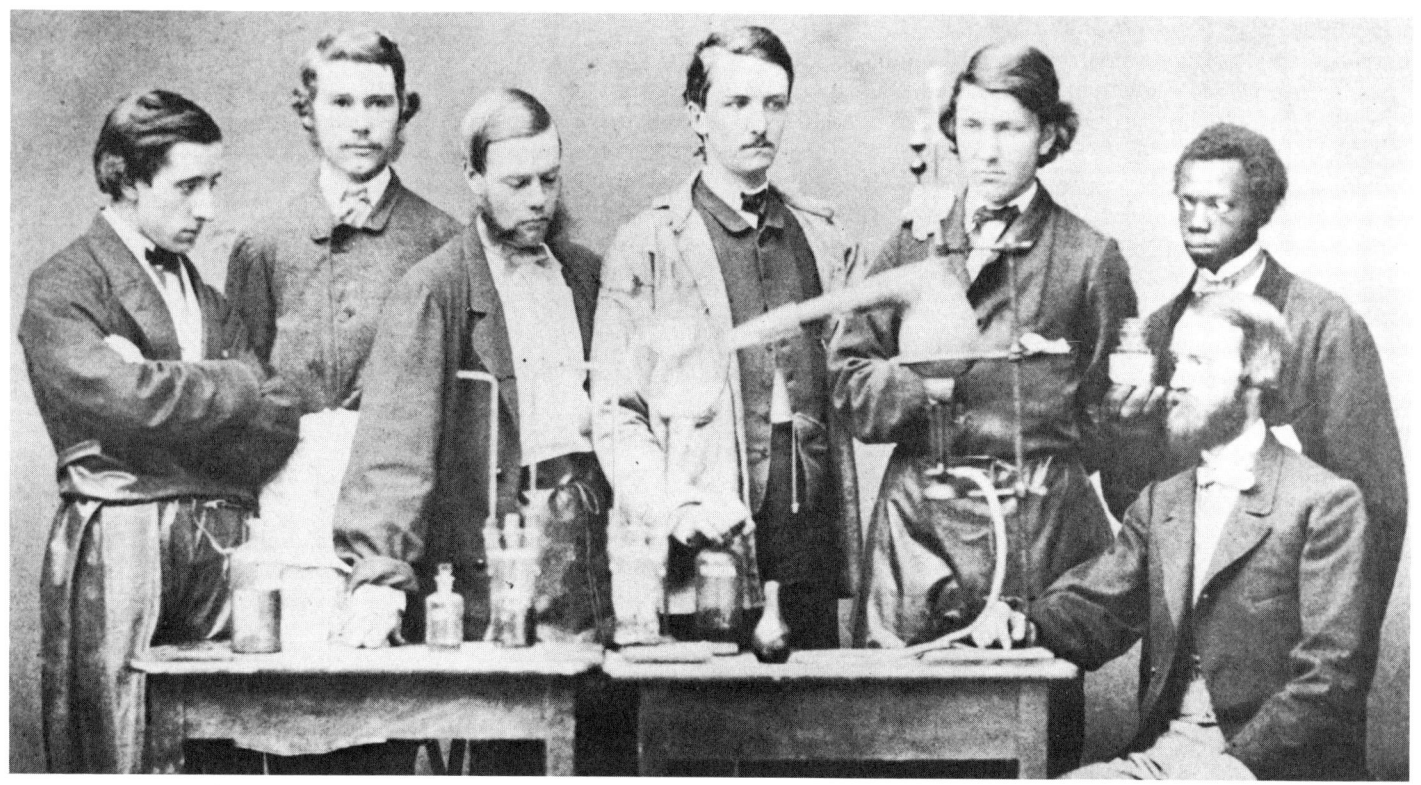

An 1857 photograph of chemistry students at Columbia College. Charles A. Joy (seated), a student of Friedrich Wöhler's, joined the Columbia faculty that year to begin regular instruction in practical chemistry. From the E. F. Smith Memorial Collection.

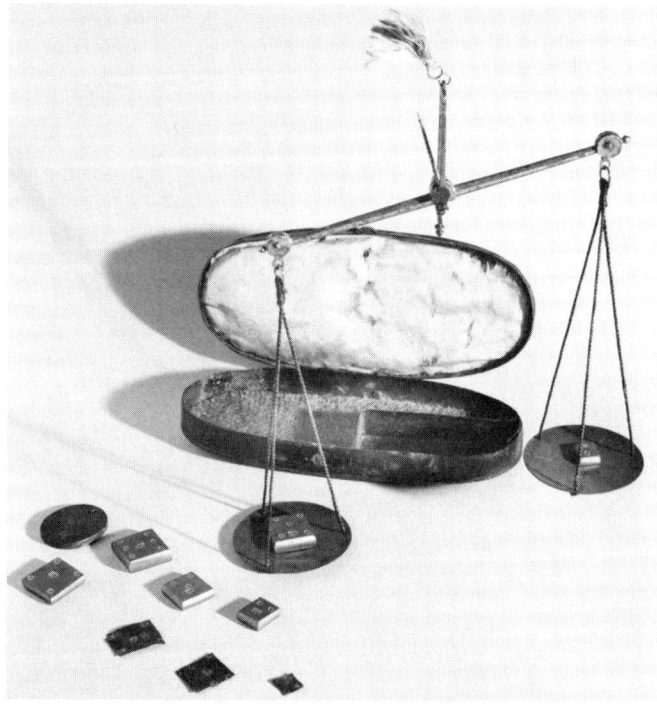

Joseph Priestley's pocket field balance, brought to America in 1794. From the E. F. Smith Memorial Collection.

A1 **Abbot, Stephen.**
Business records, 1785-86.
1 envelope.
In Essex Institute collections (Salem, Mass.)
Resident of Salem, Mass. Receipts of a trading business in Salem dealing chiefly in rum, butter, molasses, sugar, tobacco, cloth, notions, and tea; a few papers (1785) relating to a potash works in Salem; a bill advertising soap boiling; and a tax list (1786) for the town of Salem, when Abbot was collector.

A2 <u>Abbott, George Alonzo, 1874-1973.</u>
Papers, 1930-59. ca. 5 ft.
In University of North Dakota Library (Grand Forks) (214)
Chemistry professor and chairman of Dept. of Chemistry, University of North Dakota. Texts of lectures and radio addresses.

A3 <u>Abel, John Jacob, 1857-1938.</u>
Papers, 1857-1939. - 63 ft.
In Johns Hopkins Medical Institutions, Alan M. Chesney Medical Archives (Baltimore, MD).
Professor of Pharmacology at Johns Hopkins School of Medicine. - Correspondence, diaries, manuscripts of writings, notebooks, documents, clippings, and pictures.
Unpublished name index of correspondence in the library.
Access permitted to accredited scholars upon written application.
Deposited by Dr. Abel's sons, 1940.

A4 **Academy of Medicine of New Jersey.**
Records, 1775-1968. - 16 ft.
The New Jersey Historical Society (Newark).
Records of the Academy from 1911; records of several other local and state medical associations; lectures T. Gaillard gave on gynecology at the Columbia University College of Physicians and Surgeons during 1872 and 1873; the pharmacopoeia of Gertrude M. Watson, ca. 1898-99; and a casebook of Charles L. Ill concerning caesarian section.
Published guide available.

A5 Academy of Science of St. Louis.
Records, 1856-1958. 300 items and 3 reels of microfilm.
In University of Missouri-St. Louis, Library.
In part, microfilm (negative) made in 1971.
Minutes of the academy's boards of directors and trustees, proceedings of its council, and copies of its journal, Transactions. Subjects represented include American Indians, anthropology, biology, botany, chemistry, entomology, physical phenomena, physics, scientific societies, and surveying.
Information on literary rights available in the repository.
Gift of the academy, 1971.

A6 <u>Acheson, Edward Goodrich, 1856-1931.</u>
Papers, 1872-1968. ca. 13,000 items.
In Library of Congress, Manuscript Division (Washington, D.C.)
Electrochemist, engineer, and inventor. Correspondence, diaries (1906-31), biographical material, including typescript of autobiography entitled A Pathfinder: Discovery, Invention and Industry (1910), financial papers, laboratory notebooks relating to Acheson's experiments and inventions, newspaper clippings, and other material, chiefly 1899-1930. Includes 2 v. of court proceedings concerning a case (1894) involving Acheson's Carborundum Company, and research materials of his biographer, Raymond Szymanowitz. Correspondents include Acheson's sons, Edward, Jr., George, John, and Raymond Acheson, John P. Deringer, Thomas A. Edison, Alfred E. Hunt, John S. Huyler, Andrew W. Mellon, Edward L. Nichols, Walther Rathenau, William Acheson Smith, Edmund C. Sprague, Bakewell and Bakewell, Electric Smelting and Aluminum Company, and the Electrochemical Society.
Finding aid in the repository.
Information on literary rights available in the repository.
Chiefly gift of Acheson's children, Howard A. Acheson and Margaret A. Stuart, 1972.

A7 <u>Adams, Mark F., b. 1914.</u>
Papers, 1943-1962. - Ca. 50 items.
In Washington State University Library (Pullman).
Chemical engineer. - Correspondence, articles, drafts, notes and other papers. Subjects include water analysis, manganese production, and clay analysis.
Gift of M. F. Adams, 1974.

A8 <u>Acheson, Edward Goodrich, 1856-1931.</u>
Papers, 1875-1878, 1892. - 0.1 cu. ft.
In National Museum of American History Archives Center (Washington, DC).
Inventor and manufacturer. - Acheson is noted for his design of mathematical tables for tank gauging, and his discovery of carborundum in 1892. His later work dealt mainly with improved techniques for the manufacture of graphite. Collection includes a record of notes and experiments, ca. 1875-1877; a volume of manuscript tables for tank gauging, 1878; and a report on the first chemical analysis of carborundum, 1892.
Unpublished finding aid available.

A9 Adams, Roger, 1889-1971.
Papers, 1812-1971. - Ca. 23 ft.
In University of Illinois Archives (Urbana).
Professor and head of the Chemistry Department, University of Illinois. - Correspondence with family, foundation officials, businessmen, officials of professional societies, scientists, and former students; personal papers relating to family, education, and travel; speeches; reports; mss. of writings; tape-recorded interviews on the Chemistry Department and financial support of research and graduate work; photos; printed matter, and other papers. Includes correspondence and research on the chemistry of marijuana; documents about service in the Chemical Warfare Service (1917-1918) and National Defense Research Committee (1941-1946) and as scientific adviser in Germany (45-1946) and Japan (1947-1948); and material relating to elections, appointment of personnel, research, awarding of grants, securities investments, and Adams' associations with Abbott Laboratories, Alfred Sloan Fund, New York City, American Academy of Arts and Sciences, American Association for the Advancement of Science, American Chemical Society, American Institute of Chemists, American Philosophical Society, Battelle Memorial Institute (Columbus, OH), Council for Agricultural and Chemurgic Research, Coca Cola Company, Cosmos Club, (Washington DC), E. I. du Pont Company, Illinois Board of Natural Resources and Conservation, International Sugar Research Foundation, International Union of Pure and Applied Chemistry, National Academy of Sciences - National Research Council, National Science Foundation, Otto Haas Trust Fund, Robert Welch Foundation, Sloan-Kettering Cancer Research Institute, and the periodicals Organic Reactions and Organic Synthesis. Correspondents include Wallace Carothers, James Conant, Robert Robinson, Ernest Volwiler, and Richard Willstatter.
Unpublished finding aid in the repository.

A10 **Adams Sugar Refinery**, *Boston*.
Records, 1868.
2 v.
In Harvard Business School, Baker Library.
Letter books of Adams Sugar Refinery and its successor, Stanwood and Co.
Gift of Robert Storer, 1953.

A11 **Adams White Lead Company, Baltimore, Maryland**.
Papers, 1868-1886. - 6 boxes.
In Maryland Historical Society (Baltimore, MD).
Manufacturer of white lead, a primary constituent of paint. - Correspondence relating to the manufacture and sale of white lead, financial records, factory reports, time sheets for the labor force and agreements with representatives of other firms.
Unpublished inventory in the repository.

A12 **Adelberg & Raymond**.
Records, 1858-75. 1 box.
In New York Public Library.
Correspondence, reports on iron, coal, gold, silver, and other mineral deposits in the United States and Canada, geological surveys, and other records of an assay office in New York City.
Purchase, 1938.

A13 Aeronautics: The Lammot du Pont, Jr. Collection of Aeronautics.
Papers, 1903-44. - Ca. 2900 items.
In Hagley Library (Greenville, DE) (Acc. # 676).
Materials concerning the aircraft industry largely centering around the activity of Edward C. Worden (1875-1940) during World War I as chairman of the committee on airplane coating of the National Research Council (1916), as chief of the airplane wing coating section, Bureau of Aeronautics, Washington, D.C., 1916-18; but including later papers of Worden and the Worden Laboratory and Library at Milburn, N.J., and of his associate Leo Rutstein and the Rutstein Laboratory at Newark, N.J. Rutstein, a chemist associated with Worden, 1911-35, directed the Rutstein Laboratory from 1936 to 1953 and was a civilian consultant with Military Intelligence during World Wars I and II. The papers include a number relating to activities of Philip Drinker (1894-1972), chief of the Chemical Division of the Technical Section, United States Air Service, and also to work of the noted chemists, Camille Edouard Dreyfus (1878-1956) and his brother Henry Dreyfus (1882-1944), and their experiments and discoveries involving cellulose compounds. The work of the Chemical Division included extensive research concerning dope, fabric, water, cellulose acetate recovery, solvent recovery, gasoline, paints, oils, varnishes and all equipment for aircraft. There were many reports made for the benefit of the Air Service and these with others are found in the files.
Described more fully in the Supplement to A Guide to Manuscripts in the Eleutherian Mills Historical Library (1978), pp. 1-6; and in the original Guide (1970), pp. 703-704.

A14 **Agricultural Chemistry Lecture Notebooks.**
1863; 1901-02; 1906. - 4 items.
In Pennsylvania State University Libraries (University Park).
Notebook of C. Alfred Smith, containing lecture notes in agricultural chemistry under Dr. Pugh, 1863; notebook of Norman G. Miller, containing lecture notes in agricultural chemistry under Dr. Frear, 1901-02; and notebooks of C. O. Myers, containing lecture notes in agricultural chemistry under Dr. Frear, 1966.

A15 **Ainsworth, Frederick Crayton, 1852-1934.**
Correspondence, 1901-28.
225 items.
In Library of Congress, Manuscript Division.
Army officer. Letters relating mainly to smokeless powder, investigations of rifle bore corrosion, and a new or improved composition for the cleaning and lubricating of gun and rifle barrels. Correspondents include Edward Cathcart Crossman (b. 1881) and Arno Carl Fieldner (b. 1881)
Bequest of Mr. Ainsworth, 1936.

A16 **Air Force units (lower echelon), attached units, and stations overseas.**
Records, 1917-45.
In United States Air Force Historical Research Center (Maxwell Air Force Base, AL).
Periodic histories of lower echelon units including depots, and platoon, squadron and company levels of the Army Air Force in World Wars I and II. Included in these histories are accounts of chemical warfare units, chemical warfare tactics and techniques, training, World War I gas warfare casualties and chemical warfare equipment.
Open to researchers under restrictions of the repository.
Catalog card index in the repository.

A17 **Alabama Pharmaceutical Association.**
Records, 1917-1981. - ca. 15 ft.
In Auburn University Archives (Ala.).
Founded 1881, affiliated with American Pharmaceutical Association, 1971. - Correspondence, officers' reports, history files, ledgers, cashbooks, daybooks, notebooks, subject files, convention programs, publications, sound and video recordings, and photos. Includes material on American Pharmaceutical Association conventions, drugstores, laws and legislation, Association of Retail Druggists, awards, schools of pharmacy, women in pharmacy, and VA Hometown Pharmacy Program.
Deposited by Jon Bargainer, executive director, 1981.
Access restricted.
Finding aid in the repository.

A18 **Albany Billiard Ball Company.**
Records, 1869-1973. - 6 linear ft.
In National Museum of American History Archives Center (Washington, DC).
Plastics company, founded in 1858. - John Wesley Hyatt was one of the company's founders and the American inventor of celluloid. These records include documents relating to Hyatt; legal documents, 1871-1966; bank books; patents; patent certificates; patent assignments; memorandum and articles of association; union contracts; balance sheets; profit and loss accounts; journals and cash books; ledgers; time books; invoices; photographs; and correspondence to and from the company, 1881-1973.
Unpublished finding aid available.

A19 **Albis Company, Portland, Me.**
Records, 1935-50. 1 v.
In Maine Historical Society Library (Portland)
Articles of association, certificate of organization, subscription list, corporate bylaws, and minutes of directors' meetings, of a cosmetics and toiletries manufacturing firm.
Information on literary rights available in the repository.

A20 **Alchemy Collection.**
17th-18th centuries. - 7 items.
In Library of Congress, Manuscript Division.
Chiefly 17th and 18th century copies of earlier works. Includes: Traite du lapis philosophes ([17--] in French, 431 p.); Liber mutus ([ca. 1790?] in French, 43 p. with hand-colored plates); [Summa sapientia Dei] ([17--] in German, originally by John of Padua, 122 p.); Miraculum mundi ([17--] in German, 28 l.); Il segreto libro di Artefio ... che tratta dell' arte occulta e della pietra filosofale ([16--] in Italian, 55 p.); Traite de la nature metaux ([ca. 1790?] in French, 125 p.); Liber sapientae (1745, in German, by Joseph Antoni Maichelbackh, 87 l.).
Gift (in part) of Mrs. H. Carrington Bolton, 1912-1914.
Described in the Catalogue of Latin and Vernacular Alchemical Manuscripts in the United States and Canada by W. J. Wilson (Bruge, 1939), items 58-64.

A21 **Alden (L. H.) and Company,** *Aldenville, Pa.*
Records, 1829-83.
75 v.
In Historical Society of Pennsylvania collections.
Records of a leather-tanning company which cover the life of a one-industry town from 1848 to 1883, including store records, wage contracts and payments, factory accounts, grist mill accounts, etc. Includes a few records (1829) of the library at Windham, N. Y.
Purchase, 1949.

A22 Alexander, F. W., ca. 1850-1900.
Notebook, 1855-56. - 1 v.
In Maryland Historical Society Library (Baltimore).
Notes on chemical, English, and literary subjects. Also some drawings.

A23 **Alexander, Jerome**, 1876-1959.
Correspondence, 1908-51. 1 box.
In New York Public Library.
Chemist. Correspondence with American and European scientists, educators, and others, relating to Alexander's work in colloidal chemistry and other scientific and personal matters.
Gift of Alexander Jerome, 1954.

A24 Alexander, John H., 1812-1867.
Papers, ca. 1800-1883. - 6 boxes.
In Maryland Historical Society Library (Baltimore).
Chemist and physicist. - Correspondence, memos, official records, and personal writings related to Alexander's interests and activities. Included are items on: George's Creek Coal and Iron Co.; the Episcopal Church of Maryland diocese; weights, measures, and inventions; maps and drawings, glossaries; life insurance tables; copy and manuscripts for several books; treatise on leveling and on mathematical instrument dictionary of weights and measures.
Gift, May 10, 1920.

A25 Alexander, Samuel.
Lecture notes, 1812-1813. - 2 items.
In Dickinson College Library, Archives and Special Collections (Carlisle, PA).
Dickinson College student. - Alexander's notes on the chemical lectures of Thomas Cooper, 1812 (74 page manuscript); and Alexander's "Notes on Chemistry" from Professor Thomas Cooper, 1812-1813 (103 page bound manuscript).

A26 Altshuler, Henry Irving, 1898-
Papers, 1913-68. ca. 6 ft.
In University of Wyoming, American Heritage Center (Laramie)
Forms part of the repository's Western History Research Center collection.
Mining engineer. Professional files, reports, printed material, including maps, and other papers, relating to metallurgy, mining technology and techniques, labor and union matters, and various mines and mining companies, in the Western Hemisphere.
Unpublished finding aid in the repository.
Gift of Mr. Altshuler, 1969, 1974.

Aluminum Company of America. A27
Correspondence, 1927-47.
91 items.
In State Historical Society of Wisconsin collections.
Photocopies.
Letters (1927-31) relating to the discovery of fluorine as a component of public water supplies by H. V. Churchill, chief chemist of the Aluminum Company of America, in 1931. Much of the correspondence deals with fluorine as a cause of mottled enamel. A letter from Dr. Frederick S. McKay (1947) recalls the history of the discovery and its importance in establishing a means to control dental decay.

Amalgamated Leather Companies, Wilmington, A28
Delaware.
Records, 1898-1967. - Ca. 18,300 items, including 400 vols.
In Hagley Library (Greenville, DE) (Acc. # 973, 979, 983).
Leather manufacturers for the shoe and kid glove market from the late 19th century until 1966. - A full range of administrative, financial, production, and sales records including records from the files of William C. Blatz, Jr. and Peter Blatz which deal with tanning formulae, tanning methods and procedures, laboratory reports and patents.
Described more fully in the Supplement to A Guide to Manuscripts in the Eleutherian Mills Historical Library (1978).

Amelung family. A29
Papers, 1784-1953. - Ca. 200 sheets.
In Maryland Historical Society Library (1543) (Baltimore).
Photocopies (mostly positive) made in 1967 from originals in the possession of Mrs. W. Robert Milford, of Baltimore, MD.
Letters, many from Germany, to Mr. W. Robert Milford, authority on glass, relating to the glassmaking Amelung family, its genealogy, and the founding of the New Bremen and Fredericktown glassworks in Maryland. Includes a family genealogy and a few papers of John Frederick Amelung (d. 1798), glassmaker of Frederick Co., MD.
Open to investigators only with the permission of Mrs. Milford.

American Association of Colleges of A30
Pharmacy.
Records, 1871-1885, 1900-1971. - 45 boxes, including 89 v.
In State Historical Society of Wisconsin (Madison).
Professional organization which sets educational standards and clarifies educational objectives for the pharmaceutical profession.- Records include correspondence, committee records, financial records surveys, and other document types.
Gift of the Association, Silver Spring, Maryland, 1970-1973.
Unpublished finding aid available.

A31 American Chemical Society.
Records, 1940-59. 18 ft. (ca. 14,000 items)
In Library of Congress, Manuscript Division (Washington, D. C.)
Correspondence, minutes, reports, contracts, and other records, chiefly 1950-55. Includes material relating to bequests and donations, licensing clinical chemists and supervisors in various States, fellowships and scholarships, patents, professional and economic status legislation, scientific manpower legislation, standardization and safety studies, the publication of the periodical "Chemical Abstracts," and listings of accredited educational and training institutions.
Unpublished finding aid in the Library.
Open to investigators under restrictions accepted by the Library.
Gift of the society, 1969.

A32 American Chemical Society. Detroit Section.
Papers, 1907-1960. - 13 boxes.
In Detroit Public Library, Burton Historical Collection (Michigan).
Correspondence, minutes of meetings and papers reflecting various section activities.

A33 American Chemical Society. Division of Cellulose, Paper and Textile Chemistry.
Records, 1919-85. - Ca. 6 linear feet.
In University of Pennsylvania Libraries, E.F. Smith Collection (Philadelphia).
Records documenting divisional governance 1919-1984; the Payen Award, 1962-1985; Technical program abstracts, 1922-1951; other technical programs; a 2 volume history of the division, 1920-1974; and other documents and photographs.
Unpublished inventory at the repository.

A34 American Chemical Society. Division of Chemical Education.
Records, ca. late 1930s - 1981. - 8 boxes.
In University of Pennsylvania Libraries, E.F. Smith Collection (Philadelphia).
Minutes, reports, correspondence, programs, conference materials, teaching materials, and committee proceedings.
Unpublished inventory available.

A35 American Chemical Society. Division of History.
1956-1981. - 2 boxes.
In State Historical Society of Wisconsin (Madison).
The collection consists primarily of correspondence, 1973-1979. The remainder of the material consists of reports, copies of conference proceedings, and information on the Dexter Award in the History of Chemistry.
Unprocessed collection.

A36 American Chemical Society. Division of Inorganic Chemistry.
Records, 1956-66. ca. 3 ft.
In State Historical Society of Wisconsin collections (Madison)
Correspondence, newsletters, minutes of meetings, committee files, and other records, relating to the administration of the division, planning of symposia and special meetings, and professional matters.
Unpublished finding aid in the repository.
Gifts of Aaron J. Ihde, University of Wisconsin Chemistry Dept., Madison, Wis., 1965 and 1969.

A37 American Chemical Society. Erie Section.
Records, 1923-1978. - Ca. 3 ft.
In Mercyhurst College Archives (Erie, PA).
Minute books, research reports, financial records, social affairs programs, publications, membership lists, news releases, correspondence, awards and publications, ACS Constitution and By-Laws, Project Interface material, other science publications relating to chemistry.
Gift of Mercyhurst College Science Department.
Unpublished finding aid available.

A38 American Chemical Society. Eastern North Carolina Section.
Records, 1954-1981. - 5 cu. ft. and 7 volumes.
In East Carolina University Library, East Carolina Manuscripts Collection (Greenville, NC).
Chapter records include minutes, correspondence, annual reports, ENCHEM newsletters, tax records, membership lists, and other materials pertaining to operation and activities of the ACS.
Described in "East Carolina Manuscripts Collection, Bulletin No. 8."
Gift of Executive Committee, E.N.C. Section of ACS, 1980.

A39 American Chemical Society. Memphis Section.
Records, 1939-1977. - 9 cu. ft.
In Memphis State University Library, Mississippi Valley Collection (Tenn.)
Correspondence, minutes and agenda (1959-1973), reports, membership and scholarship files, ledger, cashbook, constitution and bylaws, publications, and other records. Includes published issues, circulation files, and other records pertaining to Southern Chemist, a periodical published by the section.
Gift of H. Graden Kirksey, Jr., president, 1979.
Unpublished finding aid in the repository.

A40 **American Chemical Society.** *Michigan Section.*
Records, 1900–52. 50 items and 3 v.
In University of Michigan, Michigan Historical Collections.
Correspondence and other papers, mostly dealing with the support of a French orphan; minutes, treasurer's reports and other records.

A41 American Chemical Society. Minnesota Section.
Records, 1929-1983. - 4 cu. ft.; 4 boxes.
In Minnesota Historical Society (St. Paul).
General correspondence, 1962-1966; annual reports, 1948-1979; meetings of the Executive Committee, 1950-1979; Council meetings, 1964-1967; financial records, 1949-1974; membership records, 1950-1971; conventions and symposia, 1929, 1947-77; bylaws, 1936, 1948-1967; treasurer's records, 1974-1976 and 1976-1978; bylaws, Minnesota Section, 1975-1977; charter and other papers, American Chemical Society, 1978; programs and abstracts, Great Lakes Regional Meetings, 1975, 1983.
Gift of Irwin L. Jacobs, 1981.
Unpublished finding aid available.

A42 American Chemical Society. Northern Intermountain Section.
Records, 1909-1924. - Ca. 200 items.
In Washington State University Library (Pullman).
Minutes and correspondence, membership lists and publications. Correspondents include section chairs George Olson and Elton Fulmer.
Unpublished finding aid available.

A43 American Chemical Society. Northwest Regional Meeting.
Records, 1967-1983. - Ca. 5 in.
In University of Washington Archives (Seattle).
Minutes, correspondence, proceedings, and abstracts.

A44 American Chemical Society. Puget Sound Section.
Records, 1947-1984. - 13 linear ft.
In University of Washington Archives (Seattle).
Correspondence, minutes, reports, programs, photographs, membership directories, and other material of the section.
Accessioned 1971-1977.
Unpublished finding aid available.

A45 American Chemical Society. St. Louis Section.
Records, 1903-69. 5 ft.
In University of Missouri-St. Louis, Library.
Correspondence of officers and committees, minutes, membership directory and records, histories of the society, printed matter, and other papers.
Gift of the section, 1974.

A46 American Chemical Society. University of Missouri Section.
Records. - 1 folder.
In University of Missouri-Columbia Library, Western Historical Manuscripts Collection and State Historical Society of Missouri Manuscripts.
Materials pertaining to the 75th anniversary of the organization.
Gift of Robert L. Wixom, March 28, 1984.
Unprocessed as of 1985. Available for research.
Unpublished finding aid available.

A47 American Chemical Society. Wisconsin Section.
Records, 1907-1966. - 4 boxes.
In State Historical Society of Wisconsin (Madison).
Correspondence, constitution and by-laws, minutes, newsletters, lecture notices, rosters, membership applications, annual and treasurer's reports, history of the section, clippings, and other records; together with newsletters, annual reports, reports of national meetings, and constitution and by-laws, from the national office. As of 1985, there are also unprocessed additions under this heading.
Gift of the section, 1967.
Unpublished finding aid available.

A48 American College of Apothecaries.
Records, 1939-1963. - Ca. 8 ft.
In State Historical Society of Wisconsin (Madison).
National society of professional pharmacists. - Correspondence, constitution and by-laws, annual reports, membership applications, minutes of conventions, conferences, and Board of Directors' meetings, administrative and financial records, files on awards and honors, scrapbook (1940-1950) of Secretary Charles Selby, and other records; together with material relating to earlier organizations, including the American Board of Apothecaries, American College of Pharmacists, Association for the Advancement of Professional Pharmacy, National Council on Pharmaceutical Practice, and Society of Pharmacy Colleagues. Most of the correspondence is that of Charles Selby for 1940-1950. As of 1985, there are also unprocessed additions under this heading.

Gift of American Institute of the History of Pharmacy via Glenn Sonnedecker, Madison, Wisconsin, 1970.
Unpublished finding aid available.

A49 American Crystal Sugar Company.
Records, 1899-1973. - 192 cu. ft., and 2 rolls of microfilm.
In Minnesota Historical Society (St. Paul).
Incorporated in 1899 as American Beet Sugar Company. - Records include minutes (several series), ledgers and journals, stock records, and annual reports, all nearly complete for the entire period. Subject files, statistical data, and labor records, as well as a variety of supplementary materials, are also present and concentrate most heavily on the 1950's and 1960's. Photographs and photo albums are included. Records of fourteen other predecessor or subsidiary companies, mainly in 1910's-1930's, are present in varying degrees of completeness.
Written permission of the company required for access to, and for publication of anything quoted from or based on, records less than twenty-five years old. Records more than twenty-five years old are unrestricted.
Transferred from corporate headquarters, Moorhead, Minnesota, and from a closed sugar factory in Rocky Ford, Colorado.
Published finding aid available.

A50 American Crystallographic Association.
Records, 1958-1973. - 25 in.
In American Institute of Physics, Niels Bohr Library (New York, NY).
Records of the American Crystallographic Association, including correspondence and publications.
Finding aid available.

A51 American Institute of Chemical Engineers.
Records, 1922; 1955-82. - Ca. 5 ft.
In University of Pennsylvania Libraries, E. F. Smith Collection (Philadelphia).
Professional association of chemical engineers founded in 1908. - Collection contains information on the history of chemical engineering departments at nearly 100 universities and colleges in the United States. Includes the survey sheets used in the gathering of historical data; newsletters; photocopies of departmental catalogs; commemorative brochures; general correspondence and material relating to the AIChE Publications Committee; and a 1955 brochure prepared by the Committee on the subject of careers in chemical engineering.
Unpublished finding aid available.

A52 American Institute of Chemical Engineers, Great Salt Lake Section.
Records, 1955-1984. - Ca. 1 ft.
In University of Utah Libraries, Special Collections Department (Salt Lake City).
Organizational material, correspondence, rosters, minutes and meeting notices, 1955-1984. Additional material expected.
Gift of Noel deNevers, 1984.

A53 American Institute of Mining and Metallurgical Engineers. North Pacific Section.
Records, 1917-1930. - Ca. 60 items.
In University of Washington Archives (Seattle).
Professional society of engineers in the fields of mining and metallurgical and petroleum engineering. - Minutes, 1924-30; reports; correspondence; guest book, 1917-30, of the organization.
Accessioned 1931.
Unprocessed.
Unpublished description available.

A54 American Iron and Steel Institute Collection.
Papers, 1768-1947. - 27 v. and 20 other items.
In Hagley Library (Greenville, DE) (1631).
Miscellany relating to the iron and steel industry in the United States including both original financial and administrative records from early iron enterprises and draft histories of various aspects of the iron and steel industry. The American Iron and Steel Institute was founded in 1908. In 1912 it absorbed the American Iron and Steel Association, records which form the bulk of this collection. The Institute functions as a research, educational and information facility for the industry, with headquarters in Washington, DC.
Unpublished inventory available.

A55 American Philosophical Society, *Philadelphia*.
Archives, 1768-
ca. 55 ft. and ca. 14,000 items.
In American Philosophical Society Library.
Correspondence with individuals, universities and learned societies throughout the world; minutes of the society and its committees; account books; catalogs and accession books of its library; publication records; and collections of papers submitted to the society on natural history, science, education, commerce and many other subjects. Important persons whose papers are included are Alexander D. Bache, Spencer F. Baird, George Bancroft, Joseph Banks, Benjamin S. Barton, William Barton, Charles Bonaparte, Daniel G. Brinton, Cadwalader Colden, Thomas Cooper, Edward D. Cope, Peter S. Du Ponceau, Pierre S. Du Pont, Andrew Ellicott, Felix Fontana, Anthony Fothergill, Benjamin Franklin, John F. Frazer, Persifor Frazer, Robert Fulton, Nicholas Fuss, Asa Gray, Joseph von Hammer-Purgstall, Robert Hare, Ferdinand R. Hassler, Ferdinand V. Hayden, John G. E. Heckewelder, Joseph Henry, John F. W. Herschel, bart., David Hosack, Thomas P. James, William Jardine, John K. Kane, William H. Keating, Lafayette, Benjamin H. Latrobe, Antoine L. Lavoisier, Isaac Lea, John LeConte, Joseph Leidy, J. Peter

Lesley, William Maclure, Humphrey Marshall, Timothy Matlack, Matthew F. Maury, André Michaux, François A. Michaux, Samuel L. Mitchill, John Morgan, Samuel G. Morton, Thomas Nuttall, George Ord, Robert Patterson, Robert M. Patterson, Charles W. Peale, Rembrandt Peale, Titian R. Peale, Henry Phillips, Joel R. Poinsett, William H. Prescott, Eli K. Price, John S. Price, Joseph Priestley, Lambert A. J. Quetelet, David Rittenhouse, Benjamin Rush, Thomas Say, Henry R. Schoolcraft, Benjamin Silliman, William Smith, Jared Sparks, Thomas Sully, Philip Syng, Charles Thompson, Charles B. Trego, Benjamin Vaughan, John Vaughan, Sears C. Walker, Noah Webster, and George B. Wood. Includes papers of persons for whom the library has separate collections.
Ca. 14,000 items cataloged individually in manuscripts catalog of the library. Many of the papers and records have appeared in the Society's various publications.
Acquired, 1769–

A56 **American Philosophical Society,** *Philadelphia.*
Papers on science, 1720–1958.
ca. 32 items.
In American Philosophical Society Library.
In part, transcripts and photocopies.
Writings on various scientific subjects, including astronomy, chemistry, agriculture, and natural philosophy, by William Alexander, Johann Beckmann, Hermann Boerhaave, the Marquis de Condorcet, Frederick Henry Cramer, Charles Gobrecht Darrach, Daniel Freechauff, Frederick Augustus Genth, Earl T. Glauert, Theophilus Grew, Paul Henri Thiry, baron J'Holbach, Jean Baptiste de Lamarck, Benoît de Maillet, William Maule, Edward Mulhern, Joseph Priestley, John Questebrune, Joseph Roberts, A. Sagey, George Gaylord Simpson, Thomas Peters Smith, Jacob Stauffer, Jethro Tull, Lucian McShan Turner, and C. L. B. Wavran. Also includes a Formulaire medical, and Problems in astronomy. Some of the material is in French or German.
Cataloged individually in manuscripts catalog of the library.
Gifts and purchases, 1773–1958.

A57 **American Physical Society. Division of High Polymer Physics.**
Records, 1943–1944. – 150 items.
In American Institute of Physics, Niels Bohr Library (New York, NY).
Correspondence and related materials documenting the formation of the Division in 1944. Materials were assembled by W. James Lyons, past chairman of the Division.
Deposited by W. J. Lyons, 1969.
Finding aid available.

A58 **American Potash Company.**
Records, 1917–20. 1 box.
In Nebraska State Historical Society collections.
General correspondence and statements to the commissioner of the Nebraska Dept. of Public Lands and Buildings, of a business firm at Antioch, Neb.
Gift of Elmer H. Mahlin, secretary of the Board of Educational Lands and Funds.

A59 **American Shale Refining Company.**
Records. – 4,000 items.
In Colorado Historical Society (Denver).
Features papers of Leonard L. Aitken, Jr. (1908–1974), lawyer and officer of the American Shale Refining Company. Also contains business records of American Shale, plus miscellaneous papers of the Colorado Shale Products Company and the Troy-American Petroleum Corporation.
Donated by Leonard L. Aitken, Jr.
Unpublished finding aid available.

American Smelting and Refining Company. A60
Baltimore plant records, 1887–1926. 29 v., 1 box, 1 folder, and 1 tube.
In Maryland Historical Society Library (Baltimore) (2360)
General ledgers containing a variety of accounts, cashbooks, copper receipts, shop drawings and specifications for machinery and equipment, and other records, of a copper refining plant operated as Baltimore Copper Smelting and Rolling Company until its acquisition by the American Smelting and Refining Company in 1923; together with maps, plats, and plans of property of the Canton Company, Canton Railroad, Canton Distilleries, and Baltimore Copper Smelting and Rolling Company, all in Baltimore, Md. Includes letter (1907) from Thomas Edison ordering copper.
Gift of R.H. Funke, Jr., manager of the plant.

American Society for Testing Materials. A61
Otis collection, 1954–61. ca. 900 items.
In Tennessee State Library and Archives.
Correspondence, information bulletins, minutes of meetings, coding data, reports covering the work in organizing the information storage and retrieval programs in the U. S. and abroad, and other papers relating to the activities of Marshall V. Otis with the American Society for Testing Materials and the Office of Critical Tables of the National Academy of Sciences. Correspondents include Robert L. Bowman (National Heart Institute), Fleur C. Byers (National Research Council), George L. Covert (Industrial Laboratory, Wilmington, Del.), Carrol E. Creitz (secretary, Committee on Spectral Absorption Data), Robert C. Hirt (American Cyanamid Co.), Raymond Hopkins (Standard Oil Co.), Robert J. Jakobsen (Battelle Memorial Institute), Wescott C. Kenyon (chairman, Standard Data Subcommittee ASTM), L. E. Kuentzel (Wyandotte Chemical Corp.), M. G. Mellon (professor of analytical chemistry, Purdue University), Robert T. O'Connor (Physics Unit, U. S. Dept. of Agriculture), Thomas V. Parke (chairman, Subcommittee on Standard Data, ASTM), R. F. Robey (secretary, Committee on Absorption Spectroscopy), E. J. Rosenbaum (chairman, Committee on Absorption Spectroscopy), Philip Sadtler (Sadtler Research Laboratories), Eugene Sawicki (Cancer Research, University of Florida), Richard E. Seeber (National Airline Division, Chemical Corp.), Charles Lea Smith and Clara D. Smith (Battelle Memorial Institute), J. M. Vandenbelt (Parke, Davis Co.), and Guy Waddington (National Academy of Science and National Research Council).
Unpublished register in the library.
Information on literary rights available in the library.
Gift of Marshall V. Otis.

American Sugar Refinery Company (San Francisco, A62
Calif.)
Records, 1879–1903. – 5 v.
In California Historical Society Library (San Francisco)
Formed in 1879 with the purchase of the Bay Refinery (established by Claus Spreckels in 1864); owned and operated by Havemeyers and Elder, New York. – Bylaws (1879, 1885), journal (1890–1895), general ledger (1890–1903), and cashbook (1890–1903).

A63 American Zinc Company.
Records, 1901-57. 800 items.
In University of Missouri-St. Louis, Library.
Correspondence, geological and mining production reports, ledgers, audits, annual reports, and other business records, of a company in Central City, Joplin, and St. Louis, Mo. Includes material on American Zinc Lead and Smelting Company, W. G. Swarts Company, Howard I. Young, iron ore, lead, metallurgy, mining, and zinc.
Gift of the company.

A64 Amory family.
Papers, 1697-1823. - 1,400 items.
In Library of Congress, Manuscript Division.
Merchant trading and shipping family of Ireland, South Carolina, and Boston. - Letterbooks, 1711-1728, and 1 letterbook, 1798-1799; correspondence, 1697-1808, from commercial clients, traders, and associates, legal documents; family correspondence, 1714-1731; and a scrapbook of financial papers, 1723-1744. The bulk of the collection, apart from letterbooks, is for the period 1802-1804, and represents the incoming correspondence of Thomas Amory (1762-1823), Boston, successor to his grandfather Thomas Amory (1682-1728) in operation of the Amory enterprises (import, shipbuilding, and distilleries).
Gift, Copley Amory, 1934-1936, and purchase, 1937-1938.

A65 Andrews, Andrew Irving, 1895-1966.
Papers, 1924-1966. - 4.3 ft.
In University of Illinois Archives (Urbana).
Professor of Ceramic Engineering at the University of Illinois. - Correspondence, published articles, ms. reports and speeches, manufacturers' literature, and other papers, relating to the properties of ceramic materials, the adherence of enamels to metals, porcelain enamels, manufacturing processes, standards and tests, ceramic-engineering education, professional associations, consulting work, and the policies, curriculum, and growth of the Department of Ceramic Engineering.
Unpublished finding aid in the repository.
Information on literary rights available in the repository.
Acquired, 1967.

A66 Andrews, Roy Chester, 1888-1955.
Papers, 1902-55.
2 ft.
In University of Oregon Library.
Professor of chemistry at the University of Oregon and teacher in Michigan, Texas, Arkansas, Washington, and Oregon. Correspondence, diaries, notebooks, and photos.

A67 Andrus, Leonard A 1883-1965.
Reports on the history, condition, and prospects of certain business firms in the Pacific Northwest, 1921-30. 1 v.
In University of Oregon Library (Eugene)
Consulting engineer. Reports of surveys of the following companies: American Telephone and Telegraph Company, Armstrong Manufacturing Company, Portland, Or., Automatic Electric Brake Company, Barnes-Lindsley Manufacturing Company, Portland, Or., Bell Telephone System, Big Creek Trout Hatcheries, Black Carbon Coal Company (Oregon and Washington), Camel Chemical Company, Portland, Or., Cascade China Company, Portland, Or., Concrete Pipe Company, Seattle, Wash., Dunthorpe Bus Service, Portland? Or., Haynes-Foster Baking Company, Portland, Or., Marshfield Hotel, Prairie Box Company, Prairie City, Or., and Prairie Power Company, Prairie City, Or.

A68 Apothecary's Notebook, ca. 1800-1864.
1 octavo v., 79 leaves (24 blank).
In American Antiquarian Society (Worcester, MA).
Unidentified apothecary's recipes for treatment of various diseases. - Kept from ca. 1800 to 1864. The last entry includes a reference to Oakham, Massachusetts and there is also an entry dated 1808 in Shirley, Massachusetts, but there are no clues as to the identity of the writers. The entries were not all written in the same hand. The pages contain detailed medical recipes for treatment of a wide variety of diseases, such as gout, cancer, diabetes, and epilepsy. The recipes appear under specific names of treatments, and sometimes include the proportions of each ingredient and the recommended dosage.
Gift of Barbara L. Wyatt, 1983.

A69 Arceneaux, Claude Joseph, 1910-1978.
Papers, 1943-1975. - 1,269 items.
In Louisiana State University, Department of Archives (Baton Rouge).
Chemist and research microscopist (Ethyl Corporation); recipient of the first President's Award of the Electron Microscopy Society of America, 1974; alumnus of LSU; founder and President for several terms of the Louisiana Society for Electron Microscopy. - The collection consists of papers relating to his career and professional interests, particularly to the activities of the Electron Microscopy Society of America and the Louisiana Society for Electron Microscopy.

A70 **Archbold, John Dustin,** 1848–1916.
Papers, 1886–1916. 2 boxes.
In Syracuse University Library (N. Y.)
Businessman. Correspondence, legal documents, memorabilia, and other papers, relating to the Waters-Pierce Oil Company, the Hearst newspapers, the Standard Oil Company, of which Archbold was president and director, and to Syracuse University, which Archbold served as president of the board of trustees.
Unpublished inventory in the library.
Gift of Mr. Adrian Archbold, 1959.

A71 Armstrong, Lyndon King, 1859-1942.
Papers, 1897-1941. 6 ft. (ca. 2500 items)
In Washington State University Library (Pullman)
Mining engineer, of Spokane, Wash. Correspondence, field notes, reports, bulletins, and maps, relating to gold and silver mining in the Pacific Northwest; correspondence and minutes of meetings of the Columbia Section of the American Institute of Mining and Metallurgical Engineers and of the Washington State Natural Resources Association; and papers presented at regional professional meetings.
Unpublished container list in the library.
Information on literary rights available in the library.
Purchase, 1956.

A72 **Arnold, Benedict,** 1741-1801.
Papers, 1761-1794. - 85 items and 2 v.
In New Haven Colony Historical Society Library (Conn.) (106).
In part, facsimiles.
Revolutionary patriot and traitor. - Correspondence, accounts, and commercial and other papers, relating chiefly to Arnold's pre-Revolutionary mercantile interests in New Haven, Conn., where he worked as a bookseller and pharmacist and was active in the West Indies trade; pharmaceutical recipes; waste book (1773-1780), including accounts from his years in Philadelphia, Pa.; inventories (1781-1786) of Arnold's confiscated estates in New Haven and Philadelphia; miscellaneous military papers, including letters; diary (1775) of Samuel Barney, kept in Maine during Arnold's expedition against Quebec; and other papers relating to Arnold.
Gift.
Finding aid in the repository.

A73 **Arnold, Frederick,** 1811–1873.
Papers, 1823–61.
64 items.
In State Historical Society of Wisconsin collections.
Chandler and soap manufacturer. Correspondence, including letters from Georg Leonhard Marck; legal records of land transfers in Milwaukee; and bills and accounts for Arnold's business. Correspondence is in German.

A74 Arrhenius, Svante August, 1859-1927.
Correspondence, 1904-25. 1 reel of microfilm (60 items)
In Library of Congress, Manuscript Division (Washington, D. C.)
Microfilm made from originals in private hands.
Swedish physicist and chemist. Primarily letters from Jacques Loeb (1859-1924), physiologist and educator, including letters from Loeb's son Robert.
Gift of Dr. Olaf Arrhenius, 1962.

A75 Asarco Smelter, Tacoma, Washington.
Records, ca. 1890-1975. - 1 box.
In Washington State Historical Society (Tacoma).
Collection contains time books, 1910-1912 and 1915-1916; newpaper and magazine articles about the Arsaco Smelter; an inventory; plant metallurgical data; organizational documents of the Asarco Smelter; and photographs, mostly 1910-1920, which trace the development of the smelter from the early 1900s to the present.
Unpublished finding aid available.

A76 Ashdown, Avery Allen, b. 1891.
Papers, 1928-1956. - 0.35 cu. ft.
In Massachusetts Institute of Technology, Institute Archives and Special Collections (Cambridge, MA).
Chemical engineer; educator. - Collection contains scrapbook of newspaper clippings and diary entries, material on Ashdown House, and biographical essays on the following M.I.T.-associated chemical engineers: Arthur Alphonzo Blanchard, William H. Carlisle, Jr., Arthur Clay Cope, James Mason Crafts, Tenney Lombard Davis, Frederick George Keyes, Gilbert Newton Lewis, Arthur D. Little, James Flack Norris, Allan Winter Rowe, William P. Ryan, and Charles William Tucker.
Finding aid available.

A77 **Ashley, William Henry,** 1778–1838.
Papers, 1778–1840. ca. 85 items.
In Missouri Historical Society collections (St. Louis)
Fur trader and U. S. Representative from Missouri (1831-37). Papers relating to Ashley's interest in the fur trade, manufacture of gunpowder, and mining; improvement in navigation on the Mississippi River in the vicinity of St. Louis, Mo.; real estate in St. Louis; politics; Indian trade; and the West. Correspondents include James P. Beckwourth, James Bridger, Robert Campbell, Thomas Fitzpatrick, David E. Jackson, Etienne Provost, Jedediah S. Smith, William L. Sublette, and Samuel Tulloch.
Gift.

A78 Assay Office, Granite Falls, Washington.
Records, 1939-40. - 1 vol.
In Granite Falls Historical Society (WA).
Records of an office apparently located in Granite Falls, listing mineral analyses and other related information.

A79 **Association of Cambridge Scientists.**
Records, 1945–46. 28 folders.
In University of Chicago Library.
Papers of an association founded in late 1945 as a response to the growing controversy over the uses of atomic energy, and whose membership was comprised mainly of persons on the staffs of Harvard University and Massachusetts Institute of Technology. The main purpose of the association was to inform the general public on scientific matters when such information was pertinent to matters of public policy, and to stir discussion among scientists themselves. Includes correspondence, newsletters, press releases, and statements of the association and other similar organizations.
Unpublished guide in the library.
Deposit, 1961.

A80 **Association of Los Alamos Scientists.**
Records, 1945–48. 5 ft.
In University of Chicago Library.
Business and financial papers, membership lists, minutes of meetings, press releases, statements, memoranda, information sheets, photos., and phonograph records of radio broadcasts of an association organized in 1945 to promote the attainment and use of scientific and technological advances in the best interests of humanity by educating the public to the nature and control of atomic energy.

A81 **Association of Oak Ridge Engineers and Scientists.**
Records, 1946–51. 9 ft.
In University of Chicago Library.
Papers (mostly 1946–48) including correspondence; administrative and information files; newsletters; publicity handouts; and educational material; together with the correspondence and papers of related and affiliated groups including the Emergency Committee of Atomic Scientists; the Independent Citizen's Committee of the Arts, Science and Professions; the National Committee of Atomic Scientists; and Oak Ridge groups prior to their amalgamation into the AORES and the Federation of American Scientists. The correspondence with the Senate Atomic Energy Committee, Congressmen, and other government officials concerns the McMahon bill, the National Science Foundation bill, and international control of atomic energy.
Unpublished guide in the library.
Gift of Herbert Pomerance, member of the Executive committee of AORES, 1959.

A82 **Association of Pasadena Scientists.**
Records, 1945-46. 1 ft. (1000 items)
In University of Chicago Library.
Correspondence, press releases, minutes of meetings, newsletters, membership data, and notices concerning internal matters and dealings with the Federation of American Scientists, of which the Association of Pasadena Scientists was a part. Represented in the correspondence are William N. Lipscomb, Gene Maun, Richard M. Noyes, J. R. Oppenheimer, Bradford Shank, David P. Shoemaker, Charles Wagner, the American Association of Scientific Workers, Fellowship of Reconciliation and the Hollywood Writers Mobilization, Independent Citizens of Arts, Sciences, and Professions, inc., National Association of Science Writers, National Committee on Atomic Information, Science Society of Washington, and several major universities. Includes letters to various legislators concerning atomic energy legislation.
Unpublished guide in the library.
Gift of James Bonner, 1962.

Association of Scientists for Atomic Education. A83
Records, 1946–48. 2 ft.
In University of Chicago Library.
Correspondence, minutes of meetings, and financial and membership records of a group devoted to educating the public to the uses and dangers of atomic power.
Unpublished guide in the library.
Information on literary rights available in the library.
Deposit, 1960.

Atkinson and Smith families. A84
Papers, 1806-1943. - 6 boxes.
In Maryland Historical Society Library (Baltimore).
James E. Atkinson (d. 1851) and Thomas Marsh Smith (1810-1877) were merchants and partners in the firm known as Baltimore Chemical Works from 1833 to 1851. - Collection consists of personal papers of the Smith and Atkinson families, as well as business papers of the Baltimore Chemical Works. The company was involved in the wholesale trade of drugs, paints, and oils as well as other merchandise. Business papers consist primarily of incoming correspondence (1837-1866) and several account books, including inventories (1851, 1856); lists of balances (1850, 1856); expense accounts (1870s); and checkbooks (1860s).
Bulk of the collection is family papers: correspondence, diaries, travel accounts, passports, school books, expense accounts, bills, receipts, and estate papers.
Gift of Matthew S. Atkinson III.
Unpublished finding aid available.

Atlas Powder Company Collection. A85
Records, 1891-1940. - Ca. 700,000 items.
In Hagley Library (Greenville, DE) (Acc. # 1516, 1528).
The firm was formed in 1912 from separated elements of the Du Pont Company in compliance with an anti-trust decree of the U.S. Circuit Court. The Company engaged in the manufacture, storage, and sale of explosives, blasting supplies, and related chemicals. In 1913 it established the Reynolds plant near Tamaqua,

Pennsylvania, where it manufactured electric exploders and blasting caps, nitric and sulphuric acid, dynamite, and blasting detonators. Its research laboratory was started there in 1916. The Reynolds plant was the center of explosives operations that were national in scope. Aside from high explosives and blasting supplies, it manufactured black powder, cellulose products, and industrial chemicals. Among its best known products was Zapon, widely used in industrial finishes for a variety of consumer merchandise. Elements of the Atlas Powder Company were absorbed by Imperial Chemical Industries-United States, Inc. in 1972. The records include those of absorbed companies and are represented by the following groups: (1) Reynolds plant production records (1915-1940) comprise the bulk of the collection; (2) Research and Development Department records (1901-32) relating to lacquers, enamels, and other finishes, cements, airplane dope, leather, cotton solutions, solvents, Zapon and other cellulose products; (3) Accounting Department records (1891-1934); (4) American Forcite Powder Manufacturing Company (Landing, NJ), employment record wage cards (1903-32); (5) Peerless Explosives Co. (White Haven, PA) stockholder records.

Described more fully in the Supplement to A Guide to Manuscripts in the Eleutherian Mills Historical Library (1978).

A86 Atomic Scientists of Chicago.
Records, 1945-1959. - 17 linear ft.
In University of Chicago Library.
Created in 1945 to promote public awareness and influence government policy concerning nuclear energy and its possible consequences. Consisting of a large contingent of University scientists who had participated in the Manhattan Project, the organization sponsored conferences, lobbied for policies, and published the Bulletin of the Atomic Scientists. It was dissolved in 1959. Subject files and correspondence concerning conferences; organization and incorporation; publications and public relations; legislation; and other international, national, university, and scientific bodies.
Unpublished guide.

A87 Atwater, Wilbur Olin, 1844-1907.
Papers, 1840 - 1943. 13 Paige boxes, 1 Hollinger box, 1 oversize box.
In Wesleyan University Archives (Middletown, CT).
-Agricultural chemist; Director of the Connecticut Agricultural Experiment Station; Director of the Office of Experiment Stations, U.S. Department of Agriculture. - Includes letterpress books, 1876-1901; incoming letters, 1881-1903; student papers; and notes on experiments. The correspondence is concerned mostly with establishment of experiment stations in the U.S., with chemical analyses carried out, with other professional and administrative duties, and with family affairs. Correspondents include American Copyright League, American Chemical Society, American Society for the Extension of University Teaching, Franklin Institute in Philadelphia, members of Congress, scientific and other periodicals, laboratories, businesses, and other institutions and agencies. Also includes a journal of a trip to Europe in 1869, speeches, articles, and pamphlets. The collection also contains letters to him from various family members, papers of his daughter Helen Woodard Atwater (1876-1947) and letters written by his son Charles Atwater (1885-1946) to family members.

The Department of Manuscripts and University Archives of Cornell University Libraries holds 39 reels of negative microfilm of that portion of the collection dealing directly with Atwood (1869- ca. 1914). Copies of the positive microfilm are held at Wesleyan University as well as at the E. F. Smith Collection at the University of Pennsylvania Libraries. The permission of the Wesleyan University Archivist is required for copying.
Unpublished finding aid available.

A88 Atwood, Mary Ann (South) 1817-1910.
Papers, 1882-1910. ca. 700 items.
In Brown University Library (Providence, R. I.)
Theosophist, of Thirsk, Eng. Letters, chiefly to Mme. Isabelle de Steiger; mss. and drafts of writings, and miscellaneous notes. Subjects include alchemy, ontology, metaphysics, mythology, neo-Platonism, heraldry, and theosophy.
Complete list of correspondents and miscellaneous mss. in the library.
Purchase, 1967.

A89 Audrieth, Ludwig Frederick, b. 1901.
Papers, 1927-1966. - 1.3 ft.
In University of Illinois, University Archives (Urbana).
Professor of Chemistry at the University of Illinois, and scientific advisor and attache at Bonn, Germany (1959-1963). - Correspondence, a report on the status of inorganic chemistry at German universities prepared for the National Science Foundation, reprints of articles, clippings, and other papers relating to Audrieth's work at Bonn, non-aqueous solvents, nitrogen chemistry, inorganic compounds, the discovery of sucaryl, Paul

Walden, University of Illinois policies and practices, and Audrieth's service at Picatinny Arsenal.
Received 1966.

A90 **Aufrichtig, Alois.**
Business records, 1888–1912.
1 v. and 1 folder.
In University of Missouri Library, Western Historical Manuscripts Collection.
Correspondence, accounts, and records of the Copper, Brass, and Sheet Iron Works Co., St. Louis, Mo., owned by Aufrichtig. Includes a patent for a water filter issued to Aufrichtig, orders and accounts of various brewing companies, and records concerning prices, specifications of custom-made brewery equipment, and advertisements of new equipment.

A91 <u>Austin family.</u>
Austin memorial collection, 1763–1887. – ca. 5 ft. and 145 v.
In Greene County Historical Society collections (Coxsackie, N.Y.)
Business and personal papers of a family in Massachusetts and Greene County, N.Y., chiefly Augustin Austin, of Sheffield, Mass., and his son, Abner Austin, merchant, of Hudson and Athens, N.Y., and paper mill operator in Greene County. Includes business diaries, farming records, papers documenting the family business enterprises at Jefferson Heights, Catskill, N.Y., and records (1813-1880s) from the paper mill.
Gift of Helen Austin Berhendt, 1981.

A92 <u>Autry, James Lockhart, 1859-1926.</u>
Papers, 1834-1925.   20 ft.
In Rice University Library (Houston, Tex.)
Lawyer, of Houston, Tex. Chiefly correspondence and legal documents relating to Autry's business activities (1904-20) with American Republics Corporation (petroleum refiners), Fidelity Trust Company, Houston, Tex., and the Texas Company. Correspondents include Joseph Stephen Cullinan, John Warne Gates, and Will Clifford Hogg. Includes letters of Autry's grandfather, Micajah Autry, about the Texas Revolution, and of his father, James Lockhart Autry, L. Q. C. Lamar, and others, about the Civil War in Vicksburg, Miss., and the Battle of Murfreesboro, Tenn.
Unpublished guide in the repository.
Information on literary rights available in the repository.
Gift of Mr. Autry's daughter and granddaughter, Mrs. Edward W. Kelley and Mrs. James Dittmar.

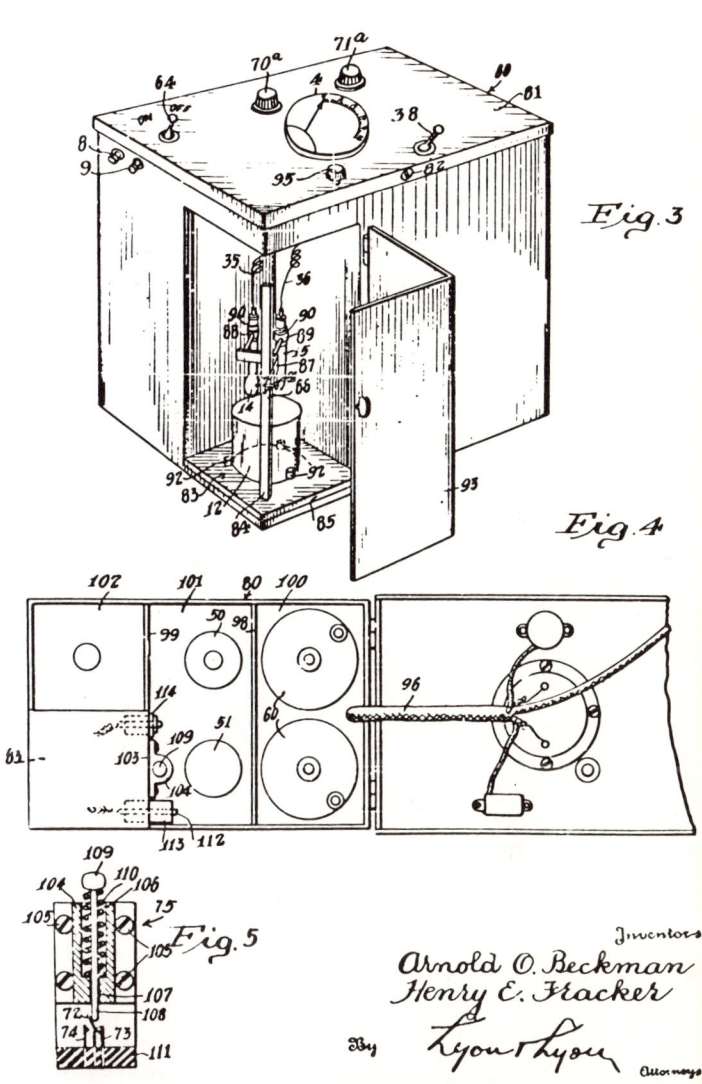

Drawings that accompanied the 1934 patent application for a pH meter by Arnold O. Beckman and Henry E. Fracker of National Technical Laboratories in Pasadena, California. From the E. F. Smith Memorial Collection.

B1 B. F. Goodrich Company, Akron, Ohio.
Records, ca. 1870-1974. - Ca. 150 boxes and unattached items.
In University of Akron, American History Research Center (Akron, OH).
Records of Akron tire and rubber company. - Collection contains general corporate records; advertising and sales/marketing information; information by and about Goodrich employees; labor relations material; personnel records; legal records and correspondence; public relations material; product information; periodicals and publications; photographs; financial and accounting records; company histories and news clippings; and research and development material, ca. 1930-1974, consisting of research notes, reports, correspondence, publicity, budget material and technical papers.
Unpublished finding aid available.
Partially restricted; consult archivist for further information.

B2 Babb, Albert L., b. 1925.
Papers, 1950-1969. - 11 boxes.
In University of Washington Archives (Seattle).
Professor of Chemical Engineering. - Correspondence, minutes, speeches and writings, conference and convention materials, notes, clippings, consulting and subject files, and other papers regarding Babb's activities as Professor of Nuclear and Chemical Engineering and Chair, Department of Nuclear Engineering, University of Washington, and member of various professional organizations.

B3 Babb, H. B.
Notebook, 1872-1875. - 190 p.
In University of Missouri-Columbia Library, Western Historical Manuscripts Collection and State Historical Society of Missouri Manuscripts.
Notebook kept from October 5, 1872 to January 18, 1875 on the cultivation of small fruits, soil, and chemistry. The back contains rolls of elementary school classes.
Received July 31, 1946.
Unpublished finding aid available.

B4 **Babbitt, Harold Eaton**, 1888-
Papers, 1922-56. 1 ft.
In University of Illinois Archives (Urbana)
Professor of sanitary engineering at the University of Illinois. Correspondence, speech notes, unpublished papers and addresses, abstracts for journals, textbooks by Babbitt, newspaper clippings, and photos, relating to sewage treatment, water purification, garbage disposal with sewage, diatomite water filters, flow of sludge, engineering in public health, and disposal of radioactive wastes in sewage.
Unpublished finding aid in the repository.
Information on literary rights available in the repository.
Acquired, 1966.

B5 Babbitt, James Aloysius, 1877-1969.
Papers, 1895-1969. - 3 manuscript boxes, and 39 volumes.
Hoover Institute Archives, Stanford University (Stanford, CA).
Mining and metallurgy. - Correspondence, lectures, reports, surveys, patents, newsclippings, sketches, and photographs. Materials pertain to economic, scientific, and technological developments in the mining and metallurgical industries of China, Japan and the Far East, with emphasis on nickel alloys.
Catalog card entry.

B6 Babcock, Stephen Moulton, 1843-1931.
Papers, 1814-1931. - 13 boxes.
In State Historical Society of Wisconsin (Madison).
Inventor of the Babcock milk test. - Personal correspondence; account books kept while a student at various schools; Cornell University classbooks; brief and irregular diary entries (1867-75); and memorandum books. Includes early family correspondence of Babcock's father Peleg B., concerning his agricultural and dairy pursuits; letters from Babcock's classmates at Tufts, Cornell, and the University of Gottingen in Germany; correspondence concerning chemical analyses and experimental dairy work; Babcock's correspondence with his wife; letters of friends and relatives; letters from organizations and individuals soliciting funds, and from faculty colleagues and fellow scientists. As of 1985, the collection also includes unprocessed additions.

B7 Babcock, Stephen Moulton, 1843-1931.
Biographical material, 1890-1948. - 22 items.
In Cornell University Libraries, Department of Manuscripts and University Archives (Ithaca, NY).
Chemist, New York State Experiment Station, Geneva; Instructor, Cornell University; Professor, University of Wisconsin; inventor of a test to determine butterfat content of milk. - Papers include typed copy of biographical sketch, pamphlet on Babcock's life, program, photograph, and clippings.

B8 **Babst, Earl D., 1870-1967.**
Papers, 1894-1967. – 23 boxes.
In University of Michigan, Bentley Historical Library, Michigan Historical Collections (Ann Arbor).
New York attorney and business executive. – Collection contains papers relating to the publication of <u>Michigan and the Cleveland Era</u>; correspondence concerning University of Michigan alumni affairs and his interest in the Michigan Historical Collections; legal materials and memorabilia concerning work as general counsel of the National Biscuit Company, and numerous scrapbooks relating to his role as president of the American Sugar Refining Company; a collection of pamphlets concerning the American Honest-Money League and the question of free silver in the election of 1896; and pamphlets of the Anti-Imperialist League.
Gift of Earl D. Babst, 1939.
Unpublished finding aid available.

B9 **Bache, Franklin, 1792-1864.**
Papers of Franklin Bache and family, 1818-60.
ca. 500 items.
In Historical Society of Pennsylvania collections.
In part, transcripts.
Physician, chemist, and educator. Copies of letters written by Bache (1833-61); historical memoranda, arranged for a lecture; and notes for "A chemistry," by Bache. Also, letters from Bache's brother-in-law, Albert Dabadie, and family letters containing comments on social events, financial matters, and political questions.

B10 **Bachmann, Werner Emmanuel, 1901-1951.**
Papers, 1924-1951. – 3 boxes.
In University of Michigan, Bentley Historical Library, Michigan Historical Collections (Ann Arbor).
Professor of Chemistry at the University of Michigan. – Includes correspondence; subject files; research notes and notebooks relating to cancer and penicillin research; teaching materials; and reprints of writings.
Unpublished finding aid available.

B11 **Bacon, Nathaniel Terry, 1857-1926.**
Papers, 1789-1954. 72 ft.
In Rhode Island College Library (Providence)
Engineer, of Rhode Island. Personal correspondence, financial papers, writings, printed material, scrapbooks, and photos (1876-1926); incomplete records (1885-1924) of several companies with which Bacon was associated, including correspondence, databooks, lab reports, and photos, of a soda alkali firm, Solvay Process Company, Syracuse, N.Y., and scattered personal papers of the following family members: Rowland Hazard (1763-1835), Joseph Peace Hazard (1807-1894), Thomas Rutherford Bacon (1850-1913), Caroline Hazard (1856-1945), Helen Hazard Bacon (1861-1925), and Leonard Bacon (1887-1954). Bacon's correspondents include industrialists Rowland Hazard and Rowland Gibson Hazard, and economist Irving Fisher.
Register published by the repository.
Gift of Professors Ronald and Martha Bacon Ballinger, 1967.

B12 **Badollet, M___ S___**
Papers, 1925-67. 5 folders.
In Vincennes University Junior College, Byron R. Lewis Historical Library (Ind.) (RHC 124)
Correspondence, speeches, and articles (1942-67) relating to asbestos and its role in everyday living, written by Badollet, former resident of Vincennes, Ind., and authority on asbestos; together with genealogy of the Badollet family.
Gift of Mr. Badollet.

B13 **Baekeland, Leo H., 1863-1944.**
Papers, 1881-1968. – 13.1 linear ft.
In National Museum of American History Archives Center (Washington, DC).
Chemist; inventor. – Collection includes student notebooks; private laboratory notebooks and journals; commercial laboratory notes; diaries; patents; technical papers; biographies; newspaper clippings; maps; graphs; blueprints; account books; batch books; formula books; order books; photographs; and correspondence to, from, and regarding Baekeland, 1887-1943.
Unpublished finding aid available.

B14 **Bahn, Gilbert Schuyler, 1922-**
Papers, 1946-73. 28 ft.
In University of Virgihia Library (Charlottesville) (10237)
Chemical engineer. Correspondence, memoranda, technical reports, and computer printouts, documenting Bahn's career with General Electric Company Thermal Power Systems Division, Schenectady, N.Y., and as thermodynamics engineer with Marquardt Corporation, Van Nuys, Calif., specializing in rapid combustion research; correspondence as editor and secretary of Combustion Institute Western States Section; and biography and resumés.
Unpublished finding aid in the repository.

B15 **Bailey, Alton Edward, 1907-1953.**
Papers, 1920-1953. – 1 cu. ft.
In Memphis State University Libraries (Memphis, TN).
Research chemist and author. – Collection contains correspondence, journal articles and other publications written by Bailey, specifications, diagrams, patents, reports, business and financial papers, chemical analyses, reprints, experimental data, and other material documenting Bailey's career as a research chemist who published and served as a consultant. Material pertaining to his associations with the Girdler Corporation in

Louisville, KY (1946-1950) and Humko, Inc. in Memphis (1950-1953) is present. There is also a small amount of memorabilia, containing information on Bailey's high school and college work, clippings, and other aspects of his personal life.
Gift of Dorothy Nicely Bailey, April 8, 1980.
Unpublished finding aid available.

B16 **Bailey, Jacob Whitman,** 1811-1857.
Papers, 1828-53. 57 items.
In U. S. Military Academy Library (West Point, N. Y.)
Professor of chemistry, geology, and mineralogy, at the U. S. Military Academy. Letters written to Bailey's mother and brother while a cadet at the academy, and while a member of various scientific societies; together with a diary kept while at West Point and on a southern botanical tour (1835). Includes a typewritten copy of a vol. entitled, "Selected letters of J. W. Bailey and Miscellaneous Papers Relating to the Bailey Family of West Point," compiled by Dr. Alfred G. Bailey.

B17 **Baltimore Chemical Works.**
Records, 1833-1907. 45 v.
In Maryland Historical Society Library (756, 756A)
Correspondence, journals (1846-68), ledgers (1833-72), receipt books (1851-63), notes and bills payable and receivable (1855-61), accounts open (1869-72), inventory books (1843-69), invoice book (1856-60), check stubs (1882-89), list of firms in Maryland, Virginia, Pennsylvania, and Tennessee, trading with Smith and Atkinson, scrapbook of newspaper clippings (1862-1907), will of Thomas U. Smith, account of Mary M. Smith's estate, and other records of a Baltimore firm engaged in the wholesale trade in drugs, paints, and oils, first run by Robert M. Smith and James E. Atkinson, then by Smith and Philip S. Chappell, and finally by Smith alone. Includes business records (1869-72) of R. M. Smith, collector of internal revenue.

B18 **Bancroft, George,** 1800-1891.
Papers, 1811-1901. (MS 62-3395)
—— —— Addition, 1851-1923. 310 items.
In Cornell University Libraries, Dept. of Manuscripts and University Archives (Ithaca, N. Y.) (2528, 2531)
In part, transcripts.
Historian and diplomat. Correspondence (1862-89) of Bancroft with his sons, John Chandler Bancroft and George Bancroft, Jr., and other relatives, concerning family matters, and with publishers, brokers, and others, concerning business and financial matters, and legal documents; papers of his sons and other family members including correspondence (1864-77) of John Chandler Bancroft and his wife, Louisa Denny Bancroft, correspondence (1881) between John Chandler Bancroft and George Bancroft, Jr., regarding the will of Abisha Learned of Oxford, Mass., letters (1894-95) to Wilder D. Bancroft from his fiancée, Katherine Bott, and her European travel diary (1880), 15 letters (1906) from British scientists to Sir William Ramsay pertaining to Wilder's suggestion that the Journal of Physical Chemistry be printed in England, and clippings (1923) relating to Wilder's experiments in rainmaking.
Permanent deposit by Warner B. Berry, 1965.

B19 Barclay family.
Papers, 1906-34. 8 ft.
In Pennsylvania Historical and Museum Commission collections (Harrisburg)
Correspondence, account books, bills, and invoices, relating to the family's business and financial interests, particularly Barclay Chemical Company, Williamsport, Pa., its plant at Lacquin, Pa., Lacquin Lumber Company, and Northwest Lumber Company, Seattle, Wash. Persons represented include W. L. Barclay and his sons, George S. and S. D. Barclay.
Unpublished finding aid in the repository.
Permanent deposit by Mrs. Stanley Barclay, Jersey Shore, Pa., 1973.

B20 Barkdull, Calvin H., 1875-1960.
Papers, 1895-1958. - 2 linear ft.
In University of Washington Archives (Seattle).
Mining engineer, mineralogist. - Correspondence, 1904-1958; financial records, 1909, 1917-1920; legal documents, 1909-1955; speeches and writings; notes; photographs; clippings; reports, 1895, 1909-1955; maps and other papers regarding Barkdull's career as mining engineer in Alaska, Washington, California and Mexico, and his activities as member of West Coast Mineralogical Association, Evergreen Rock Club and Boeing Mineralogical Society.
Accessioned 1966.
Unpublished finding aid available.

B21 Barkman family.
Papers, 1917-21. 30 items.
In University of Michigan, Bentley Historical Library, Michigan Historical Collections (Ann Arbor)
Business correspondence of Barkman Lumber Company, especially correspondence with Dow Chemical Company relating to the manufacture of bromine in the East Tawas, Mich., area; and family material, of an East Tawas, Mich., family.

B22 **Barnes, George E** 1898-
Papers, 1923-63. 25 ft.
In Case Western Reserve University Archives (Cleveland, Ohio)
Professor of hydraulic and sanitary engineering at Case Institute of Technology, consulting engineer, and expert on industrial waste and pollution problems, whose career included hydroelectric and irrigation projects in Argentina and water supply and irrigation programs in Nicaragua and Honduras. Correspondence, notes, drafts, research reports, reprints, blueprints, diagrams, and printed material

relating chiefly to Barnes' career and to projects with which he was concerned.
Inventory available in the repository.

B23 **Barron, Springall and Company,** *Dexter, Me.*
Records, 1862–76. 1 ft.
In Merrimack Valley Textile Museum (North Andover, Mass.)
Ledgers, cash books, and day book, of a company dealing in chemicals, drugs, music, and stationery.

B24 **Barrows, Horace Aurelius,** 1809–1851.
Papers, 1829–58. ca. 2 ft.
In Maine Historical Society Library (Portland)
Physician. Diaries, medical notes, church records (1814–79) of First Parish Church, Otisfield, Me., and papers relating to Barrows' making of proprietary medicines and his medical practice.
Information on literary rights available in the library.
Gift of the estate of George Barrows Turner, Portland, Me., 1964.

B25 **Bartell, Floyd Earl,** 1883–1961.
Papers, 1903–11. 200 items.
In University of Michigan, Bentley Historical Library, Michigan Historical Collections (Ann Arbor)
Professor of chemistry, University of Michigan. Correspondence and newspaper clippings concerning Bartell's professional activities.

B26 **Bartlett, Paul D.,** b. 1907.
Papers, 1934-1974. – 14 linear feet.
In Harvard University Archives (Cambridge, MA).
Professor of chemistry. – Collection contains general correspondence, 1934-1959: incoming and outgoing letters chiefly of professional interest, including letters of recommendation; general correspondence, ca. 1954-1974: letters of recommendation, research and research contracts, professional societies, Harvard University, and refereeing of journal articles; letters, 1938-1948, regarding award of Theodore William Richards Medal by Northeastern Section of American Chemical Society; correspondence, 1948-1950, with the Office of Naval Research; and correspondence and other papers, ca. 1965-1972, relating to the International Union of Pure and Applied Chemistry.
Gift of Professor Bartlett, 3/6/73 and 8/27/74.
Unpublished finding aid available.
Access restricted. Contact archivist for further information.

Barton, Benjamin Smith, 1766-1815. B27
Papers, 1789-1815. – 14 boxes and 32 vols.
In American Philosophical Society Library (Philadelphia, PA).
Physician, naturalist, Professor of Medicine at the University of Pennsylvania. – The collection reflects Barton's wide-ranging interests which included American Indians, botany, chemistry, materia medica, and rattlesnakes. Material dealing with or touching on chemistry appears in the following series: American Indian Materials (Indian dyes); Correspondence; Medical (materia medica); Miscellaneous (notes and writings on chemistry and material medica); and Volumes (notebooks on chemistry, mineralogy, and materia medica).
Unpublished finding aid available.

Bartow, Edward, 1870-1958. B28
Papers, 1870-1959. – 3.5 cu. ft.
In University of Illinois Archives (Urbana).
Professor of Chemistry; Director of State Water Survey, 1905-1917. – Subjects covered include the American Chemical Society; Bartow's autobiography; the Columbian Exposition; crystallography; history, geography, science and technology in France; University of Gottingen; history of chemistry; International Union of Pure and Applied Chemistry; Kansas; agriculture and natural resources of Mexico; mineralogy; organic chemistry; sanitary engineering; sanitation; sewage treatment; University of Iowa; University of Kansas; University of London; water quality; water supply; Illinois water survey; Williams College; World War I, agricultural and industrial production and construction; World War I, hospitals, medicine, and mortality; and World War I, supply, support and transportation.
Received in 1974.
Unpublished finding aid available.

Bartow, Virginia, 1896-1980. B29
Papers, 1908-1980. – 1.3 cu. ft.
In University of Illinois Archives (Urbana).
Professor of Chemistry. – Subjects covered by this collection include the American Chemical Society; chemical education; the Chemistry Department at the University of Illinois; history of chemistry; history of science; Iota Sigma Pi; Vassar College; and World War I, agricultural and industrial production and construction.
Received in 1980.
Unpublished finding aid available.

B30 Baruch, Bernard Mannes, 1870–1965.
   Papers, 1905–65. ca. 221 ft. and 521 v.
   In Princeton University Library (N. J.)
   Businessman and public official. Correspondence, articles, speeches, statements, clippings, memorabilia, and other papers, relating chiefly to Baruch's intervals of Government service as chairman of the U. S. War Industries Board (1918), economic advisor for the American Commission to Negotiate Peace at the Versailles Conference (1919), chairman of the Fact-Finding Commission on Synthetic Rubber (1942), U. S. representative to the United Nations Atomic Energy Commission (1946), and in the Council of National Defense (1917–18), the National Industrial Conference (1919), and the Office of War Mobilization as an advisor to James F. Byrnes and member of the War and Postwar Adjustment Unit (1943–44).
   Unpublished guide in the library.
   Gift of Mr. Baruch, 1963.

B31 Basore, Annie Terrell, 1895-1982.
   Papers, 1852-1983. - Ca. 10.5 ft.
   In Auburn University Archives (Auburn, Alabama).
   Teacher and genealogist. - Includes papers of Basore's husband, Cleburne Ammen Basore (1893-1974), Auburn University Assistant Professor of Chemistry and Chemical Engineering, Assistant Director of Engineering Experiment Station from 1937 to 1960, and head of the Chemical Engineering Department from 1939 to 1963. Cleburne Basore's papers consist of 32 file folders arranged alphabetically by subject, including correspondence, speech notes, certificates and awards, research notes, publications and personal papers. The correspondence files include 3 file folders relating to consulting work.
   Donated by Mrs. Emily Hixon Gunter, niece of Annie Basore, March 7, 1984.
   Unpublished finding aid available.

B32 Bauer, Frederick Charles, 1886–
   Papers, 1909–60. 1 ft.
   In University of Illinois, University Archives.
   Professor of agriculture at the University of Illinois. Correspondence, notes, research papers, addresses, clippings, publications, photos. and workpapers relating to agronomy, teaching, soil productivity, the Morrow plats, experiment fields, fertilizers, the McKinley Foundation and Bauer's career, including his positions in charge of soils extension, chief of experimental fields, and professor of soil fertility.

B33 Baugh & Sons Company.
   Records of Baugh & Sons Company and Baugh Chemical Company, ca. 1880-1930. 4 v.
   In Eleutherian Mills Historical Library (Greenville, Del.) (663)
   Correspondence, advertising material, estimates, and a daybook of orders and shipments, of fertilizer companies of Philadelphia and Baltimore.

Baugh & Sons Company, Philadelphia, Pennsylvania B34
   Papers, 1905-1932. - 101 items.
   In Duke University Library (Durham, NC).
   Producer and distributor of fertilizer and agricultural chemicals; founded in 1855; for many years, the leading producer in its field. Numerous branches opened around the turn of the century. Baugh Chemical Co. acquired in 1963 by Kerr-McGee Oil Industries, Inc. - Collection consists primarily of photographs and of mats for advertisements or illustrations. Also several printed advertisements. Most of the pictures relate to the company and to the use of its products. Scenes include offices, factories, docks, machinery, products, displays of products, personnel, dealers, and farms. Also numerous photographs of crops, farmers, workers, buildings, and equipment.
   Inventory of photographs is available.

Baugh Chemical Company. B35
   Records, 1801–1916. ca. 50 items.
   In Maryland Historical Society Library (1199)
   Correspondence, land deeds, and an insurance policy of a Baltimore firm.
   Gift of Arthington Gilpin, Jr., and Arthington Gilpin III, 1944.

Baumgartner, Jacob, 1830-1916. B36
   Record books, 1846-1916. - 14 v.
   In State Historical Society of Wisconsin (Madison).
   Fennimore, Wisconsin farmer. - Dairies and reminiscences, accounts, weather notes, and financial records. Diary (1846-65) records Baumgartner's experiences as an apprentice and journeyman dyer in Germany, 1844-50, his travels in America, family records, and Wisconsin local history. With the exception of a few financial records, all material is in German.

Baxter, Gregory Paul, 1876–1953. B37
   Papers, 1897–1953. ca. 2 ft.
   In Harvard University Archives.
   Instructor and professor of chemistry at Harvard. Correspondence, professional papers, and lecture notes.
   Open to investigators only upon prior application to the repository.
   Information on literary rights available in repository.

Beck, Lewis C. B38
   Papers, 1841-52. - 1 folder.
   In Rutgers University Library, University Archives Collections (New Brunswick, NJ).
   Professor of Chemistry. - Letters and

reports relating mainly to procurement of equipment and supplies for the Department of Chemistry and Natural History and the expenditure of funds provided by the Alumni Association.
Unpublished finding aid available.

39 Beck, Lewis Caleb, 1798-1853.
Papers, 1834-1852. - 2 vols., 5 fascicles.
In Rutgers University Library, Special Collections Department (New Brunswick, NJ).
Chemist; professor. - A volume of addresses on temperance; addresses on the importance of natural science, delivered before the Philosophical Society of Rutgers College, July 4, 1834; miscellaneous lecture notes on mineralogy and other scientific subjects, April 1852; notebook with Biblical citations; lecture notes on chemistry, electricity, and other scientific topics.
Unpublished finding aid available.

40 Becker, George Ferdinand, 1847-1919.
Papers, 1814-1928.
15 ft. (ca. 9000 items)
In Library of Congress, Manuscript Division.
Geologist, mathematician, engineer, and physicist. Correspondence (including family letters dating from 1814), diaries, letter books, notebooks, notes and memoranda, charts, tables, reports, articles, memorabilia, landscape sketches, blueprints, maps, and miscellaneous printed matter. Chief interest centers in Becker's professional papers covering his long service as Geologist-in-Charge, U. S. Geological Survey, during which time he conducted investigations in Nevada, southern Alaska, South Africa, the Pacific slope, and the Philippines. Other papers concern his service as geophysicist of the Carnegie Institution (1898), as U. S. representative at the Radioactivity Congress in Brussels (1910), and as president of the Geological Society of America (1914). Correspondents include Andreas Arzruni, James F. Bell, Theodore E. Burton, William Crozier, Edward S. Dana, James D Dana, Samuel F. Emmons, Archibald Geikie, Arnold Hague, Eugene W. Hilgard, Edmund O. Hovey, Henry M. Howe, Louis Janin, Waldemar Lindgren, Charles W. Merrill, Simon Newcomb, Charles S. Peirce, Chester W. Purington, Theodore Roosevelt, Henry W. Turner, Charles D. Walcott, and Robert S. Woodward.
Indexed in part. Includes a list of Becker's notes and unpublished MSS.
Unpublished finding aid in the library. Also described in the Library's Quarterly journal of current acquisitions, v. 11, no. 3 (May 1954) p. 164.
Open to investigators under restrictions accepted by the library. Information on literary rights available in the library.
The first group of papers, given by Mrs. Becker to the U. S. Geological Survey, were transferred to the library in 1952; two subsequent groups were given directly to the library by Mrs. Becker in 1953-54.
1. U. S. Geological Survey. 2. Geology—Collected works.
3. Becker family.

41 Beeman Chemical Company, Cleveland, Ohio.
Records, 1891-99. 1 v.
In Western Reserve Historical Society collections (Cleveland)
Articles of incorporation, bylaws, minutes of meetings, treasurer's reports, and other documents, relating to a company which manufactured and sold pepsin, pepsin gum, and other confections.
Gift of Alvin Good, 1941.

Belcher family. B42
Papers, 1834-85.
ca. 50 items.
In Missouri Historical Society collections (St. Louis)
Letters of William H. and Charles Belcher of St. Louis to their brother Nathan in New London, Conn., and other family correspondence, concerning family affairs and genealogy, the Belcher Sugar Refinery in St. Louis, and the 1849 cholera epidemic and fire.
Gift, 1957.

Bellinger, Frederick. B43
Papers, 1926-1978. - ca. 3 cu. ft.
Addition to: Bellinger, Frederick. Papers, 1940-1970s (MS 79-57).
In Atlanta Historical Society collections (Ga.).
Chief, Material Sciences Division, Engineering Experiment Station, Georgia Institute of Technology (Atlanta). - Correspondence, articles and publications by Bellinger, programs, certificates, newsletters, genealogy charts, scrapbooks, and other family and personal papers.

Bellinger, Frederick. B44
Papers, 1940-70's. 2 ft.
In Atlanta Historical Society collections (Ga.)
Chief of Material Sciences Division, Engineering Experiment Station, Georgia Institute of Technology, Atlanta. Published and unpublished articles, essays, and speeches on chemical engineering, civil defense, atomic weapons, radioactive fallout, and national security.

Benedum and the oil industry: oral history collection, 1951. B45
33 items.
In Columbia University Libraries (New York City)
Transcripts of tape-recorded interviews with Michael Late Benedum (1869-1959) and other persons having special knowledge of leasing and financing, geology, oil and gas production, and legal and tax problems, reflecting the oil industry from 1890 to 1950 as shown in the development of the Benedum oil interests and of his associates, especially Joe Clifton Trees. Subjects discussed include wildcatting activities of Benedum and Trees, finding of new oil and gas sources in China, Colombia, Mexico, Philippines, Rumania, and the U. S., marketing, refining, storage, transportation, conservation and the Interstate Oil Compact Commission, and the development of oil companies, including Bentex Oil Corporation, Hiawatha Oil and Gas Company, Plymouth Oil Company, and Transcontinental Oil Company. Includes impressions of Woodrow Wilson at Princeton, John Archbold, John W. Davis, E. L. Doheny, Senator Joseph F. Guffey, the Slick Research Foundation, and the Texas General Land Office.
Described, with a complete list of persons interviewed, in The Oral History Collection of Columbia University (1964) p. 134. Part of this material was used in "The Great Wildcatter," by Sam Mallison (1953).

B46 Benson, Henry K.
Papers, ca. 1907-1951. - 1.5 linear ft.
In University of Washington Archives (Seattle).
Professor of Chemistry. - Correspondence, writings, student notes, subject files, and other papers pertaining to Benson's career as faculty member; Chairman, Chemistry and Chemical Engineering Departments, UW, 1919-1947; and his activities in professional organizations.
Accessioned 1977.

B47 Berent family.
Papers, 1772-1861. 300 items.
In Leo Baeck Institute collections (New York City)
Chiefly business correspondence (1814-61) of Berent & Company, a Jewish banking house founded in Potsdam, Ger., by Samuel Bacher-Berent and his brother, Lewin (Louis) Bacher-Berent, which in 1812-13 moved to Berlin and established a sugar refinery there; together with birth certificates, citizenship papers, and other family records (1772-1830). 37 letters are from the banking house of Rothschild in Frankfurt, London, Paris, and Vienna. In German.
Purchase, 1962.

B48 Bergmann, Max, 1886-1944.
Papers, ca. 1930-45. 15 boxes.
In American Philosophical Society Library (Philadelphia)
Biochemist. Letters, reports, addresses, and lectures, relating to biochemistry and other scientific topics, the Rockefeller Institute, refugee scientists, and professional associations. Correspondents include Lawrence W. Bass, George W. Beadle, Franz Boas, James McKeen Cattell, Jaques Cattell, Alfred E. Cohn, H. D. Dakin, René J. Dubos, Albert Einstein, Simon Flexner, Paul Gyorgy, Karl Landsteiner, Irving Langmuir, Otto Loewi, Duncan A. MacInnes, John H. Northrop, Winthrop J. V. Osterhout, William J. Robbins, Peyton Rous, Fred M. Uber, Harold C. Urey, Donald D. Van Slyke, Selman A. Waksman, and Warren Weaver.
Table of contents in the library.
Information on literary rights available in the library.
Gift of the Rockefeller Institute, 1964.

B49 Bergmann, Max, 1886-1944.
Papers, 1934-44. 1 ft.
In Rockefeller University Archives, Rockefeller Archive Center (Pocantico Hills, North Tarrytown, N.Y.)
Chemist and biochemist at Rockefeller Institute for Medical Research, New York, N.Y. Correspondence, relating chiefly to the immigration process and to administrative matters, and biographical documents, memoir and funeral address by Joseph S. Fruton, bibliography, reprints (1911-45) of Bergmann's works, and photos.
Unpublished inventory in the repository.
Information on literary rights available in the repository.
Gift of the Rockefeller University administration, 1944.

B50 Berry family.
Papers, 1791-1936.
45 ft. of microfilm.
In Pennsylvania Historical and Museum Commission collection.
Microfilm made from typewritten transcripts lent by Mrs. J. L. Lockhart of Pittsburgh.
Records and background material of the Berry and Anshutz families. Includes a history of early iron furnaces in western Pennsylvania.

B51 Beverley family.
Papers, 1654-1901. - 4,788 items.
In Virginia Historical Society (Richmond).
Section one hundred and seven (5 items), comprising the student notebooks of Robert Beverley (1769-1843) (while studying mathematics at Trinity College, Cambridge University), George Clark (while studying at Brandywine Academy, Carlisle, Pa., under John Ferguson Grier and at Dickinson College, Carlisle, Pa.), James Bradshaw Beverley (while studying chemistry at Dickinson College, Carlisle, Pa.), and Jane Bradshaw Beverley (while studying Latin at Misses Lymans School, Philadelphia, Pa.).

B52 **Bienfang, Ralph David,** 1905–
Papers, 1900-45.
ca. 3 ft.
In University of Oklahoma Library.
Pharmacist and university professor. Notes relating to army pharmacy, sketchbooks and notes pertaining to plant and animal drugs, along with other papers regarding the role of pharmacy in military history.

B53 Black, Joseph, 1728-1799.
Papers, ca. 1773-1777. - 12 v; also on microfilm.
In Stanford University, Lane Library (Stanford, CA).
Professor of chemistry, University of Edinburgh. - Two sets of students' notes: "The substance of a course of chemical lectures as delivered by Dr. Black"; and "Lectures on chemistry delivered in the University of Edinburgh."
Finding aids available.

B54 Black, Joseph, 1728-1799.
Lectures, 1773-74. - 2 v.; 21 cm.
In National Library of Medicine (Bethesda, MD).
Chemistry lectures, Edinburgh, 1773-74. Lecture 118 is dated May 2, 1774.

B55 Black, Joseph, 1728-1799.
Lectures, ca. 1785. - 1 v.
In Library of Congress, Manuscript Division.
Professor of Chemistry and Medicine, Edinburgh, Scotland. - Contemporary copies of lectures on chemistry by Professor Black, with the name of John Bacon on first page.
Purchase, 1935.

56 Blackburn, Sam.
Papers, 1937-45. 1/2 ft.
In Wayne State University, Walter P. Reuther Library, Archives of Labor and Urban Affairs (Detroit, Mich.)
Correspondence, transcripts, resolutions, and clippings, concerning the relationships between the United Auto Workers, Society of Designing Engineers, and Federation of Artists, Engineers, Chemists, and Technicians.

57 Blackman, Elmer Ellsworth, 1865(?)-1942.
Papers, 1897-1938. - Ca. 1500 items.
In Nebraska State Historical Society Collections (Lincoln).
Lincoln archaeologist. - Correspondence, legal and financial files, memorandum books, essays and reports, genealogical data on the Blackman family, and memorabilia; relating to Blackman's business activities in the Hydrozo Waterproofing Company at Kansas City, Missouri, (1911-1917), and at Lincoln, (1922-1924); his invention of a waterproof paint; the Nebraska State Board of Agriculture; Nebraska State Horticultural Society; and Quivera Historical Society, of which Blackman was a founder. Correspondents include Jeremiah Behm, Robert W. Furnas, William A. Gwyer, Charles A. Holmes, J. Sterling Morton, James H. Noteware, F. W. Samuelson, Philip K. Slaymaker, Moses Stocking, John Taffe, A. W. Touzalin, Albert Watkins, Oliver T. B. Williams, and S. P. Wisner.

58 Blake, William Phipps, 1826-1910.
Papers, 1850-1910.
5 ft. (275 items)
In Arizona Pioneers' Historical Society collections (Tucson, Ariz.)
Mineralogist and university professor. Notebooks containing records of Blake's worldwide scientific trips and pencil and ink sketches. Includes entries relating to Arizona, California, New Mexico, and Texas; and sketches of San Xavier del Bac Mission and Pilots Knob.
Unpublished calendar in the repository.
Gift, 1959.

59 Blatchford, Seward, & Griswold.
Papers, 1841-1910. - 6 linear feet.
In Massachusetts Institute of Technology, Institute Archives and Special Collections (Cambridge).
Law firm. - Firm developed from two partnerships, one started in New York City by Richard Blatchford in 1819 and another upstate at Auburn by William Henry Seward and Elijah Miller in 1823. BS&G assisted most of the prominent American inventors of the nineteenth century with patent application and patent extension proceedings. The collection includes handwritten copies of patent applications and letters patent, correspondence, notes, newspaper clippings, drawings, printed court records, court notices, bills of complaint, briefs, witness lists, depositions, subpoenas, affidavits, transcripts of testimony, and patent assignments. Includes material on the following patents: gutta percha, 1844-1860; rubber, 1846-66, involving Charles Goodyear; oleomargarine, 1873-75; illuminating gas, 1881-84; gunpowder, 1889-90; artificial silk, 1896-98; macadamite, 1899; and sodium manufacture, 1873-75.
Gift of the New York City law firm Cravath, Swaine & Moore, May 1954 and March 1955.
Unpublished finding aid available.

Blomquist, Alfred Theodore, 1906-1977.   B60
Progress reports, 1943-1965. - 2 cu. ft.
In Cornell University Libraries, Department of Manuscripts and University Archives (Ithaca, NY).
Professor of Chemistry, Cornell University, 1941-1971; consultant to B. F. Goodrich Company for 25 years; author of Organic Chemistry, 1948; editor of a series of monographs on organic chemistry. - Collection contains 29 volumes of reports (carbon copies) written for B. F. Goodrich Company by Blomquist and others.

Bodansky, Meyer, 1896-1941.   B61
Papers, 1916-1949. - 1.75 linear ft.
In University of Texas Medical Branch, Moody Medical Library (Galveston).
Biochemist and pathologist. - The collection reflects Bodansky's research activities as a biochemist and pathologist. Includes correspondence with fellow scientists, reprints, manuscript drafts, photographs, and material relating to Bodansky's work with committees formed to assist in the settlement of physicians fleeing the Nazi occupation of Europe.
Donated to the History of Medicine and Archives Department of the Moody Medical Library by Bodansky's daughter, Mrs. William N. Roddy, on December 27, 1978.
Unpublished finding aid available.

Boettcher, Charles, 1852-1948.   B62
Papers, 1870-1969. ca. 10,000 items.
In State Historical Society of Colorado collections (Denver)
In part, photocopies.
Businessman, of Colorado. Correspondence, business records, newspaper reports, scrapbooks, photos, memorabilia, and taped interviews (with transcripts) of people who knew and worked with Boettcher and his son, Claude K. Boettcher (1875-1957), who carried on and expanded his father's business interests. Most of the collection consists of Boettcher's correspondence as an executive

in the companies he directed or controlled. Includes personal and family material and minute books of Capitol Life Insurance Company, Cement Securities Company, Great Western Sugar Company, Ideal Cement Company, and of subsidiaries owned by these cement companies. Business interests represented include banking, beet sugar refining, portland cement, and electrical, mining, and transportation enterprises.
    Calendar published by the repository in 1969.
    Access to family records restricted until 1995.
    Acquired from various sources.
    The repository also has 4 reels of microfilm from which photocopies of the correspondence were made.

B63 Bogert, Nicholas I Marsellus, *b.* 1842.
    Papers, 1852-1939. 10 items and 9 v.
    In Rutgers University Library (1447, 1471, 1521, and 1672)
    Clergyman of the Reformed Church in America in New York City and New Brunswick, N. J. Correspondence (1879-1939); texts of lectures attended (1865-67) at the New Brunswick Theological Seminary on church history and pastoral theology; texts of lectures attended (1862-63) at Rutgers College on Roman and Greek history, the English language, and chemistry; other lecture notes, a pastor's register, and other papers. Includes a journal (1852-93) of Bogert's father, William Bogert, concerned largely with religious experiences and meditations. Correspondents include Mark Hopkins and Theodore Roosevelt.
    Gift of Julia T. Bogert.

B64 Bolton, Henry Carrington, 1843-1903.
    Papers, 1873-1887. - 1 box.
    In Library of Congress, Manuscript Division.
    Chemist; co-founder of American Chemical Society. - Writings, notes, extracts, etc., on uranium compounds, symbols in alchemy, practical chemistry, and other topics.
    Gift of Mrs. Henrietta Irving Bolton, 1912, and transfer from the Library's Cataloging Division, 1931.

B65 Boltwood, Bertram Borden, 1870-1927.
    Papers, 1890-1932. 3 ft.
    In Yale University Library (New Haven, Conn.)
    Professor of radiochemistry at Yale University. Correspondence, laboratory notebooks (1892-1926), lectures on radioactivity, and scientific papers and reports, including 4 papers by Lord Ernest Rutherford and The Life of Radium written by Ellen Gleditsch. Correspondents include A. S. Eve, Hans Geiger, Ellen Gleditsch, Otto Hahn, Howard A. Kelly, J. C. McLennan, Stefan Meyer, Ernest Rutherford, Frederick Soddy, Robert W. Wood, and miners, prospectors, and chemical manufacturers.

    Unpublished register in the library.
    Information on literary rights available in the library.

Book of Secrets.   **B66**
    Collection of chemical, alchemical and medical recipes, ca. 1650-84. - 1 v (ca. 451 l.); 21 cm.
    In National Library of Medicine (Bethesda, MD).
    German "Rezeptbuch:" manuscript, in various hands, dealing with chemical analysis of gold and silver, processing of metals and minerals, production of numerous materials, and cures for plague, gout, renal stones, consumption, and other ailments.

Booth, James Curtis, 1810-1888.   **B67**
    Papers, 1833-52. - ca. 9 inches.
    In University of Pennsylvania Libraries, E.F. Smith Collection (Philadelphia).
    Considered the first professional consulting chemist in the U.S. - Analytical notebooks, numerous lecture manuscripts for chemistry courses and on topics in industrial chemistry, and a manuscript diary of his trip to Europe in 1834-35.
    Cataloged by item on cards in the repository.

Bouldin, Thomas Tyler, 1813-1891.   **B68**
    Student notebook, 1831-1834. - 136 p., 7 3/4 in. holograph.
    In Virginia Historical Society (Richmond).
    Bound volume. - Concerns the study of chemistry and mathematics at the University of Virginia.
    In the Bouldin family papers, 1737-1960.

Bowman, Henry.   **B69**
    Papers, 1846-1929.
    221 items and 11 v.
    In Cornell University Library, Collection of Regional History and University Archives (64, 566)
    Chiefly records relating to a tannery business in Elmira, N. Y. Includes correspondence (1855-70) between Bowman and leather dealers in Boston, Philadelphia, and Chicago; letters (1855) on the hide market in Michigan; accounts, a diary (1849) describing a trip to Chicago by way of the Great Lakes, and records of a fire department and of a mutual protective society. Francis Bowman is also represented.

Boyd, James, 1904–   **B70**
    Papers, 1941-59. 5 ft.
    In Harry S. Truman Library (Independence, Mo.)
    Educator, Army officer, Government official, and mining corporation executive. Correspondence, memoranda, speeches, reports, orders, news releases, and other papers relating to Boyd's activities as an Army officer during World War II, Director of the U. S. Bureau of Mines, Administrator of the Defense Minerals Administration, and Chairman of the National Science Foundation Advisory Commission on Mineral Research.
    Open to investigators under restrictions accepted by the library.
    Information available on literary rights in the library.
    Gift of Mr. Boyd, 1961.

B71 Boyd, Overton F.
Papers, 1915-1925; 1938; - 23 items; 1 mss. v.; 2 printed v.
In Louisiana State University, Department of Archives (Baton Rouge).
Professor; alumnus of LSU; son of LSU President Thomas D. Boyd; sugar technologist at the Imperial College of Tropical Agriculture in Trinidad, British West Indies. - Two broadsides (in Spanish), 1916, pertain to the military control of the Dominican Republic by the United States during the administration of President Wilson. A memorandum book, 1915-1916, contains formulas for sugar processes and miscellaneous notes. Printed volumes include "Proceedings of the Sixth Congress of the International Society of Sugar Cane Technologists," held at LSU, Baton Rouge, in 1938.

B72 Bradley, Joseph P., 1813-1892.
Papers, ca. 1836-1936.
ca. 5000 items.
In New Jersey Historical Society collections.
Associate Justice, U. S. Supreme Court. Correspondence, diary, journal of law cases, photos., speeches, legal documents, genealogical notes, printed matter, and other papers. Material on Bradley's private law practice includes documents about Bell Telephone, Charles Goodyear, Morse telegraph, and other patent cases; railroad litigation involving the Camden and Amboy and other railways; and the Delaware and Raritan Canal. Papers on Bradley's Supreme Court career include those dealing with Alaska seal, Guiteau appeal, Slaughterhouse, Legal tender, and other cases. Includes papers of his son, Charles Bradley, and letters of Chester A. Arthur, J. J. Astor, George Bancroft, George Bayard, James G. Blaine, Mary H. Bradley, Vincent L. Bradley, Simon Cameron, Alex. G. Cattell, Roscoe Conkling, George H. Cook, Jay Cooke, George T. Curtis, David Davis, A. L. Dennis, William M. Evarts, Stephen Field, Hamilton Fish, F. T. Frelinghuysen, Theo. Frelinghuysen, Melville W. Fuller, Horace Grey, William M. Harlan, Benjamin Harrison, Joseph Henry, George F. Hoar, William Hornblower, John P. Jackson, James G. King, Lucius Q. C. Lamar, Martha J. Lamb, S. P. Langley, Arthur MacArthur, A. W. Markley, Weir Mitchell, William Nelson, Cortland Parker, William Pennington, Benj. Perley Poore, Theo. G. Randolph, Charles H. Reed, Carl Schurz, John Sherman, B. Silliman, Ellen L. Stanton, R. H. Stevens, John P. Stockton, Robert Stockton, Robert F. Stockton, M. R. Waite, Marcus L. Ward, Stephen Wickes, and J. G. Wilson.
Indexed in MSS. index; list available.
Gift of Charles B. Bradley and Joseph G. Bradley, 1950.

B73 Brakeley, John H 1816-1897.
Correspondence, 1835-50. 6 folders.
In Rutgers University Library (1740)
Methodist clergyman, of Wilmington, Del., and Pennington, N. J. Correspondence relating to the collection and exchange of mineral specimens. Some of the letters were received by Brakeley while a student at Lafayette College.
Cataloged in the library.
Gift of Gretchen M. Smith.

B74 Brasch, Frederick E., 1875-1967.
Papers, 1745-1963. - 104 boxes.
In Stanford University Library, Department of Special Collections (Stanford, CA).
History of science bibliographer; chief of Scientific Collections at the Library of Congress. - Manuscripts written or collected by Brasch concerning the history of science, especially science in colonial America, and related subjects. Includes correspondence of T. J. J. See, Albert Einstein, and others.
Finding aid available.

B75 Bratton family.
Papers, 1779, 1859-1953. (MS 66-1354, 72-1261)
—— —— Addition 2, 1817-1960. 239 items.
In University of South Carolina, South Caroliniana Library (Columbia)
Personal and genealogical correspondence, estate and guardianship papers of the Bratton and Rainey families, medical accounts and prescriptions, recipes for paints, lists of policyholders of the Southern Life Insurance Company, and other papers of a York County, S. C., family. Includes correspondence relating to the efforts of John S. and J. Rufus Bratton to return from Canada, Civil War hospitals, and Sherman's march through Georgia.

B76 Bray, William Crowell, 1879-1946.
Papers, ca. 1890-1945. - 2 linear ft.
In University of California, Berkeley, Bancroft Library.
Chemist, Professor at UC Berkeley. - Correspondence: letters written to Bray and copies of letters by him (includes a letter from Bray to Professor Mel Gorman, May 19, 1944, cataloged separately); papers and articles; notes; reprints and other materials concerning his career as professor at UC Berkeley.
Finding aid available.

B77 Brays, William H.
Formula book, 1862-1864. - 1 item.
In State Historical Society of Wisconsin (Madison).
Brays' formula book, April 16, 1862-1864, containing numerous medical recipes.
Gift of Mrs. J. B. Noble, Waukesha, Wisconsin, 1951.
Unpublished description available.

B78 Brazel, David.
Recorded oral interview, 1980.
In University of Texas at Arlington, Library, Department of Special Collections.
Union organizer for the International Union of Operating Engineers and other unions in Texas. - Discusses organizing efforts, strikes, and other aspects of union activities with workers in oil refining, petrochemicals, rubber production and other areas related to industrial chemistry.

B79 Brewer, William Henry, 1828-1910.
Papers, 1830-1927. - 26 linear ft.
In Yale University Library, Department of Manuscripts and Archives (New Haven, CT).
Professor of Chemistry; taught chemistry at Ithaca Academy, 1850-1851, 1852; Ph.D. from Yale, 1852; taught at Ovid Academy, 1852-1855; Professor of Natural Science at Washington and Jefferson College, 1858-1860; First Assistant to California State Geological Survey, 1860-1864; Professor of Agriculture at Yale Sheffield Scientific School, 1864-1903; helped organize and run the Connecticut State Experiment Station; member of the U.S. Forestry Commission, and served on other government commissions in subsequent years, 1895-1897; Professor of Forestry at Yale, 1900-1903. - Collection contains correspondence, diaries, notebooks, lectures, articles, essays, genealogical materials, photographs, and other papers.
Gift of Arthur and Henry Brewer, 1931.
Unpublished finding aid available.

B80 Brewster, Carl Milton, 1881-1961.
Papers, 1902-1958. - Ca. 400 items.
In Washington State University Library (Pullman).
Chemist and educator. - Memoirs, correspondence, photographs, clippings, and publications of C. M. Brewster.
Gift of C. M. Brewster, 1960.

B81 Brinegar, Thomas P
Papers, 1912-35.
150 items.
In Arizona Pioneers' Historical Society collections (Tucson, Ariz.)
Journal (1918) of a trip made by Brinegar through southern Arizona, assay and analysis certificates (1912) for mining companies, MS. of Nuggets of precious metals (a listing of the gold nuggets found in Arizona), and a folder relating to the Alamos Cacharama group of mines, Alamos, Sonora.

B82 Brinton, Clement Starr, 1875-1963.
Papers relating to the iron industry, 1889-1957.
ca. 800 items.
In Eleutherian Mills Historical Library (Greenville, Del.) (692)
Chemist, of Haddonfield, N. J. Correspondence, articles, notes, printed matter, topographical maps, photos, glass lantern slides, mineral samples, and other research data, relating to iron works, primarily in Pennsylvania. Includes articles by Thomas D. Cope of the University of Pennsylvania and material on John Fritz, ironmaster at Bethlehem, Pa., the American Iron and Steel Institute, Bethlehem Steel Company, Cornwall Furnace (Lebanon Co., Pa.), Eagle Furnace (Pa.), Elizabeth Furnace (Lancaster Co., Pa.), Franklin Blast Furnace (N. J.), Hopewell Furnace (Berks Co., Pa.), Lukens Steel Company, Martha Furnace (N. J.), Mont Alto Furnace (Franklin Co., Pa.), Oxford Furnace (N. J.), Pine Grove Furnace (Pa.), Principio Iron Works (Md.), Ringwood Iron Mines (N. J.), Rock Grove Furnace and Forges (Centre Co., Pa.), Saugus Iron Works (Mass.), Speedwell Iron Works (N. J.), Trenton Iron Works (N. J.), and Alan Wood Iron and Steel Company (Pa.). Correspondents include R. D. Billinger, K. Braddock-Rogers, Stewart Huston, John C. Long, Wyndham Miles, Harvey Moore, G. R. Pullinger, W. P. Valentine, T. T. Watson, and Melvin J. Weig.

B83 Brookhaven National Laboratory.
Program proposals, 1947. - 49 p. photocopies.
In American Institute of Physics, Niels Bohr Library (New York, NY).
Proposed programs for the departments of physics, chemistry, biology, medical research and engineering and miscellaneous materials including budgets and map.
Finding aid available.

B84 Brookhaven National Laboratory. Office of the Director.
Records of Philip Morse and Leland Haworth, 1946-1961. - 20 reels microfilm, 16 mm. pos. and neg.
In American Institute of Physics, Niels Bohr Library (New York, NY).
Primarily records of Haworth's tenure. Files of outgoing letters from the Directors' correspondence, memoranda and committee reports relating to Associated Universities, Inc., the corporate holder of the BNL contract; Atomic Energy Commission policy and administrative materials; materials relating to the activities of internal committees including the Cosmotron Failure Investigating Committee; records of the scientific and non-scientific departments; records documenting the relations to outside agencies such as industries and universities; and materials relating to Leland Haworth's activities outside the laboratory, some of which pertain to national scientific planning.
Microfilmed in 1981 from original records at Brookhaven National Laboratory.
Finding aid available.

B85 Brotherton, William H
Papers of William H. and James Brotherton, 1803-1910.
137 items.
In Duke University Library.
Correspondence, indentures, and business papers of William H. and James Brotherton, of North Carolina and Tennessee. About 60 letters from William Brotherton relate his experiences as a private in the Confederate Army, describing camp life, field activities, desertions from the ranks, and prisoners. They were written from the vicinity of Richmond, Fredericksburg, and Orange Court House, Va. The remainder of the letters are from James Brotherton, who moved fom North Carolina to east Tennessee, and concern the distilling of whiskey and brandy, and Ku-Klux activities near Lynchburg, Tenn., in 1868.
Indexed in part.
Acquired 1942 and 1958.

Brown, H. Clifford.
    Chemistry laboratory notebook, 1920. - 111 p.
    In University of Missouri-Columbia Library, Western Historical Manuscripts Collection and State Historical Society of Missouri Manuscripts.
    Student. - Notebook describes experiments, lists equations, etc.
    Received July 31, 1946.
    Unpublished finding aid available.

Brown, Harry Fletcher, 1867-1944.
    Papers, 1893-1900. 22 v.
    In Eleutherian Mills Historical Library (Greenville, Del.) (708)
    Chemist. Correspondence, reports, memoranda, charts, and drawings, as chief chemist of the Chemical Dept., Torpedo Station, U. S. Naval Station, Newport, R. I.; together with 4 vols. entitled "Manufacture of Pyro-cellulose," containing reports signed by Brown and giving details of manufacture and chemical examination of pyro-cellulose for use in the manufacture of smokeless powder at the Naval Torpedo Station (1896-1900).

Brown, Herbert C., b. 1912.
    Papers and publications, 1935-1985. - Ca. 480 linear ft.
    In Purdue University, Department of Chemistry, (West Lafayette, IN).
    Nobel laureate in Chemistry, 1979. - Of its total volume, ca. 150-200 feet of the Herbert C. Brown Collection are manuscript materials. Included are research records; reports and summaries written by Brown's students and co-workers; original research notebooks; correspondence arranged topically and covering all aspects of Brown's career (Lectures, Major Lectures, Awards, Nobel Award, Consulting, etc.); reprints of Brown's 966 publications; Ph.D. and M.S. theses written by Brown's graduate students; and commemorative tributes.
    The remainder of the collection contains copies of Brown's five books; his personal Library; portraits, photographs, and other memorabilia.
    Arranged and labeled.

Brown, Matthew, 1766-1851.
    Letter book, 1804-18.
    1 v.
    In Cornell University Library, Collection of Regional History and University Archives (1420)
    Physician and merchant, of Rome, N. Y. Letters pertaining to the purchase and sale of foodstuffs, drugs, hardware, and other merchandise, as well as ashes for the manufacture of potash, the payment and collection of debts, tax payments and land sales, a project for establishing a glassmaking concern near Rome, N. Y., and related business matters. Correspondents include Seth Capron, Henry Franklin, Benjamin Huntington, Samuel Leggett, Robert Troup, John Westworth, and Benjamin Wright.

Brown, Rachel Fuller Brown, 1898-1980.   B90
    Notebooks. - 2.5 linear ft.
    In University of Chicago Library.
    Ph.D., Chemistry, 1933. Lecture notes from courses in bacteriology, organic chemistry, and inorganic chemistry (1920-26) taught by Sarah Branham, William Friedrich, Edwin O. Jordan, Ben Nicolet, Mary Rising, Hermann Schlesinger, and Julius Stieglitz, among other. Reading notes. Laboratory experiment notebooks.
    Unpublished guide available.

Brown, William Robinson, 1875-1955.   B91
    Papers, 1907-57. 6 ft.
    In Yale University Library (New Haven, Conn.)
    Forester and businessman, of Berlin, N. H. Business correspondence, articles, speeches, galley sheets of Brown's book, Our forest heritage: a history of forestry and recreation in New Hampshire (1958), photos, forestry studies, and Congressional bills. Includes papers relating to the Brown Company, Berlin, N. H. (manufacturer of paper products, chemicals, and other wood-using products), American Forestry Association, New Hampshire Disaster Emergency Council, New Hampshire Forestry Commission, North East Forest Research Council, Society for the Protection of New Hampshire Forests, Society of American Foresters, and the Timber Lands Mutual Fire Insurance Company.
    Unpublished register in the library.
    Information on literary rights available in the library.

Browne, Charles Albert, 1870-1947.   B92
    Papers, 1895-1945.
    17 ft.
    In Library of Congress, Manuscript Division.
    In part, photocopies and transcripts (typewritten)
    Chemist, food technologist, and writer on science. Correspondence, reports, addresses, notes, and MSS. of articles. Includes 8 volumes reflecting Browne's laboratory research, and his journals and records of scientific trips and observations in foreign countries. Part of the material relates to the New York Sugar Trade Laboratory and the International Society of Sugar Cane Technologists.
    Unpublished finding aid in the library.

Browne, Charles Albert, 1870-1947.   B93
    Papers, 1896-1947. 68 v. and 9 boxes.
    In New York Public Library.
    Chemist, food technologist, and writer on science. Journals (1926-45, 41 v.) describing Browne's travels in U.S., Europe, Near and Far East, and his social and business routine in Washington, D.C.; digests of chemical and agricultural data; published and unpublished writings; together with correspondence and photos.
    Unpublished finding aid in the repository.
    Gift, 1963.

Browne, E. Wayles, Jr., 1909-1980.   B94
    Papers, 1937-1979. - 43 ft.
    In University of Wyoming, American Heritage Center (Laramie).
    Economist. - Correspondence, manuscripts, photographs, printed material pertaining to Dr. Browne's career as a government economist and a consultant on oil and gas matters to public interest groups.

Includes material on natural and synthetic rubber, a subject on which Dr. Browne was an authority.
Gift of Dr. Barbara Moulton Browne, 1982.
Unpublished finding aid available.

B95 Brownell, Lloyd Earl, 1915-
Papers, ca. 1952-56. 1 ft.
In University of Michigan, Bentley Historical Library, Michigan Historical Collections (Ann Arbor)
Professor of chemical and nuclear engineering and director of Fission Product Laboratory, University of Michigan. Professional correspondence and reports, largely concerning Brownell's interest in the use of radiation for preserving food.

B96 Bruson, Herman A., 1901-1981.
Papers, ca.1920s-1980. - ca. 16 cu ft.
In University of Pennsylvania Libraries, E.F. Smith Collection (Philadelphia).
Pioneering industrial polymer chemist. - Correspondence, research notes, research reports, patent files, and company publications from Bruson's years at such companies as Goodyear, Rohm and Haas, Industrial Rayon, and Olin Mathieson Chemical. Most of the collection is from the post-1948 era.
Access restricted. Contact repository for further information.

B97 Buckeye Steel Castings Company, Columbus, Ohio.
Records, 1883-1977. ca. 8 ft.
In Ohio Historical Society collections (Columbus) (MSS 662, MIC 129)
In part, 1 reel of microfilm (negative).
Records documenting administration, finance, products, technology, sales, and labor, of a company begun as an iron foundry and now a division of Buckeye International, producing plastics and microtechnology. Early records include minutes, notebook of Samuel P. Bush, financial records, and employee statistics; later records document product development, patent activities, and labor negotiations.
Unpublished inventory in the repository.
Gift of Buckeye International, Inc., 1979.

B98 **Bulkeley, Gershom, 1636-1713.**
Papers, 1661-1721.
ca. 2 ft.
In Hartford Medical Society Library.
Clergyman and physician, of New London and Wethersfield, Conn. Notebooks containing medical and chemical notes, accounts and invoices, notes on Thomas Willis, sermons, and poetry by Bulkeley and his younger brother Peter. Many of the notes are in Greek and Latin or in a code or shorthand devised by Bulkeley.
Indexed in the library's card catalog.
Open to investigators under library restrictions.
Bequest of Gurdon W. Russell, 1909.

Bulletin of the Atomic Scientists. B99
Records, 1945-1952. - 16.5 linear ft.
In University of Chicago Library.
Established in 1945 as the journal of the Atomic Scientists of Chicago to provide a forum for the discussion of the responsibilities of scientists and the implications of nuclear energy. General subject files include correspondence and manuscripts concerning: the Atomic Energy Commission; Brookhaven, Oak Ridge, and Los Alamos laboratories; J. Robert Oppenheimer; Michael Polanyi; Charles E. Merriam; and others. Clippings files (1946-51), hearing transcripts (1949), Atomic Energy Commission press releases (1947-51), and miscellaneous material on civil defense programs.
Unpublished guide available.

Bunsen, Robert Wilhelm Eberhard, 1811-1899. B100
Papers, 1892. - 1 item and 1 v.
In Duke University Library, (Durham, North Carolina).
Chemist. - Photocopy of a holograph of R. W. E. Bunsen's laboratory notebook. In German script; identified as being from Heidelberg at that time.
Unpublished finding aid available.

Burden Iron Company, Troy, N.Y. B101
Records, 1818-1935. 33 boxes and 155 packages.
In New York State Library, Manuscripts and Special Collections (Albany)
Correspondence, costbooks, receipt books, ledgers, vouchers, check stubs, payroll and furnace records, inventory books, scrapbooks, and other records.
Unpublished finding aid in the repository.
Gift of Burden Iron Works, Troy, N.Y., 1940.

Burt, Le Van Merchant, 1875-1971. B102
Family papers, 1860-ca. 1924. ca. 2 ft. and 11 ft. of microfilm.
In Cornell University Libraries, Dept. of Manuscripts and University Archives (Ithaca, N.Y.) (2327)
Papers pertaining to Civil War service of Burt's father, Capt. Franklin Burt (1841-1924), of Corbettsville, N.Y., with the 89th Regt., New York Volunteers, and Burnside's Expedition to North Carolina, including correspondence from family, friends, and fellow officers, diaries (1861-63), and other papers, relating to conditions during the war, veterans' pensions, reminiscences, and a search for his brother, Lt. Albert C. Burt, lost in the war. Papers pertaining to Franklin Burt's business and political activities include business letters and documents, and clippings, relating to Calvert Tannery, Mannington, W. Va.,

owned with his father, Friend H. Burt, Burt Oil Company (W. Va. and Ky.), Frank Burt Mines and Milling Company (Colo.), the Masons of West Virginia, and the Prohibition Party in West Virginia. Other material includes biographical data on Friend H. Burt, temperance speeches and related matter (1885-89), and Le Van Merchant Burt's autobiography (dictated ca. 1919-ca. 1924).

Business history, administration, and education: oral history interviews, 1973-75. 19 items.
In University of California, Berkeley, Bancroft Library.
Transcripts of tape-recorded interviews relating to Cutter Laboratories (pharmaceutical firm), immunization research, Levi Strauss Company, certified public accounting, moving and storage business, mining geology, University of California governance and teaching, marketing theory and practice, and social consequences of business decisions. Persons interviewed include Harry J. Cooper, David L. Cutter, E. A. Cutter, Jr., Robert K. Cutter, John F. Forbes, Ernest T. Gregory, Ewald T. Grether, Milton Grunbaum, Peter Haas, Walter Haas, Sr., Walter Haas, Jr., James Edward Hammond, Anson Herrick, Daniel E. Koshland, Sr., C. W. Lagoria, Harry Lange, Harvey R. Lyon, Donald McLaughlin, and Howard M. Winegarden.
Catalog and card index in the university's Regional Oral History Office, which conducted the interviews, retaining the original tape recordings and depositing the transcripts in the Bancroft Library. Each transcript also is indexed.
Information on literary rights available in the repository.
Photo-offset duplicates have been made. The Library of the University of California, Los Angeles, also has a copy of the transcripts. Information on the location of other copies is available from the Regional History Office.

Butters, Charles, 1854-1933.
Papers, ca. 1894-1933. 1 carton.
In University of California, Berkeley, Bancroft Library.
Mining engineer and metallurgist. Correspondence, invitations, memorabilia, scrapbooks, and photos, relating to Butters' career, pertaining to his experiences in South Africa with the Reform Committee of the Transvaal and Dr. L. S. Jameson; newspaper clippings concerning the Nicaraguan insurgent, Sandino, and U. S. intervention; and to mining interests in California, Salvador, and Mexico; and to the University of California and Class of 1879. Part of the collection has been designated the Charles and Jessie Butters Memorial Collection.

Bancroft, Wilder Dwight, 1867-1953.
Papers, ca. 1895-1953. 15 ft.
In Cornell University Libraries (Ithaca, NY).
Physical chemist. — Bancroft taught physical chemistry at Cornell from 1895 to 1937, founded Journal of Physical Chemistry in 1896, and was president of the Electrochemical Society in 1905 and 1919, and of the American Chemical Society in 1910. Papers consist of research reports and materials, miscellaneous notebooks and catalogues, pamphlets on various topics, magazines, biographical material on Bancroft and others, miscellaneous personal material, lectures, lecture books and notebooks, material relating to the Journal of Physical Chemistry, and 15 boxes of correspondence.
Unpublished finding aid available.

Stephen M. Babcock (1843-1931) and his original butterfat tester. See entry B6. From the E. F. Smith Memorial Collection.

C1 **C. Wakefield and Company (Bloomington, Illinois).**
Records, 1875-1915. - 7 linear inches.
In Illinois State Historical Library (Springfield).
Patent medicine company established in 1846 by brothers Zera (d. 1848) and Cyrenius Wakefield (1815-1885). Manufactured and sold drugs and medicines wholesale and retail; specialized in cough syrups, liver pills, tonics, and blackberry balsam. - Business records include a journal, records of invoices, commissions and "worm capsule" sales, day book, cash book and ledger, and samples of promotional material advertising Wakefield's Blackberry Balsam. Also includes handwritten book of formulas for home remedies, tonic pills, salves, etc. Collection also contains records of the Bloomington and Rosita Mining Company, of which Cyrenius Wakefield was President. Records of this coal company include a payroll book and minute book.

C2 **Caldwell, George Chapman,** 1834-1907.
Papers, 1853-97. (MS 62-3892)
—— Addition, 1855-1903. 1 box and 6 v.
In Cornell University Library, Collection of Regional History and University Archives (14/8/583, 21/13/411)
Notebooks and lecture notes (1868-1903); together with Caldwell papers collected by Anna Elsbree, consisting of personal correspondence with friends and relatives and letters from families trying to locate soldiers hospitalized during the Civil War.

C3 **Caldwell, George Chapman,** 1834-1907.
Papers, 1853-97.
2 ft.
In Cornell University Library, Collection of Regional History and University Archives (14/18/271)
Professor of chemistry. Diaries, lecture notes, personal accounts, copies of Caldwell's Ph. D. thesis, photos., and pamphlets. Includes diaries of Mr. and Mrs. Caldwell and of George W. Chamberlain, Caldwell's son-in-law, a professor of architecture at Cornell University; and notes kept by Caldwell on Louis Agassiz's Lowell lectures and the Cambridge geology lectures.

C4 **California Wine Association.**
Records, 1894-1936. - 10 v.
In California Historical Society Library (San Francisco)
Worldwide wine concern founded in 1894, operating in California, with headquarters in San Francisco and New York; liquidated ca. 1941.- Minutes of executive committee meetings (1903-1924), minutes of directors' meetings (1894-1934), annual reports, and other financial records (1896-1921).

C5 **California wine industry: oral history interviews,** 1971-75. 21 items.
In University of California, Berkeley, Bancroft Library.
Transcripts of tape-recorded interviews relating to winemaking, grape growing, prohibition, immigrant contributions to California life, and business and agricultural history. Persons interviewed include Leon D. Adams, Maynard A. Amerine, Harry Baccigaluppi, Philo Biane, Sydney Block, Burke H. Critchfield, Andrew G. Frericks, Ernest Gallo, Maynard A. Joslyn, Horace O. Lanza, Louis M. Martini, Louis P. Martini, Otto E. Meyer, Harold P. Olmo, Antonio Perelli-Minetti, Louis A. Petri, Jefferson E. Peyser, Lucius Powers, Victor Repetto, Edmund A. Rossi, Arpaxat Setrakian, Robert Setrakian, Brother Timothy, Carl F. Wente, Ernest A. Wente, and Albert Winkler.
Catalog and card index in the university's Regional Oral History Office, which conducted the interviews, retaining the original tape recordings and depositing the transcripts in the Bancroft Library. Each transcript is also indexed.
Partially restricted.
Information on literary rights available in the repository.
Photo-offset duplicates have been made. The Library of the University of California, Los Angeles, also has a copy of the transcripts. Information on the location of other copies is available from the Regional Oral History Office.

C6 **Callahan, Patrick Henry,** 1866-1940.
Papers, 1908-39. 1 box.
In University of Notre Dame Archives.
Army officer, manager and president of the Louisville Varnish Company. Correspondence with prominent businessmen and business executives and other papers.

C7 **Camden, Johnson Newlon,** 1828-1908.
Papers, 1845-1908.
45 v. and ca. 97 boxes.
In West Virginia University Library (7)
U. S. Senator and businessman. Correspondence, business records, maps, and other papers relating to Camden's ventures in the purchase of land in the 1850's; his activities in oil production and refining, 1860-75, as president of the Camden Consolidated Oil Co. and the Baltimore United Oil Co.; and his development of railroads, including the Ohio River Railroad, the West Virginia and Pittsburgh Railroad, and the Monongahela River Railroad, 1879-92. Correspondents include John D. Alderson, Edward R. Bacon, Jonathan M. Bennett, John Brannon, Gideon D. Camden, John S. Carlile, William E. Chilton, E. W. Clark, Grover Cleveland, John J. Cornwell, John K. Cowen, Henry G. Davis, Stephen B. Elkins, James G. Fair, Charles J. Faulkner, Jr., H. M. Flagler, A. Brooks Fleming, Frederick T. Frelinghuysen, A. H. Garland, John W. Garnett, Nathan Goff, Arthur P. Gorman, C. W. Harkness, E. S. Harkness, N. W. Harkness, Abram S. Hewitt, Jed Hotchkiss, C. P. Huntington, H. M. Hutchins, M. E. Ingalls, James K. Jones, John E. Kenna, John T. McGraw, James M. Mason, 2d, Charles F. Mayer, Joseph S. Miller, Wesley Mollohan, O. H. Payne, Charles Pratt, John D. Rockefeller, William Rockefeller, H. H. Rogers, William Salamon, Samuel Spencer, George W. Thompson, Sr., William P. Thompson, W. G. Warden, Clarence W. Watson, Jere Wheelwright, William W. Whitney, William L. Wilson, and Henry A. Wise.
Gift of the trustees of Mr. Camden's estate, 1931-34.

C8 **Campbell, William,** 1876-1936.
Papers, ca. 1900-1925. - 4 boxes.
In Columbia University, Rare Book and Manuscripts Library (New York City).
Geologist and metallurgist; Columbia faculty member. - Most of the papers are of a personal nature, including ca. 100 items of incoming correspondence, Campbell's World War I naval records, financial

records and memorabilia. A few miscellaneous letters related to his academic career; also some notes and daily calendar memoranda. Included is a holography laboratory notebook of the British metallurgist, Saville Shaw, with typescript transcripts of contents along with notes by Campbell.
Transferred from Columbiana, 1965.
Unpublished finding aid available.

Candee, Leverett, 1795-1863.
Papers, 1843-64. ca. 1 ft. (150 items)
In New Haven Colony Historical Society collections (Conn.)
President of L. Candee & Company, New Haven, Conn., manufacturers of rubber shoes and boots. Business correspondence, articles of incorporation, real estate transactions, records of bad debts and promissory notes, advertising cards, and other papers, chiefly relating to the company.
Gift.

Carded Woolens Manufacturers Association.
Records, 1909-28. 140 items and 4 v.
In Cornell University Libraries, Dept. of Manuscripts and University Archives (Ithaca, N.Y.) (2412)
Correspondence, pamphlets, and other publications, and scrapbooks, collected by Edward Moir as president of the association, chiefly relating to the organization's fight for a tariff on wool based on value of the goods, and to manufacture of dyes and duty on dyestuffs. Includes articles of association and bylaws (1864-1908) of the National Association of Wool Manufacturers, who were opposed to such a tariff. Correspondents and persons represented include Samuel Dale, F. R. Marshall, and Herman A. Metz.
Gift, 1964-65.

Cardon (A.) and Company.
Records, 1815-30. ca. 1560 items.
In Eleutherian Mills Historical Library Greenville, Del.) (500, 501)
Correspondence, daybooks, ledgers, accounts, cashbooks, delivery books, bankbooks and check stubs, bills, receipts, orders promissory notes agreements, bank accounts, notice of dissolution, and other records of a tannery on the Brandywine near Wilmington, Del. Persons represented include James Antoine Biderman (1790-1865), Alexandre Cardon de Sandrans (1787-ca. 1837), and Joseph Charles Dalmas (1777-1859).
In part, described in A Guide to the Manuscripts in the Eleutherian Mills Historical Library, by John B. Riggs (1970) p. 716-717.
Gift of the E. I. du Pont de Nemours and Company, 1971.

Carpenter, Walter Samuel, 1888- C12
Business records, 1926-48. ca. 13,570 items.
In Eleutherian Mills Historical Library (Greenville, Del.) (542)
Business executive, of Wilmington, Del. Papers relating to Carpenter's career with E. I. du Pont de Nemours & Company as chairman of the Finance Committee and as president.
Access restricted until Mr. Carpenter's death.

Carson family. C13
Papers, 1861-1895. - 7 ft.
Forms part of: Arnoldus Vander Horst V family papers, 1682-1944 (12-194/276)
In South Carolina Historical Society collections (Charleston)
Correspondence, European travel diaries, instructions by William F. Miller on painting and photographic techniques, and other papers, of Caroline Petigru Carson, artist, of Charleston, S.C., and wife of W.A. Carson (d. 1856), of Dean Hall plantation, Berkeley County, S.C., including ca. 200 letters written to her father, James Louis Petigru, and her sons, William A. and James Petigru Carson, from New York City (1861-1872), Italy, and elsewhere in Europe (1872-1893). Correspondents include Boston attorney Edward Everett, Countess Gianotti of Rome, Gen. J.J. Pettigrew in Petersburg, Va., during the Civil War, Eugene Schuyler of the U.S. legation in Budapest, New York City attorney Clarence A. Seward, and George W. Wurts, U.S. legation at St. Petersburg. Papers of James Petigru Carson, inventor, engineer, and mineralogist, include journal (1868) of U.S. Geological Expedition west of Cheyenne City, Wyo.; journal (1870) of the Darien Expedition for Panama Canal surveys; correspondence relating to blast furnace designed for Spang, Chalfant & Company, Pittsburgh, Pa., and to the New Croton Aqueduct (N.Y.); engineering notebooks (1881-1883); plans for rice fields and trunks at Dean Hall plantation; material on experiments with explosives and a process for clarifying sugar; research material for his Life, Letters and Speeches of James Louis Petigru; and correspondence of James Louis Petigru and William Drayton and William Elliott, and Petigru's daughters, Caroline Carson, Susan Petigru King Bowen, and Mrs. Jane North, proposals for Petigru's tombstone inscription, list of books in his library, and letters of condolence at his death.

Carter, Herbert Edmund, b. 1910. C14
Papers, 1935-1970. - 1.6 ft.
In University of Illinois Archives (Urbana).
Professor and head of the Department of Chemistry, University of Illinois. - Correspondence notes, minutes, reports, reprints, travel vouchers, and brochures. Includes material relating to carcino-

genesis, food additives, and Carter's services as a member of the Faculty Advisory Committee to the Illinois Board of Higher Education and of the National Research Council's Food Protection Committee.
Information on literary rights available in the repository.
Acquired, 1965, 1968.
Unpublished finding aid available.

C15 Carter & Scattergood, Philadelphia, Pa.
Records, 1834-1910. 53 items.
In Eleutherian Mills Historical Library (Greenville, Del.) (376)
Correspondence, daybooks, ledgers, laboratory books containing records of wages, materials, apparatus, and processes, production tables, bills, receipts, and other papers, of a chemical manufacturing company. Includes a memorandum of the history of the firm by John Elliott and correspondence among Samuel Allison, Jr., John Carter, George M. Haverstick, Joseph Scattergood, and certain Virginia firms.
Described in A Guide to the Manuscripts in the Eleutherian Mills Historical Library, by John B. Riggs (1970) p.717.

C16 Case Institute of Technology. Department of Chemistry and Chemical Engineering.
Records, 1891-1965. - Ca. 1 linear ft.
In Case Western Reserve University Archives (Cleveland, OH).
Laboratory and lecture notes, correspondence, brochures, reports, department history, and ledgers; relating to curriculum, research, students, faculty, and equipment.

C17 Case Western Reserve University. Department of Chemistry.
Records, 1962-1974. - Ca. 2 linear ft.
In Case Western Reserve University Archives (Cleveland, OH).
Correspondence, newsletters, brochures, department history, and annual reports; relating to grants, equipment, financial aid, and faculty.

C18 **Catlin, Amos Parmalee,** *b.* 1823.
Business records, 1849-71.
ca. 2600 items.
In Henry E. Huntington Library (San Marino, Calif.)
Correspondence and business papers of Catlin, who came to California in 1849 and engaged in mining operations at Mormon Island, later forming the Natoma Water and Mining Co. and the American River Water and Mining Co. Includes papers of Palmer and Day, assayers, who were business associates of Catlin.
Purchased from Frank A. Guerney, 1947.
Inventory of the papers in the library.

Celluloid Company. C19
Records, 1890-1893. - 1 item.
In National Museum of American History Archives Center (Washington, DC).
A one volume record of experiments conducted by John H. Stevens and Frank C. Axtell for The Celluloid Company, 1890-1893. Most of the experiments deal with the development, use, and manufacture of pyroxyline solvents. The entries were apparently transcribed from other records.

Celluloid Corporation. C20
Records, 1892-1935. - 0.85 linear ft.
In National Museum of American History Archives Center (Washington, DC).
Celluloid company; originally called Celluloid Manufacturing Company. Founded in 1872. Products included plates for artificial teeth; lacquer; nonflammable "safety film"; and Lumarith, a thermoplastic. - Records include trade catalogues; price lists; notebooks; promotional literature; patents; a salesman's kit, including samples; photographs and prints of plant buildings, personnel, celluloid molds and by-products.
Unpublished finding aid available.

Century of Progress International Exposition. New York World's Fair Collection, 1931-1940. C21
42 linear feet.
In Yale University Library, Manuscripts and Archives Division (New Haven, CT).
Printed matter and publicity material related to the Century of Progress International Exposition in Chicago (1933-1934) and the New York World's Fair (1939-1940), including miscellaneous administrative records of the New York World's Fair, e.g., the annual report for 1938, organizational charts, etc.
Items brought together by Yale librarian Sherman Kent. Much of the New York World's Fair administrative material was contributed by its director of research, Frank Monaghan, also an Assistant Professor of History at Yale.
Finding aid available.

Chamot, Emile Monnin, 1868-1950; Fred Hoffman Rhodes, b. 1889. C22
"The Development of the Department of Chemistry and of the School of Chemical Engineering at Cornell," 1868-1957. - 44 items.
In Cornell University Archives (Ithaca, NY).
The final copy (1 vol. typescript) of

this history, begun by Professor Chamot and completed by Professor Rhodes, also contains Professor Rhodes' personal reminiscences of chemistry professors Wilder D. Bancroft, Thomas R. Briggs, Arthur W. Brown, George W. Cavanaugh, Emile M. Chamot, Louis M. Dennis, and William R. Orndorff; also, typescript drafts of this history and five pamphlets concerning chemistry and chemical engineering department history.
Restricted.

C23 Chandler, Charles Frederick, 1836-1925.
Papers, ca. 1864-1915. - 228 boxes.
In Columbia University, Rare Books and Manuscripts Library (New York City).
Industrial and public health chemist; professor. - Chandler was one of the most notable American chemists of the period. His interests were very wide, ranging from mineralogy and water to dyes, gas, sugar, electricity, photography, rubber, cement, aluminum, electrometallurgy, food, soda, soap, ceramics, pharmacy, tanning, and sewage. He had an active career as a consulting chemist; was a founder of the American Chemical Society; and was a member of many clubs and public service agencies in New York City. The collection is an extensive information file compiled by Chandler while at Columbia for his personal use. It contains clippings, brochures, pamphlets, charts, graphs, memoranda, and notations on chemistry and other scientific subjects.
Gift, 1937.
Unpublished finding aid available.

C24 Chapman, M____ J____
Papers, 1895-1922. 1 ft. (ca. 1200 items)
In Washington State University Library (Pullman)
Securities and real estate agent, of Pullman, Wash. Correspondence, documents, and printed material, relating to the Realty Company, Boston, Mass.; Athabaska Oil and Asphalt Company, Spokane, Wash.; Chacamax Land Development Company, Seattle, Wash; and the Chester Harvey Company, Spokane, Wash., representing the Texas Wonder Pools Oil Company. Includes chemistry notes (1920) of Janet Chapman.
Unpublished container list in the library.
Information on literary rights available in the library.

C25 **Chase, Warren Tinker**, 1860-1940.
Papers, 1880-1916. ca. 1 ft.
In Nebraska State Historical Society collections.
Physician and druggist, of Loup City, Neb. Customer accounts, druggist account books, and pharmacy notebooks.
Gift of the estate of Dr. Chase's daughter, Mrs. R. E. Dale.

Chedsey, William R., b. 1887. C26
Papers, 1913-1960. - 6 ft.
In University of Illinois Archives (Urbana).
Professor of Mining Engineering at the University of Illinois. - Correspondence (1934-1943), unpublished papers and biographical data (ca. 1942-60), mineral land and mine surveying notes (1913-37), "Coal preparation" extension course materials (1959-1960), "Anti-submarine tactical training course" materials (1943), mining publications (1931-1958), stock-holders quarterly and annual reports, and personal financial records (1940-1960). Subjects include mining, metallurgy, and petroleum engineering.
Information on literary rights available in the repository.
Acquired, 1967.
Unpublished finding aid available.

Chemurgy Project, 1941-1953. C27
Records, 1941-53. - .5 cu. ft.
In University of Nebraska - Lincoln Archives.
University research project commissioned by the state legislature, aimed at developing industrial raw materials from agricultural products and by-products. - Correspondence, reports, and minutes of meetings located in three separate record groups (2/11/5; 2/14/2; 8/1/2).
Inventories available at the repository.

Chester-Kent, Inc. St. Paul, Minnesota. C28
Papers, 1889-1971. - 39 feet.
In Minnesota Historical Society (St. Paul).
Manufacturer of Adlerika laxative and other health aids. - Correspondence, minutes of directors' meetings, sales and advertising files, financial records, pamphlets, clippings, and photographs relating to the company's business in the United States and Canada, fair trade laws, and issues of Federal regulation of the drug industry. Collection also contains a small quantity of the personal papers of the Carl Weschcke family, owners of the company.
Gift of Carl Weschcke in several lots.
Unpublished finding aids available.

Chicora Mining and Manufacturing Company, C29
Charleston, South Carolina.
Records, 1870?. 1 volume.
In Duke University Library, Special Collections Department (Durham, NC).
Minutes of the organization of the corporation and subsequently of the board of directors. The firm was organized to mine phosphates, earths, marls, rocks, and minerals and to manufacture chemicals, acids, and fertilizers.

C30 **Choate family.**
Papers, 1680–1911. ca. 5 ft.
In Essex Institute collections (Salem, Mass.)
Diary (1905) of Anna N. Choate (b. 1870), of Salem, Mass.; estate papers, assault and battery case papers, and other legal papers of David Choate (1796–1872), teacher and justice of the peace, of Essex, Mass.; account book (1789–1828) of George Choate, farmer, of Essex, Mass.; diary-account books (1867–87) of George Francis Choate (1822–1888), lawyer, of Salem, Mass.; correspondence with chemical and mining firms, other correspondence, plans, scientific reports, research reports, papers on metallurgical processes and patents, business reports, and family papers of Parker Cogswell Choate (b. 1862), mining engineer, of Salem, Mass., Ogden, Utah, and Portland, Or.; account books (1771–1811) of Thomas Choate (1751–1830), farmer and blacksmith, of Ipswich, Mass.; and correspondence, deeds, legal papers, estate papers, bills, school compositions, and other papers relating to family property in Salem, Mass., and to the family of George F. and Abby P. Choate.
Unpublished finding aid in the repository.

C31 <u>Christman, Adam Arthur, b. 1895.</u>
Papers, 1977–1982. – Ca. 0.5 ft.
In University of Michigan, Bentley Historical Library, Michigan Historical Collections (Ann Arbor).
Physiological chemist; faculty member of the University of Michigan Medical School, Department of Biological Chemistry. – Collection contains a paper entitled "The Department of Biological Chemistry, 1922-1955," containing a yearly summary of departmental activities and staff changes, lists of masters students, and sketches of all doctoral students including their present activities; correspondence containing biographical information on doctoral graduates of the department; and Christman's memoirs, describing life in Ann Arbor and at the university from his arrival in 1922.

C32 <u>Chymia.</u>
Papers, 1946–1951. – Ca. 6 linear ft.
In University of Pennsylvania Libraries, E. F. Smith Memorial Collection (Philadelphia).
Papers document the establishment and publication of <u>Chymia</u>, an annual volume of studies in the history of chemistry. The <u>Chymia</u> project was inspired by the resources of the E. F. Smith Memorial Collection, an internationally known resource for the study of the history of chemistry, chemical engineering, and the chemical industry, built around the personal library of Edgar Fahs Smith (1854-1928), University of Pennsylvania professor and administrator. The ACS Division of History of Chemistry co-sponsored the <u>Chymia</u> project.
The collection is made up of correspondence, photographs, page proofs, and reviews of various <u>Chymia</u> issues.

C33 **Claiborne, Herbert A**
Papers, 1935–57.
ca. 1000 items.
In Virginia Historical Society collections.
Correspondence, notes, architectural drawings, MSS. of Claiborne's works on colonial architecture and construction, other papers, and photos. Includes MS. and proof of Comments on Virginia brickwork before 1900 (1957), notes and MS. of Some paint colors from four eighteenth century Virginia houses (1948), correspondence and notes concerning Some colonial Virginia paint colors (19151), material relating to the restoration of Gunston Hall, Fairfax Co., Va., and speeches on Virginia and South Carolina architecture.
Gift of Mrs. Claiborne, 1957.

C34 <u>Clark, George Lindenberg, 1892-1969.</u>
Papers, 1914-1968. – 3.6 ft.
In University of Illinois Archives (Urbana).
Professor of Chemistry, University of Illinois. – Correspondence, mss. of writings, biographical material, research papers, and reprints, relating to the editing of encyclopedia articles on microscopy, spectroscopy, and X-rays; Clark's consultant work for the Delco-Remy Division of the General Motors Corporation, General Electric Company, Ohio Oil Company (later Marathon Oil Company), Parker Pen Company, Jos. Schlitz Brewing Company, and the U. S. Army Chemical Corps, Ordnance Corps, and Signal Corps; administration of the University of Illinois and its Analytical Chemistry Division; doctoral candidates and teaching; professional activities; lectures and symposia; crystallography, spectroscopy and X-ray diffraction research. Includes a tape-recorded interview (1956) with Clark, discussing aircraft metals and rubber research, courses, laboratories, and X-ray studies.
Information on literary rights available in the repository.
Acquired, 1969-1970.
Unpublished finding aid available.

C35 <u>Clark, John Dustin, b. 1882.</u>
Papers, 1902-1965. – 7 ft. (ca. 5000 items)
In University of New Mexico Library (Albuquerque).
Professor of Chemistry at the University of New Mexico. – Official correspondence, classroom notes and materials, and personal files.
Information on literary rights available in the library.
Gift of Mr. Clark, 1969.

C36 <u>Clark, William Mansfield, 1884-1964.</u>
Papers, ca. 1903-64. – Ca. 7,000 items (6 ft.), 24 notebooks.
In American Philosophical Society Library (Philadelphia, PA).
Biochemist. – A variety of material relating to his research, mainly at Johns

Hopkins University, and to his life. He did his primary work on the oxidation-reduction potentials of organic systems, and there are drafts of manuscripts and correspondence, especially with James B. Conant and Barnett Cohen, relating to this topic. Of particular note are Clark's Cutter Lectures at Harvard in 1930.

His participation in professional organizations is reflected in correspondence with the Society of American Bacteriologists as well as with the Journal of Biological Chemistry, the latter as seen in the Stanley Benedict and Rudolph Anderson letters. An important series is with Eric G. Ball for 1937-38, when Ball worked in Otto Warburg's lab in Berlin. There is material relating to Clark's lectures and papers, and there are 4 student notebooks (physical chemistry, optics thermodynamics/physics), and 23 notebooks, 1941-53. Significant correspondence is with: R. Keith Cannon, Walter B. Cannon, Jacques Loeb, Joseph Needham, F. Lee Rodkey, Donald D. van Slyke, Hubert B. Vickery.

Unpublished finding aid available.

C37 Clark University. Department of Chemistry.
Records, 1946-79. - 11 linear ft.
In Clark University Archives (Worcester, MA).
The records from 1946 to 1979 represent departmental correspondence and grants. In addition to these files there is one box of lecture materials for a course in the history of chemistry taught by Professor Benjamin Merigold from ca. 1910 to 1943, with one lecture by Professor Charles Kraus, later of Brown University. Also four informal histories of the department prepared at various dates early in the century by Merigold and Professor Jesse L. Bullock.

C38 Clarke, Frank Wigglesworth, 1847-1931.
Papers, 1873-1921. - 0.33 linear ft.
In Smithsonian Institution Archives (Washington, DC).
Geological chemist and professor. - Clarke collaborated with the Smithsonian Institution on atomic weight research. He also served as Chief Chemist of the U.S. Geological Survey until 1925. Papers consist chiefly of incoming correspondence from other scientists, concerning Clarke's Smithsonian publications and atomic weight research. Also included are computations and tables on the atomic weight of elements and compounds; a budget proposal for the U.S. Geological Survey; memorabilia; and a photograph of Clarke. Portions of the correspondence are in German and French.
Unpublished finding aid available.

C39 Clarke, Frank Wigglesworth, 1847-1931.
Diaries, 1865-1931. - 30 v.
In Library of Congress, Manuscript Division (Washington, DC).
Physicist, chemist, and geologist. - Diaries covering the years 1865-1931, but containing the greatest details for the early period of his life, especially the college years.
Gift of Grace C. Clarke, 1938.

C40 Clarke, Hans Thacher, 1887-1972.
Papers, ca. 1903-73. - Ca. 3,500 items (3 ft.).
In American Philosophical Society Library (Philadelphia, PA).
Biochemist. - Includes correspondence, reports, notes and notebooks. Clarke studied chemistry at University College, London (1896-1905), worked for the Eastman Kodak Co. in Rochester (1914-69), and was a professor of biological chemistry at the College of Physicians and Surgeons, Columbia University (1928-56). Among other research work, he was involved in the production of penicillin in the U.S. His participation in the following organizations is documented: American Philosophical Society, American Chemical Society, American Otological Society, and the American Society of Biological Chemists. His chemistry research is detailed in 13 notebooks, 1903-71 (the 1903 volume is on photographic chemistry and processes). There is also 1 volume on the clarinet, which Clarke played expertly. The personal and family correspondence is principally with Mrs. Dorothy Clarke Middleton and Mrs. Agnes Helfreich. Correspondents of note include: James Bryant Conant, Vincent du Vigneaud, Alfred Edwards Emerson, Joseph Stewart Fruton, William John Gies, Sir Julian Huxley, Esmond Long, J. Murray Luck, Henry Allen Moe, Erwin Planck, William Shockley, Samuel Smiles, Warren M. Sperry, Alfred Walter Stewart, Sir Geoffrey Taylor, Merle Tuve.
Presented by Mrs. Clarke, 1973.
Unpublished finding aid available.

C41 Clauss, Julius A 1886-1973.
Papers, 1908-60. 5 ft.
In University of Michigan, Bentley Historical Library, Michigan Historical Collections (Ann Arbor)
Steel industry engineer, vice president in charge of engineering, Great Lakes Steel Company, Ecorse, Mich.; and chief of steel plant facilities for Steel Division, U.S. War Production Board, during World War II. Correspondence, writings, professional papers, files relating to Clauss' interest in Association of Iron and Steel Engineers, and paper detailing production of first Bessemer steel at Wyandotte, Mich.
Unpublished finding aid in the repository.

C42 Cleaveland, Parker, 1780-1858.
Papers, 1795-1858. - 6 ft. (ca. 1800 items)
In Bowdoin College Library (Brunswick, ME).
Professor of Chemistry, Mineralogy, and Natural Philosophy at Bowdoin College. - Correspondence (chiefly scientific), meteorological journal, mineralogical, chemical, and various scientific notes, and extensive revisions for a third edition of Cleaveland's book, "An Elementary Treatise on Mineralogy and Geology" (1822). Principal correspondents include Brewer, Stevens & Cushing of Boston; George W. Carpenter; Cummings & Hilliard of Boston; Hilliard, Gray & Co. of Boston; Reuben D. Mussey, Benjamin Silliman, William Sweetser, Benjamin Vaughan, John D. Wells, and J. M. Wightman.
Alphabetical and chronological indexes to the correspondence in the library.
Gift and purchase, from various sources.

C43 Cloke, John B., b. 1897.
Papers, 1929-1968. - 7.5 linear ft.
In Rensselaer Polytechnic Institute Archives and Dept. of Special Collections (Troy, NY).
Professor of Chemistry. - Correspondence, reports, laboratory notes, and publications spanning the career of John B. Cloke as chemist and professor at RPI.
Gift of Howard and Betty Cloke Curtiss, 1983.

C44 **Coblentz, William Weber,** 1873–
Papers, 1884-1960.
4 ft. (ca. 350 items)
In Library of Congress, Manuscript Division.
Physicist and author. Correspondence (1893–1960), journal kept on the solar eclipse expedition to Sumatra and the return through Europe (1926), MS. and annotated printed copies of two books, Man's place in a superphysical world (1954), and From the life of a researcher (1951), biographical material, memorabilia, financial records (1884–1905), scientific notebooks concerning stellar and planetary radiation (1914–26) and psychic phenomena (1910–15), photos., and printed copies of Coblentz's works, frequently annotated (1903–53). Bulk of the material relates to Coblentz's pioneer work in the fields of infrared spectroscopy, and to the application of radiometry to astronomical problems. Correspondents include Cleveland Abbe, Charles G. Abbot, George E. Hale, Dayton C. Miller, August H. Pfund, and William R. Whitney.
Information on literary rights available in the Library.
Gifts of Mr. Coblentz, 1952–60.

C45 Cohn, Mildred, b. 1913.
Papers, 1957-80. - Ca. 12,000 items (19 linear ft.).
In American Philosophical Society Library (Philadelphia, PA).
Collection includes correspondence, research data (20 large loose-leaf notebooks), reviews of grant proposal manuscripts, and student recommendations (restricted). Her career prior to about 1960 is not documented in great detail: Ph.D., Columbia University, 1938 (physical chemistry); research associate in biochemistry (George Washington Univ., 1937-38; Cornell Univ., 1938-46; Washington Univ., 1946-60, Harvard Medical School, 1950-51). The correspondence in the collection dates primarily to the 1960's-1970's, when she worked at the University of Pennsylvania (Professor of Biophysics and Physical Biochemistry, 1961-78; Benjamin Rush Professor of Physiological Chemistry, 1978-   ), and at the Institute for Cancer Research, Philadelphia. The letters in the collection are professional in content, discussing biochemistry, molecular biology, nuclear magnetic resonance (NMR) experiments (with significant material on the NMR Facility for Biomedical Studies at Carnegie-Mellon University). The letters also concern lectures and participation in conferences and symposia, and there is much relating to recommendations for students and colleagues for positions and promotions. There is little material, however, reflecting Dr. Cohn's interest and involvement with the issue of women in science. Correspondents include: Paul D. Boyer, Aksel A. Bothner-By, Carl F. Cori, Gerty Cori, Marianne Grunbert-Manago, Hans A. Krebs, Lafayette Noda, Irwin A. Rose, Harold C. Urey, Vincent du Vigneaud.
Presented by Dr. Cohn, 1982.

C46 **Coleman family.**
Buckingham collection, 1832-1930.
55 ft. (462 items)
In Lebanon County Historical Society collections (Lebanon, Pa.)
Letter books and business records. Includes records (1832–1931) of the Cornwall Charcoal Furnace, which was puchased from the Grubb family by Robert Coleman in 1798, the iron ore banks connected with it remaining in the Coleman family until 1918 when the Bethlehem Steel Co. began acquiring the various properties. Other records include letter books and business papers of Percy and Ann Alden, the Alden estate (1892–1915), M. C. Buckingham, Cold Spring Furnace, the Colebrook estate (1875–1915), Colebrook Farm, Colebrook Furnace (1881–94), Colebrook Mill, Bird Coleman furnaces, Robert Coleman (1878–92), R. W. Coleman and his heirs, Sarah E. Colemen (1871–88), Sarah H. Coleman (1881–82), William Coleman (1850–73). Conewago Lake boat records, Cornwall Anthracite Farm, Cornwall Anthracite Furnaces, Cornwall Anthracite Store, Cornwall estate, Cornwall Farm, Cornwall Iron Co., Cornwall Mill, Cornwall Reading Room Association, Cornwall Railroad Company, Emaus Furnace, Fairview Farm, E. W. and W. C. Freeman, Lochiel Furnace, H. and V. L. Meigs and Company, North Cornwall Furnace, North Lebanon Railroad Company, Robesonia Furnace, Robesonia Iron Company, Speedwell Farm, Speedwell Forge, and miscellaneous items.
Open to investigators under library restrictions.
Gift of Hon. William C. Freeman, 1947.

C47 Collected Scientific Manuscripts.
Papers, 13th-20th centuries. - 200 boxes.
In National Museum of American History Archives Center (Washington, DC).
The collection contains manuscripts which document the history of science, including works by Aristotle, Boethius, and

others. In particular, several manuscripts by Isaac Newton on various subjects, including alchemy and chemistry; the correspondence of Benjamin Silliman; letters, clippings, and other items relating to Joseph Priestley; and a collection of addresses which represent, in part, the work of Antoine Henri Becquerel and Marie Curie.
By appointment only.
Unpublished finding aid available.

C48 Collection of scrapbooks relating to medical subjects, 1891-1976. 18 items.
In University of Texas Medical Branch, Moody Medical Library (Galveston)
Scrapbooks of Galveston County Medical Assistants Society (1958-61), John Sealy Hospital (1949-53), Nursing Alumnae Association (1895-1964), Resiterns (1961-74), Sigma Xi (1949-76), University of Texas Medical Branch (1900-20, 1935, 1941-42, 1942-49), University of Texas Medical Branch Pathology Dept. (1891-1942), University of Texas Medical Branch Pharmacology and Toxicology Dept. (1922-64), and University of Texas Medical Branch Surgery Dept. (1891-1942). Includes 2 v. on yellow fever in New Orleans (1897) and yellow fever in Texas (1905).

C49 Collier, Peter, 1835-1896.
Letters, 1881-82. 58 items.
In University of Wisconsin Library, Rare Books Dept. (Madison)
Chemist. Letters from Collier at the U. S. Dept. of Agriculture to Benjamin Silliman, Jr., in New Haven, Conn., dealing with various phases of sorghum investigation being carried out by Collier.

C50 Collins family.
Papers of the Collins and Marshall families, 1761-1838. 10,000 items.
In Historical Society of Pennsylvania collections.
Forms part of the Daniel Parker papers.
Correspondence and business records of the two families, to whom General Parker was connected by marriage. Includes papers of Parker's father-in-law, Zaccheus Collins, who died in 1831, of Stephen Collins, father of Zaccheus, of Christopher Marshall, Jr., 1740-1806, father-in-law of Zaccheus, and of Elizabeth Marshall. Zaccheus Collins papers consist of correspondence with his family and with others including Constantine Samuel Rafinesque, Charles Lee, Edmund J. Lee, R. B. Lee, R. H. Lee, and William Lee; business records, including papers relating to Pennsylvania lands and correspondence relating to trade with India, letters and legal opinions (1831) by William Rawle in relation to Collins' estate, and accounts of Daniel Parker as executor. Stephen Collins' papers contain old bonds and deeds (1761-95), a bankbook, a letter book, and a group of letters from Robert Hampden Pye (1778-79) Papers of Christopher Marshall, Jr. consist of a diary, a waste book, a ledger (1775-97) of the firm of Christopher and Charles Marshall, Philadelphia drug and paint manufacturers, and an account book (1811) of his estate. Miscellaneous papers consist of Elizabeth Marshall's ciphering book (1782); inventory of the estate of Thomas Paschall (1796), Christopher Marshall executor; account of the estate of Ann Collins, a minor (1807-15); a packet of Daniel Parker letters relating to his venture in the horse-breeding business (1818-38) and a book entitled War and peace register and regulations, 1814-21, containing copies of orders, regulations, etc. issued by the War Dept.
1. Real property—Pennsylvania. 2. Drug trade—Philadelphia. 3. Paint industry and trade—Philadelphia. 4. U. S.—Comm.—India. 5. India—Comm.—U. S. 6. Horse breeding—Pennsylvania. 7. U. S. Army—Regulations. I. Marshall family. II. Parker, Daniel, 1782-1846.

Columbia University. Department of Chemistry. C51
Records, 1897-1919. - 17 boxes.
In Columbia University, Rare Books and Manuscripts Library (New York City).
Records of the Chemistry Department, chiefly files from Alexander Smith's tenure as Chairman (1911-1919). Files consist mostly of correspondence.
Unpublished finding aid available.

Conant, James Bryant, 1893-1978. C52
Papers, ca. 1893-1978. - ca. 60 ft.
In Harvard University Archives (Cambridge, MA).
Professor of Organic Chemistry; President of Harvard University, 1933-1953. - Professional and personal papers (not including his official papers as President of the University). Professional papers include material dealing with Conant's courses in natural sciences; papers relating to his inauguration as President at Harvard; speeches and articles; subject files; correspondence; reports; and notebooks.
Particular topical areas included are Conant's work with the Baruch Rubber Survey (1942); a scientific notebook regarding contacts with European chemists and ideas for experiments, 1921; a journal relating scientific encounters and ideas in Germany, 1925; a file of correspondence and notes, 1916-50 regarding: algenic acid problems, Aromatic Chemical Company, Du Pont Company, Conant's organic chemistry text, the Organic Division of ACS, petroleum work, and notes on readings about cyclopropane compounds.
Personal material includes diaries of James S. Conant (father); miscellaneous material from elementary and high school; scrapbooks of clippings, reports, speeches, photographs, certificates, honorary degrees, and the like. Conant miscellany includes bibliographies, cassette recording of memorial service, Ph.D. theses about Conant, remembrances, books, and academic garb.
Researchers are advised to review the entry under "Harvard University. Office of the President." for Conant's official presidential papers.
Access restricted. Contact the archivist for further information.
Shelflist available at repository.

Condit family. C53
Papers, 1761-1917. ca. 110 items.
In Rutgers University Library (1076 and 1077)
Papers of the Condit family, of Essex Co., N. J., including a ciphering book, tuition bills, grocery accounts, and other papers of Aaron M. Condit; letters received of Oliver H. Condit; papers of Jemima D. (Tomkins) Halstead, of the Thomas Ogden estate, of Jabez Pierson, and of Aaron Tomkins; 19th cent. dyeing and tanning formulas, and other papers.

C54 Connecticut. State Library, *Hartford.*
Connecticut manuscripts.
ca. 500 items.
In Connecticut State Library.
Correspondence, diaries, and business, family, and historical papers. Includes letters and papers of Prudence Crandall, Charles Goodyear, Lydia Sigourney, Jonathan Trumbull, and Noah Webster.
Gift and purchase, various dates.

C55 Conservation Council of Virginia.
Records, 1968-74. ca. 2 ft.
In University of Virginia Library (Charlottesville) (9883)
Correspondence, minutes, financial statements, bylaws, reports, and policy statements, of the council, together with correspondence and publications of member groups and special committees. Chiefly relates to Virginia legislation affecting the environment and State agencies concerned with environmental control, particularly State Water Control Board, Air Pollution Control Board, and Governor's Council on the Environment. Includes material on strip mining, water quality control, land development, preservation of scenic, historic, and environmentally unique areas, and controversial dams, bridges, highways, resorts, and electric power plants.
Gift, 1973-75.

C56 Consolidated Virginia Mining Company.
Records, 1859-1921. 16 ft.
In University of Nevada Library.
Correspondence, mine operations reports, assay reports, ore and bullion statements, mill records, annual reports, bylaws (1921), equipment specifications and performance records, payroll books, account books, deeds, court cases, patent applications and protests, agreements, contracts, resolutions, and other legal records of a silver mining company in Virginia City, Nev. Includes records of the California Mining Company, the Consolidated California and Virginia Mining Company, and the Consolidated & California Shaft, which was jointly operated by the Consolidated Virginia Mining Company and the California Mining Company.
Unpublished guide to the collection in the library.
Gift of the company, ca. 1929.

C57 Cook, George Smith, 1819-1902.
Papers, 1845-61. 55 items.
In Library of Congress, Manuscript Division.
Pioneer photographer, of Charleston, S. C., and Richmond, Va. Business correspondence, 4 account books (1845-54, 1859-64), and a photo. The account books include references to the mixing of chemicals and the processing of photographic plates.
Purchased from Richard S. Wormser, 1940; transferred from the Library's Prints and Photographs Division, 1951.

C58 Cooke, Josiah Parsons, 1827-1894.
Papers, 1850-94. 4 v.
In Harvard University Archives.
Professor of chemistry and mineralogy at Harvard University. Lecture notes.
Information on literary rights available in the repository.

Coolidge, Albert Sprague, 1894-1977. C59
Papers, ca. 1950-1966. - 0.7 linear ft.
In Harvard University Archives (Cambridge, MA).
Lecturer in chemistry. - Collection includes correspondence and other papers relating to scientific research, especially correspondence with Philip B. Lorenz, 1954-1966; also some miscellaneous correspondence and scientific notes and notebooks. In addition, one file of reprints and pamphlets.
Gift of Mrs. A. S. Coolidge, via Chemistry Library, August 11, 1977.
Unpublished finding aid available.
Access restricted. Contact archivist for further information.

Cooper, Hewitt & Company, Ringwood, New Jersey. C60
Records, 1833-1907. - 233.6 lin. ft.
In Library of Congress, Manuscript Division (Washington, DC).
Business papers of the iron works and glue factory at Ringwood, New Jersey, of Peter Cooper and Abram Hewitt.
Finding aid available.

Cooper, Thomas, 1759-1840. C61
Letters, 1806-1838. - Ca. 50 items.
In University of Pennsylvania Libraries, E. F. Smith Collection (Philadelphia).
Pioneer American chemist. - Correspondence with Benjamin Silliman, Thomas Jefferson, and Martin Van Buren, among others. Some are facsimiles of originals in the Library of Congress.
Individually cataloged on cards in the repository.

Coosaw Company. C62
Records, 1892-1913. - 1 ft.
In South Carolina Historical Society collections (Charleston) (23-157/160)
Phosphate mining company based in Charleston and Beaufort, S.C., organized 1892.- Letter book, minutes, stock certificates, dividend books, lists of stockholders (1912), and bylaws. Correspondents include James Adger, phosphate inspectors T.J. Cunningham and S.W. Vance, company superintendent D.H. Lopez, South Carolina state comptroller general, state treasurer, and Smythe, Lee & Frost. Principal stockholders include Andrew M. Adger, James Adger & Company, Edward Austen, Alexander Brown, A.T. Smythe, and C.C. Wylie.

C63 Cori, Carl Ferdinand, 1869-1984.
Papers, 1919-1984. - 8 ft.
In Washington University School of Medicine Archives (St. Louis, MO).
Biochemist, pharmacologist, and university professor. - Correspondence, notes, manuscripts, and card files relating to Cori's career as a researcher and teacher at the State Institute for the Study of Malignant Diseases (Buffalo), Washington University (St. Louis), and Harvard University. Includes some correspondence addressed to Gerty T. Cori (1896-1957), biochemist and first wife of Carl F. Cori.
Gift of Anne Cori, 1985.

C64 Cornell University Medical Center. Department of Pharmacology.
Papers, 1931-1978. - 5 in.
In New York Hospital-Cornell Medical Center, Medical Archives (New York, NY).
Annual reports, correspondence, biographical data on faculty members, bibliographies, etc.
Unpublished finding aid available.

C65 Cottrell, Frederick Gardner, 1877-1948.
Papers, 1900-1951. - 1000 items.
In Library of Congress, Manuscript Division (Washington, D.C.)
Chemist and inventor. - Diaries, notebooks, notes, scientific papers, writings, and miscellany, chiefly 1907-1940, relating primarily to Cottrell's work and daily routines as a research chemist and scientist. Includes papers relating to patents and to his associates, P.H. Royster and Farrington Daniels.
Unpublished finding aid in the repository.

C66 Cowles, Calvin Josiah, 1821-1907.
Papers, 1773-1941. 72,000 items.
In University of North Carolina Library, Southern Historical Collection (Chapel Hill) (3808)
Merchant and assayer of the U. S. Mint, of Wilkes Co., N. C. Correspondence and family and business accounts, relating to Cowles' business activities including his merchandising in roots and herbs, land speculation, mining, and railroad development in western North Carolina, Kansas, and South Dakota, and his service as assayer of the U. S. Mint at Charlotte, N. C. (1869-85). Includes campaign papers of his son, Charles Holden Cowles, Republican Congressman (1908-10).
Unpublished description in the library.
Gift of Hamilton C. Horton, 1968.

C67 Cowles, Calvin Josiah, 1821-1907.
Papers, 1817-85. 12 ft. (ca. 23,000 items)
In North Carolina State Office of Archives and History collections (Raleigh)
Merchant, postmaster, and assayer of the U. S. Mint, of Elkville and Wilkesboro, Wilkes Co., N. C. Business and personal correspondence, accounts, memoranda, Confederate quartermasters' records, catalogs, advertisements, and other papers. Includes ca. 15,000 letters reflecting Cowles' career as a merchant, chiefly in the roots and herbs trade, land speculation, and assayer of the U. S. Branch Mint at Charlotte, N. C., and his Unionist and Republican sympathies during Civil War and Reconstruction; ca. 400 letters from Cowles' father, Josiah Cowles, giving family news and advice; and letters referring to elections, North Carolina Constitutional Convention of 1868; impeachment of Gov. William W. Holden, Gap Creek copper mine, Ashe Co., N. C., U. S. Army in the West, secession, and economics, politics, merchandising, and other conditions during the Civil War. Correspondents include Andrew Cowles, Calvin D. Cowles, Henry Cowles, and William W. Holden.
Unpublished description in the repository. Also described in Guide to Private Manuscript Collections in the North Carolina State Archives (1964) p. 98-99.
Gift of Mrs. Cowles, 1914-36.

C68 Cox, Joseph Warren.
Papers, 1829-59. 117 items.
In Rutgers University Library (659)
Inventor, of Malden, Mass. Correspondence, receipts, notes on experiments and inventions relating to rubber manufacture in New Brunswick, N. J., and elsewhere, and miscellaneous papers. The bulk of the collection dates from 1852 to 1859. Places represented include Cambria Co., Pa.
Gift of Herbert M. Waldron.

C69 Cox, Talton L L
Papers, 1858-1918.
108 items.
In Duke University Library.
Civil servant, of Randolph Co., N. C. Personal and business correspondence and other papers, many of which relate to Cox's service as assistant assessor of internal revenue in the 6th division of the 5th district of the U. S. Internal Revenue Service in North Carolina. Letters relating to surveys and operations of distilleries, including a brandy distillery. Letters of the Civil War period contain references to Federal prisoners in Richmond. Postwar correspondence includes discussions of such public topics as the Fourteenth amendment, President Grant, the campaign of 1868, State and National politics, the Republican and Democratic Parties, and President Wilson's appeal to the American people to buy bonds during World War I.
Card index in the library.
Acquired 1949 and 1959.

C70 Craig family.
Papers, ca. 1733-1956. ca. 6 ft. (7500 items)
In Minnesota Historical Society collections (St. Paul)
Correspondence, diaries, genealogical material, MSS. of books, articles, sermons, and tracts, notes, pamphlets, clippings, and other papers, of Austin Craig (1824-1881), pertaining to his activities as minister of the Christian Church and president of Antioch College and the Christian Biblical Institute, and his friendship with Horace Mann; of Austin Craig, Jr. (1872-1949), concerning his law practice in Oregon, teaching career in the Philippine Islands, at the University of the Philippines and the University of Manila,

travels in Europe and the Orient, and his research and writings on religion, Japan, and World War II, José Rizal, the history of the Philippines, Filipino-American relations, and other subjects; of Adelaide (Craig) Snyder (1870-1967) and her husband, Harry Snyder (1867-1927), chemist at Cornell University and the University of Minnesota, including his research and writings on soils, nutrition, wheat, and flour milling; and of Mary Adelaide (Churchill) Craig, wife of Austin Craig, Sr., their children, Josephine Craig and Moses Craig, and Eunice Malvina Churchill. Correspondents include Henry W. Bellows, Fausto V. Carlos, S. J. Coffin, Edmund Day, Livingston Farrand, David Felt, Horace Greeley, Edward Everett Hale, Francis Burton Harrison, Thomas Hill, E. G. Holland, Charlotte (Garrigue) Masaryk, Lucretia Mott, Eulogio B. Rodriguez, and Clinton D. Smith.

Unpublished inventory in the repository.

Information on literary rights available in the repository.

Gifts of Adelaide (Craig) Snyder, 1952, and of her estate, through Mary H. Powell, Minneapolis, 1967.

C71 **Craig, Isaac,** 1742?-1826.
Papers, 1766-1907.
15 ft. (ca. 5000 items and 77 v.)
In Carnegie Library (Pittsburgh)
Merchant, manufacturer, farmer, and Army officer. Military orders and payrolls as Colonel at Fort Pitt, logistical records (1791-1801) as deputy quartermaster and commissary of supplies; records of Craig, Bayard & Co., of a distillery, a glass works, and other of Craig's business enterprises; records as stamp agent at the time of the Whiskey Insurrection; private account books and ledgers (1767-1816); records of his farm on Montour's Island, in the Ohio River; and other papers, the majority of which are of the period 1780-1800. Includes papers of Henry Knox Craig, 1791-1869, an Army officer, and of Neville B. Craig, 1787-1863, a lawyer, editor, publisher and author.
Unpublished finding aid in the library.
Open to investigators under library restrictions.
Deposited by Theodore Diller, 1911, given by his family, 1957.

C72 **Craig, Locke,** 1860-1925.
Papers, 1880-1924.
125 items and 1 v.
In Duke University Library.
Lawyer, legislator, and political leader, of Asheville, N. C. Personal correspondence and other papers, in part relating to Craig's activities as member of the North Carolina Legislature (1898-1903) and Governor of North Carolina (1913-17) Includes copies of political and religious speeches, and one on Freemasonry; and a chemistry notebook kept by Craig while a student at the University of North Carolina. The correspondence includes recommendations from university professors; letters from his mother; letters from Craig to his sons, Carlyle and Arthur, at the U. S. Naval Academy, Annapolis; and a letter from Carlyle Craig describing a voyage to the Azores and the Island of Fayal. Other correspondents include J. W. Bailey, Kemp P. Battle, H. G. Connor, A. W. Mangum, W. J. Peele, J. C. Pritchard, Woodrow Wilson, F. D. Winston, and G. T. Winston.

C73 <u>Craig, Lyman Creighton,</u> 1906-1974.
Papers, 1935-74.   28 ft.
In Rockefeller University Archives, Rockefeller Archive Center (Pocantico Hills, North Tarrytown, N.Y.)
Biochemist at Rockefeller University. Personal and professional correspondence, mss. of writings, addresses, lectures, biographical material, bibliography, collected reprints, laboratory notes, photos, and other papers. Includes correspondence relating to American Chemical Society and American Society of Biological Chemists and files on development of techniques of countercurrent distribution for separation and identification of compounds, National Institutes of Health study section (1963-72), National Institute of Mental Health (1971-74), National Institute of Arthritis, Metabolism, and Digestive Diseases (1965-74), NMR Consortium (1968-74), and Gordon Research Conferences (1952-73). Correspondents include Walter Jacobs, students, and colleagues.

Unpublished finding aid in the repository and in Rockefeller University Archives in New York City.

Gift of Mrs. Rachel P. Craig, 1975.

C74 **Crawford, Fred Lewis,** 1888-1957.
Papers, 1925-53. ca. 4 ft., and 25 v.
In University of Michigan, Michigan Historical Collections.
Sugar processor and U. S. Representative from Michigan. Correspondence, speeches, and other papers relating mainly to personal and financial affairs (especially Crawford's Texas ranch) and political and congressional affairs in the 1930's; papers relating to his activities as a member of the Insular Affairs Committee of the House of Representatives; scrapbooks of clippings and photos.; and papers of the Michigan Sugar Company. Correspondents include Wilber Brucker, Leonard Hall, George A. Malcolm, and relatives of Crawford.

Gift of Mrs. Crawford, 1964.

C75 <u>Crews, J. M.</u>
Chemistry lecture notes, 1872. - 1 v.; 25 cm.
In Auburn University Archives (Auburn, AL).
Student. - Contains notes from eighty-one lectures in general chemistry taken by Crews while a student at Southern University, Greensboro, Alabama in 1872.
Transferred from Ralph Brown Draugon Library, Auburn University, 1969.

C76 <u>**Crisp, Lucy Cherry,**</u> 1899-1977.
Papers, 1794-1972. - ca. 6500 items.
In East Carolina University Library, East Carolina Manuscript Collection (Greenville, N.C.).
In part, transcripts of diary (1860-1865) of Delha Mabrey of Edgecombe County, N.C.; and of early letters of George Washington Carver.
Director, North Carolina State Art Gallery (Raleigh), and Florence Museum (S.C.). - Correspondence, diaries, speeches, daybooks, financial and legal papers, literary mss., publications, notes, newspaper clippings, pamphlets, and photos, relating to Crisp's career, her relationships with artists and with scientist George Washington Carver, whose biography she wrote, other writings by Crisp, and USO Club, Greenville, N.C. Correspondents include Charles Baskerville, George Washington Carver, Lloyd C. Douglas, William Coffield Fields III, Frank Porter Graham, Charles Sylvester Green, Robert Lee Humber, Charles Edward Jefferson, Douglas Ellsworth Lurton, David Ulrey McDowell, Lewis Mumford, Gorham B. Munson, Louis Orr, Hobson Pittman, and Charles Kenneth Sibley.

Gifts of Miss Crisp, 1970 and 1973; and of Richard H. Crisp, Greenville, N.C., 1978-1979.

Finding aid in the repository.

C77 Cross, Paul C., b. 1907.
Papers, 1934-1967. - 2.5 in.
In University of Washington Archives (Seattle).
Professor of Chemistry; head of Chemistry Department at the University of Washington, 1949-61. - Correspondence, 1934, 1937-67; speeches, 1961-63. Major correspondents were Edgar Bright Wilson; James R. Killian, Jr.; Mellon Institute; U.S. Naval Ordinance Laboratory; and the Chemistry Department of the University of Washington.
Gift of Mrs. Cross, April 21, 1981.
Open for research for undergraduates.
Unpublished finding aid available.

C78 Crowell family.
Papers, ca. 1873-1975. - 1 v. and 2 boxes.
In Woods Hole Historical Collection (Mass.).
Chiefly papers (ca. 1873-1893) of Azariah Foster Crowell (1846-1918), chemist with Pacific Guano Company, including personal letters; financial records; estate papers of his father, Prince Sears Crowell (1813-1881) founder of the company; company records; letters (1881-1882) representing Isaiah Spindel & Company, fish company in Woods Hole, and Spencer Fullerton Baird, U.S. Commissioner of Fish and Fisheries; and plans for Crowell House, now owned by Woods Hole Oceanographic Institution. Other family members represented include Crowell's children, Azariah Foster Crowell, Jr. (1879-1945), Polly Lauraetta Crowell (1886-1969), and Prince Sears Crowell (1881-1972). The latter's papers include recollections of Woods Hole; material relating to Spiritsail, fishing and transportation boat; history of Pacific Guano Company; and family Bible with births, marriages, and deaths from 1846-1975.
Register in the repository.

C79 Cullen, William, 1710-1790.
Lectures, ca. 1756. - 2 v.; 22 cm.
In National Library of Medicine (Bethesda, MD).
Chemistry lectures, Edinburgh, ca. 1756.

C80 Cullity, Bernard D., 1917-1978.
Papers, ca. 1945-1978. - 33 linear ft.
In University of Notre Dame Archives, (Notre Dame, IN).
Professor of Metallurgical Engineering and group leader in the World War Two Manhattan Project. - Collection includes Cullity's correspondence with colleagues, students, Notre Dame administrators, government agencies, and corporations; research notes and reports; manuscripts of his articles and textbooks; scientific journal reprints; grant applications and supporting documents; and material relating to the courses he taught. Much of the collection documents Cullity's research on projects supported by the Atomic Energy Commission, the Office of Naval Research, and other agencies and corporations. His major research interest was in X-ray diffraction of metals.
Unpublished finding aid available.

C81 Culver, Stephen Berry, 1841-1902.
Diary, 1855-1902. 17 items and 7 v.
In University of North Carolina at Chapel Hill, Library, Southern Historical Collection (3992)
Mining and chemical engineer, clerk in the Naval Office, New York, N.Y., and resident of Sandy Hill, N.Y. Intermittent and retrospective entries, reflecting Culver's interest in work, Methodist church, family affairs, and national and local social, cultural, and political affairs.

C82 Curie, Marie (Sklodowska) 1867-1934.
Papers, 1920-34.
182 items.
In Columbia University Libraries.
Polish-French chemist and physicist. Chiefly letters, cards, and telegrams to Marie Mattingly Meloney, beginning from the time arrangements were being made for Mme. Curie to come to the United States to receive a gift of a gram of radium for experimental use in the Curie Institute, and relating to personal, biographical, and scientific matters. Includes the original French draft of Mme. Curie's address of acceptance of the radium gift, an English draft of the address, and an article by Mme. Curie giving her impressions of America.
Gift of William Brown Meloney, Jr., 1956.

C83 Curtis & Brother Company, Newark, Del.
Records, 1823-1942. ca. 2000 items and 40 v.
In Eleutherian Mills Historical Library (Greenville, Del.) (394)
Correspondence, journals, ledgers, cashbooks, bill books, invoice book, customer orders, cost sheets, freight records, time books, roll sizes book, machine record book, record of materials and supplies, bylaws, agreements, judgment bonds, and other papers, of a paper manufacturing firm. Includes letters pertaining to a Federal Reserve loan (1932-35) and deeds and other papers concerning real estate at Newark, Del.
Described in A Guide to the Manuscripts in the Eleutherian Mills Historical Library, by John B. Riggs (1970) p. 723.

C84 Curtis, Charles Elbert, 1862-1958.
Student notebook, 1883. - 1 v.
In Cornell University Libraries, Department of Manuscripts and University Archives (Ithaca, NY).
Chemistry notebook, January 13-June 5, 1883.

C85 **Curtis, Harry Alfred, 1884–1963.**
Papers, 1912–63. ca. 11 ft. (ca. 2000 items)
In University of Tennessee Library (Knoxville)
Chemical engineer distinguished in industry, government service, research, and education. Business and personal correspondence, research notes, published and unpublished articles, talks, and book reviews, printed technical material, an account of Curtis' career, and photos. Correspondents include Frederick G. Cottrell, Albert Gore, Estes Kefauver, Trygve Lie, Harcourt A. Morgan, and Harry S. Truman.
Unpublished guide in the library.
Open to investigators under library restrictions.
Information on literary rights available in the library.
Gift of Mrs. Curtis, 1964.

C86 **Cushing Refining Company,** *Cushing, Okla.*
Records, 1927–45.
239 items.
In University of Oklahoma Library.
Correspondence and records. Includes directives and regulations issued by the Petroleum Administration for War.

C87 **Cutler, Manasseh, 1742–1823.**
Papers, 1783–1837.
1 envelope.
In Essex Institute collections (Salem, Mass.)
Clergyman and U. S. Representative from Massachusetts. Correspondence, meteorological diaries (1783–1837), 4 botanical notebooks, a chemistry notebook, a pharmacopeia, a notebook entitled A description of animals of North America taken from actual observation by Manasseh Cutler, 1786, and another entitled Astronomical recreations performed at Yale College 1763 by Manasseh Cutler. Seven letters (1803–05) from Cutler while he was in Congress to Francis Low of Hamilton, Mass., tell of the Louisiana Purchase resolution and other Congressional business, and describe the trial of Justice Samuel Chase (1741–1811) before a court of impeachment.

C88 Cutler, Thomas Robinson, 1844-1922.
Biography (edited copy of ms.). - 1 v. (182 p.) bound; appendices; endnotes; index.
In University of Utah Libraries, Special Collections Department (Salt Lake City).
Utah businessman, prominent developer of Utah's beet sugar industry, and Latter Day Saints church leader. - This biography, prepared by Cutler's grandson, Jesse R. Smith, records his personal and business history with emphasis given to his role as Vice President and General Manager of Utah-Idaho Sugar Company. Contains penned in remarks and suggestions.
Gift of Jesse R. Smith.
Unpublished finding aid available.

Dabney family.
Papers, 1716–1945.
12 ft.
In University of North Carolina Library, Southern Historical Collection (no. 1412)
Business and personal correspondence, and legal, financial, and military papers of five generations of the Dabney family of Virginia. Beginning in the 1830's the papers are chiefly of Robert Lewis Dabney, Presbyterian clergyman, teacher, author, and Confederate chaplain, who was associated with Hampden-Sydney College and the Union Seminary of Virginia; of his brother; of the James Morrison family into which Dabney married in 1848; and of his son Charles William Dabney, scientist, educator, and author, who held State and Federal Government positions in the field of chemistry and agriculture and who was president of the University of Tennessee (1887–1904) and of the University of Cincinnati (1904–20). Robert Dabney's papers include letters to his wife during the Civil War and papers regarding his biography of Thomas J. Jackson. Charles Dabney's papers reflect his wide range of professional activities and public service in addition to his specific institutional positions and writings. They relate to such topics as international affairs, national defense and world peace, education in the South and in Mexico, agricultural and scientific education, mineral resources in Alabama, Texas, and North Carolina, and Dabney family history. Correspondents include William James Battle, Wallace Buttrick, Frank R. Chambers, Philander P. Claxton, Clarence H. Dodd, E. L. Doheny, J. D. Eggleston, Parke P. Flournoy, Jr., James D. Hoskins, John McLaren McBryde, Alexander J. McKelway, James Morrison, R. Hall Morrison, Edward Pearson Moses, Robert C. Ogden, Walter Hines Page, George F. Peabody, Henry S. Pritchett, Edward T. Sanford, Benjamin M. Smith, Edwin Willits, A. F. Woods, and R. B. Woodworth.
Indexed description in the library.
Gift of Mrs. Alexander Thomson and Mrs. John W. Ingle, Jr., 1947.

D'Alelio, G. Frank, 1909–80.
Patent documents, 1966–1976. – 2.5 in.
In University of Notre Dame Archives, (Notre Dame, IN).
Professor of Chemistry and industrial consultant. – The collection includes correspondence, patent records, legal documents, and descriptions of D'Alelio's inventions. He held over 400 patents and was noted for his discovery of polymers capable of withstanding high temperature while remaining processable at reasonable temperatures and pressures.
Access restricted; researchers should apply to the University Archivist.

Dane, Dana, and Company, *Boston.*
Records, 1851–59.
9 ft.
In Harvard Business School, Baker Library.
Letters, accounts of ships' expenses, and accounts of cargoes bought and sold by the firm in connection with its far eastern trade. Includes papers of the Boston Sugar Refinery, a subsidiary.
Described in List of business manuscripts in Baker Library, compiled by R. W. Lovett (2d ed., 1951)
Gift of Charles H. Taylor.

Daniels, Farrington, 1889–1972.
Papers, 1912–1975. – 50 linear ft.
In University of Wisconsin Archives (Madison).
Professor of Chemistry. – Collection includes 48 boxes of correspondence, 1912–1972, with the ACS, Monsanto Chemical Company, Linus Pauling, and others, about solar energy, grant proposals, the Perkin Medal, and other topics; 4 boxes of publications, including research reports, articles, and copies of Daniels' textbooks; 9 boxes of departmental files, 1926–72, including lecture notes, exams, course evaluations, correspondence, and other material; 49 boxes of subject files, 1918–1971, on such topics as atomic power and rocket fuels; and 20 boxes of biographical files, containing family correspondence, memorabilia from Daniels' student days, newsclippings, diaries, scrapbooks, cassette tapes, and other materials.
Received in 1970 and 1972.
Unpublished finding aids available.

Darcy, Timothy Johnes. D5
Papers, 1811–1900. ca. 320 items and 15 v.
In Rutgers University Library (0.998 and 662)
Correspondence, bills, legal papers (relating to collections and suits by Darcy as attorney-in-fact for various creditors), accounts, monthly balance sheets, memos, and other papers for 1811–64; dockage records for railroad ties, accounts of Julius G. Parish of sales of a chemical compound and other drugs, diary of Eliza Darcy and household accounts, and papers (1847–1900) of John S. Darcy. Places represented include Passaic Co., N. J., and Friendship, N. Y.

Darley family. D6
Papers, 1875–1970. 10 ft.
In University of Colorado Libraries, Western Historical Collections (Boulder)
Papers (chiefly 1950's–60's) of a Colorado family. Includes correspondence, sermons (1876–1916), articles, mss. of books and articles, photos, and other papers, of George Marshall Darley (1847–1917), relating to his career as a Presbyterian minister in several Colorado mining towns and to his interest in life in mining camps; correspondence, photos, and other papers, of Darley's three sons, George Sinclair Darley, minister, of Georgetown and Alamosa, Colo., Ward Darley, Sr., who was involved in sugar beet manufacture and railroading, and William Marshall Darley, of the U. S. Forest Service in southwestern Colorado; and genealogical information.
Guide published by the repository.
Gift of Ward Darley, Jr., 1969.

Darrach, William, 1796–1865. D7
Diaries, 1830–38.
2 v.
In Historical Society of Pennsylvania collections.
Philadelphia physician and teacher of medicine. One diary (1830–38) contains Darrach's notes on pharmacology and his professional activities; the other (1832–37) reflects his interest in Presbyterian Church affairs, prayer meetings, religious revivals, and doctrinal controversies.
Gift of Alfred Darrach, 1926.

Darrow, Fritz Sage, 1882–*ca.* **1929,** *collector.* D8
Darrow collection of Helmontiana.
ca. 5 ft.
In Western Reserve University.
In part, transcripts.
Correspondence, articles, literary MSS., and transcriptions, concerning Jean Baptiste van Helmont, Flemish chemist, physiologist, and philosopher; and his son, Franciscus Mercurius van Helmont, physician and occult scientist. Other papers relate to Anne Viscountess Conway, Henry More, Manasseh Ben Joseph ben Israel, Gottfried Wilhelm Freiherr von Leibniz, and the Quakers.
Described in two unpublished studies: Darrow collection of Helmontiana at Western Reserve University, by J. H. Hanford; and Baron Francis Mercury van Helmont material, arr. and described by I. R. Watts.
Research access restricted.
Information on literary rights available in the library.
Purchased from the Darrow estate, 1930.

D9 Davis, Aaron, 1819-1903.
Papers, 1806-1903. 33 items and 2 v.
In University of North Carolina at Chapel Hill, Library, Southern Historical Collection (3788)
Harness and patent medicine manufacturer and inventor, of Newark, N.J. Miscellaneous papers, including notebooks of recipes for rubberized coatings and varnishes for harnesses.

D10 **Davis, Edward Wilson, 1888-1973.**
Papers, 1883-1973. - ca. 9 ft.
In Minnesota Historical Society collections (St. Paul)
Superintendent and director, University of Minnesota Mines Experiment Station (1918-1955), and developer of the commercial process for extracting iron ore from taconite. - Correspondence, articles, reports, technical papers, mss. of writings, histories, patents, newsletters, speeches, slides, photos, and scrapbooks, relating to Davis' role in taconite development and to Minnesota's iron ore industry. Includes lake and water pollution studies; technical data on taconite production and composition; reports, maps, and related papers of the Mesabi Iron Company; data on taconite taxes; correspondence and publicity materials on the taconite amendment to Minnesota's constitution (1964); and background materials and drafts of Davis' book, Pioneering With Taconite (1964). Letters, reports, and newsletters from Reserve Mining Company concern its role in taconite development, its plant at Silver Bay on Lake Superior, and disposal of taconite tailings. A few files contain materials on underwater archeology and the history of northern Minnesota, and genealogical notes on the Davis and related families.
Gifts from various sources, 1958-1974.
Unpublished inventory in the repository.

D11 **Davis, John William, 1873-1955.**
Papers, 1873-1955.
ca. 100,000 items.
In Yale University Library.
Lawyer and diplomat. Correspondence, diary, and reminiscences. Covers Davis's boyhood, his five-year term as Solicitor General of the United States, his service as Ambassador to Great Britain, 1918-21, the campaign of 1924 in which he was the Democratic presidential candidate, and his law practice and public affairs, 1921-55. Correspondence also concerns his activities as president of the Council on Foreign Relations, president of the English-Speaking Union, president of the American Bar Association, 1922-23, director of the Atchison, Topeka and Santa Fe Railway, director of the National Bank of Commerce in New York, and director of the United States Rubber Company. The material is related to the library's Frank L. Polk, Edward M. House, and Henry L. Stimson collections.
Unpublished register in the library.
Open to investigators under restrictions accepted by the library.
Information on literary rights available in the library.
Gift of Julia Davis Healy, 1961.

D12 **Davis, Marguerite, 1887-1967.**
Papers, n.d. - 3 folders.
In Racine County Historical Museum Library (Racine, WI).
Davis was assistant in biochemistry to Elmer V. McCollum and co-discoverer with him of vitamins A and B at the University of Wisconsin in 1913; she also founded nutrition laboratories there and at Rutgers University. - Collection contains personal correspondence, family data, a pamphlet she wrote, clippings, and other papers. Also includes information about the work of her father John Jefferson Davis, who was the first director of the University of Wisconsin herbarium.

Davis, T. M., *firm, Cambridge, Mass.* D13
Records, 1820-80.
3 ft.
In Harvard Business School, Baker Library.
Account books, price lists, and notes on experiments of the soap factory of T. M. Davis (known before 1866 as the firm of Eliphalet Davis)
Described in List of business manuscripts in Baker Library, compiled by Robert W. Lovett (2d ed., 1951)
Gift of W. Porter Adams, 1936.

Davis, William Hammatt, 1879- D14
Papers, 1934-59. 26 boxes.
In State Historical Society of Wisconsin collections (Madison)
Patent attorney, labor mediator, and public official, of New York. General correspondence (1941-57); correspondence relating to Davis' speeches and articles (mainly 1945-57); files of letters, reports, memoranda, press releases, and other material concerning his participation in many State, national, and international organizations and projects, particularly those relating to the role of government in industrial relations, and reference files compiled by Davis on atomic energy, labor arbitration, the Taft-Hartley bill and other legislation affecting labor, and wage and price stabilization. Organizations represented include the Atomic Energy Labor Relations Commission, the Atomic Energy Labor Relations Panel, the Atomic Energy Patent Advisory Panel, the board of directors of Sydenham Hospital in New York City, the Citizens' Committee for the International Labor Organization, the Commission on Industrial Relations of Great Britain and Sweden, the Labor Committee of the Twentieth Century Fund, the National Defense Mediation Board, the National Recreation Association, the New York City Housing Authority, the Special Commission for Rubber Research of the National Science Foundation, the United Nations Mediation Study, and the War Labor Board. Correspondents include Bernard M. Baruch, Charles A. Beard, Percy W. Bidwell, John Brophy, James A. Brownlow, Cyrus S. Ching, Evans Clark, Alfred E. Cohn, Henry Steele Commager, Norman Cousins, J. Frederick Dewhurst, John T. Dunlop, James A. Farley, Nathan P. Feinsinger, Thomas K. Finletter, Adrian S. Fisher, Ralph E. Flanders, Edwin B. Fred, Lloyd K. Garrison, Clinton S. Golden, William Green, Ira A. Hirschman, Harold L. Ickes, Eric A. Johnston, William M. Leiserson, David E. Lilienthal, Isador Lubin, Brien McMahon, Arnaud C. Marts, Arthur S. Meyer, Wayne Morse, Philip Murray, Robert R. Nathan, Reinhold Niebuhr, Casper W. Ooms, Walter P. Reuther, Daniel C. Roper, Bernard M. Ruml, Alexander Sachs, David Sarnoff, Sumner H. Slichter, W. Stuart Symington, George W. Taylor, Norman M. Thomas, Maurice J. Tobin, Harry S. Truman, Alan T. Waterman, Carroll L. Wilson, Charles E. Wilson, W. Willard Wirtz, and Edwin E. Witte.

D15 Davy, Sir Humphry, 1778-1829.
Correspondence, 1803-1822. - 35 items.
In American Philosophical Society Library (Philadelphia, PA).
English natural philosopher. - Chiefly correspondence with Alexander John Gaspard Marcet on chemistry, with references to Sir Joseph Banks, John Eric Berger, Johan Jakob Berzelius, Jean Baptiste Biot, and others; a few letters to and from John Bostock, Thomas Cooper, John Wilson Croker, Giovanni Fabroni, and Henry Penneck.

D16 Day, Jesse Erwin, d. 1935.
Papers, 1928-1935. - 0.5 linear ft.
In Ohio State University Archives (Columbus).
Professor of Chemistry. - Reprints of Professor Day's publications and correspondence, 1928-1935. Reflects his main research interest, the catalytic combustion of carbon.

D17 Day, Stephen Delevan, 1883-1960.
Papers, 1917-42. ca. 200 items.
In Nebraska State Historical Society collections (Lincoln)
Aviator, of Nebraska and Texas. Correspondence, certificates, and printed material, relating to Day's military service as a World War I aviator, aviation in the U.S. in the 1920's, and development of oil pipeline coating processes during the 1930's and 1940's.
Information on literary rights available in the repository.
Gift of Louise A. Nixon, Lincoln, Neb., 1971-72.

D18 Day, Walter, 1894-
Papers, 1933-62. ca. 2 ft.
In Wayne State University, Archives of Labor History and Urban Affairs (Detroit)
Union official. Correspondence, minutes, and papers concerning chemical workers, and United Mine Workers disputes with Berry Brothers and Acme Paint Company in Detroit. Correspondents include Sen. Homer Ferguson.
Unpublished finding aid in the repository.
Information on literary rights available in the repository.
Deposited by Mr. Day, 1964.

D19 Dean, Reginald S., 1897-1961.
Papers, 1927-1961. - 21 linear ft.
In Harry S. Truman Library (Independence, MO).
Government official and metallurgical engineer. - Collection includes correspondence, memoranda, studies, patent files, reports, notes and publications documenting Dean's career as Chief Engineer, Metallurgical Division, U.S. Bureau of Mines, 1929-1942; Assistant Director, U.S. Bureau of Mines, 1942-1946; and private consulting metallurgist, 1946-1961.
Gift of Mattie M. Dean, 1976.
Open to researchers under restrictions accepted by the Library.
Unpublished shelf list available.

D20 De Barr, Edwin.
Papers, 1891-1947.
525 items.
In University of Oklahoma Library.
University professor. Diaries, chemical research reports, notes on early Oklahoma mineral industries and mineral springs, other papers relating to the early history of the University of Oklahoma, and photos. of early campus scenes.

D21 Debye, Peter Josef William, 1884-1966.
Lectures on quantum theory, 1914-1918. - 1 reel microfilm.
In American Institute of Physics, Niels Bohr Library (New York, NY).
Chemist. - Probability and statistics with applications to specific heats, spectral and atomic models, and "Neuere Ergebnisse der Quantentheorie."
Finding aid available.

D22 Debye, Peter Joseph Wilhelm, 1884-1966.
Papers, ca. 1959-1966. - 0.1 cu. ft.
In Cornell University Libraries, Department of Manuscripts and University Archives (Ithaca, NY).
Professor of Chemistry, Cornell University; Nobel laureate in Chemistry. - Letters from Debye to Sherry K. Hathaway, and bibliographic material relating to Debye.

Delaney, James J., b. 1901.
Papers, 1952-1978. - 19 boxes (18.5 linear ft.).
In Syracuse University Library (Syracuse, NY).
Delaney, member of the 79th, 81st-87th (7th District, NY) and 88th-95th (9th District, NY) Congresses, dedicated his major efforts to education and regulation of chemical additives to food. The collection includes files, printed material, hearing notes, legislative bills, and other information relative to these subjects.
Unpublished finding aid available at the repository.
Gift, 1979.

D24 Delaware industrial miscellany, 1773-1950. 127 items.
In Eleutherian Mills Historical Library (Greenville, Del.) (various accessions)
In part, transcripts (typewritten), photocopies, and microfilm (negative and positive) made from originals in various repositories and privately owned.

Correspondence, account books, diaries, notebooks, petitions, receipt books, estate papers, census records, and other business papers, relating to flour milling, machinists, foundries, and other manufacturing and industrial interests in Delaware. Names represented include Jacob Broom (1752-1810), Bush & Lobdell, Wilmington, Del., Samuel Canby (1751-1832), Jessup & Moore Paper Company, Rockland, Del., Thomas Lea & Son, Wilmington, Del., Miller Brothers, Henry Clay, Del., J. Morton Poole Company, Wilmington, Del., John J. Phillips, New Castle Co., Del., Joseph Robinson, Wilmington, Del., Joseph Tatnall (1740-1813), and Horace Holden Thayer, Jr. (1878-1959), marine engineer and naval architect.

A composite collection consisting of small groups of papers or single items listed and described separately in A Guide to the Manuscripts in the Eleutherian Mills Historical Library, by John B. Riggs (1970).

Acquired from various sources.

D25 De Milt, Clara, 1880-1958.
Papers, 1891-1953. - 1,196 items.
In Tulane University Library, Special Collections (New Orleans, LA).
Appointed Assistant Professor in the Newcomb College Chemistry Department in 1925 and department head in 1926, de Milt later joined the Tulane graduate school faculty where she served as Chairman of the Chemistry Department until 1949. - Collection contains correspondence, notes, lectures, manuscripts and reprints of her works, her doctoral dissertation, course material, photos, and other papers. Includes letters of Wolcott Gibbs and Ann Hero; notes, lectures, manuscripts, and reprints relating to de Milt's work on organic chemistry and the history of chemistry and science; 2 manuscripts of a translation done by Hero and de Milt; papers written about de Milt after her death; and a funeral announcement for George Urbain.
Unpublished guide available.

D26 Derick, Clarence G., b. 1883.
Papers, 1965-1967. - 0.1 cu. ft.
In University of Illinois Archives (Urbana).
Chemist and professor. - Subjects include Chemical Abstracts; chemical manufacturers; courses; curriculum; dyestuff industry; organic chemistry; Schoellkopf Research Laboratories; spectrographics analysis; and spectroscopy.
Received in 1967.

D27 Detroit Society for Coating's Technology.
Records, 1923-70. 7 boxes.
In Detroit Public Library, Burton Historical Collection (Mich.).

Correspondence, minutes, bylaws, membership records, committee reports, printed material, and photos, of an organization founded as Detroit Paint and Varnish Production Club.
Unpublished finding aid in the repository.

Deutch, John M., b. 1938. D28
Papers, 1966-1980. - 1.3 cu. ft.
In Massachusetts Institute of Technology, Institute Archives and Special Collections (Cambridge, MA).
Chemist, educator. - Includes minutes, reports, and correspondence of the Committee on M.I.T. Research Structure and Committee on Education Division at M.I.T., grant proposals, notes, and memoranda. Also correspondence consisting mostly of letters of recommendation and applications for post-doctoral research postion.
Box list available in the Institute Archives.
Access fully restricted; consult the Institute Archivist for further information.

Dexter, Henry Bowers, 1827-1907. D29
Papers, 1845-1957. 7 ft.
In Rhode Island Historical Society collections (Providence)
Businessman, of Pawtucket, R. I. Personal and business correspondence and other papers of Dexter, his partner and friend, George Howard Clark (d. 1891), and members of the Dexter family. Includes Clark's diary (1845-55), containing descriptions of his social and cultural life and travels in Europe and elsewhere; records of the Pawtucket Cardboard Company which became the Rhode Island Cardboard Company; correspondence and business records of Ray Potter (1795-1858), former owner of the Pawtucket Cardboard Company; material relating to the Pennsylvania oil wells and attached machine shop operated by Clark and Dexter (known as Central Petroleum Company and Central Machine Works) and to their interests in Nevada mining; letters (1916-17) of Dexter's grandson, Albert Stearns, describing his work as an engineer at E. I. du Pont de Nemours; and other papers of the children of Dexter's daughter, Kate (Dexter) Stearns.
Gift, 1972.

Dexter, Simon Newton, 1785-1862. D30
Papers, 1793-1896. 29 boxes and 2 reels of microfilm.
In Cornell University Library, Collection of Regional History and University Archives (753)
Merchant, banker, cotton and woolen manufacturer, and New York State Canal Commissioner. Correspondence, accounts, orders, receipts, notes, contracts, and other papers relating to Dexter's business and political activities. Many of the papers relate to the Oriskany Manufacturing Company, a woolen manufacturing firm. Other papers relate to Dexter's activities as commissioner for the Erie Canal, to local and State politics, and to various business and industries such as dry goods, cotton manufacturing, sheep

raising, ironmongery, a distillery, a tannery, wholesale and retail merchants, railroads, and banks. Includes material concerning Thomas R. Gold and his family; Porteus B. Root as a contractor for the Black River Canal and a builder of plank roads; the Birmingham Manufacturing Company; Black River Woolen Company; Detroit Farmers' and Mechanics' Bank; Dexter Manufacturing Company; Dexter Village Society; Elgin Manufacturing Company; Galena and Chicago Rail Road Company; Oneida Manufacturing Company; Rock Bottom Factory, Stowe, Mass.; and Williams Woolen Company. Correspondents include E. Bacon, Edward Bates, Francis Bicknell, John Brinckerhoff, E. Capron, Seth Capron, A. J. Center, William M. Cheever, James Cooper, Oliver Culver, John Devereux, Nicholas Devereux, J. W. Doolittle, William Durer, Guy Foote, James Goss, Francis Granger, P. L. Green, Ephraim and Roswell Hart, Hiram Holcomb, C. L. Holley, Myron Holley, Alfred Hovey, O. Hungerford, Benjamin S. Huntington, S. P. Germain, Alexander B. Johnson, Hiram Ketchum, E. Kirby, G. A. Lansing, Wm. M. Larrabee, Sam Lawrence, Daniel Lee, Edwin C. Litchfield, T. W. Lockwood, E. T. Throop Martin, Simon Matteson, Charles Moseley, Isaac Northrup, George W. Patterson, John B. Pease, George T. Perry, Jonas Pratt, Thadeus Pomeroy, Benjamin Poor, Josiah Porter, Henry S. Randall, B. W. Raymond, R. J. Renwick, P. V. Rogers, James Satterlee, Peter B. Schenck, Alexander Seward, William H. Seward, Charles and Henry Seymour, C. S. Shurman, Derick Sibley, Theodore Sill, John Slade, Ambrose Spencer, J. A. Spencer, Theodore Spencer, J. D. Steele, B. F. Stillman, William M. Tallman, Comfort Tyler, Stephen Van Rensselaer, Abraham Varick, Charles D. and William Walcott, A. J. Waldron, Ezra Weeks, John A. and W. B. Welles, Edward A. Wetmore, Andrew White, John Williams, Samuel Woodsworth, and R. L. Yarwood.

Described in Reports of the curator and archivist, Collection of Regional History, Cornell University (1948-50) p. 30-33 and (1950-54) p. 34.

D31 D'Ianni, James D., b. 1914.
Papers, 1945-1947. - 2 in.
In University of Akron, American History Research Center (Akron, OH).
Polymer chemist; Charles Goodyear Medalist (ACS Rubber Division); instructor at University of Akron, 1941-46. - One folder of technical reports and correspondence pertaining to the early days of alfin rubber.
Gift of James D. D'Ianni.

D32 Dill, Marshall, 1882-1961.
Papers, 1920-1961. - 13 v. and 4 boxes.
In California Historical Society Library (San Francisco)
Civic leader, businessman, and president of 1939-1940 Golden Gate International Exposition, of San Francisco, Calif.- Correspondence and reports from the San Francisco grand jury of 1935-1938, of which Dill was a member, including letters from Mayor Angelo Rossi to Dill, regarding Edwin N. Atherton's report and the grand jury's investigation of corruption and graft in the San Francisco Police Dept.; Dill's speeches as president of San Francisco Chamber of Commerce (1939-1940); correspondence as member and chairman of San Francisco Housing Authority (1940-1943), dealing with low cost housing and housing for defense workers; notes from Alien Enemy Hearing Board of the Northern District of California (1941-1942); Dill's speeches for Roger D. Lapham's 1943 campaign for mayor; reports regarding San Francisco Municipal Railway, prepared for California Public Utilities Commission while Dill was commissioner (1947); business correspondence from Dill-Crosett, Inc., an import-export company Dill established in 1906, revealing his recognition of the potential market for American products in the Far East and his business acumen in making Dill-Crosset the West Coast's largest exporter of American chemicals and dyes to Japan by World War II; 13 scrapbooks; and 2 photo albums kept by Dill.

Dodson, William Daniel Boone, 1871-1950. D33
Papers, 1947-49. 1 box.
In University of Oregon Library (Eugene)
Journalist, salesman, and employee of the Portland Chamber of Commerce. Correspondence relating to Federal policy on Columbia River development and the establishment of the Willamette Valley Wood Chemical Company in Springfield, Or., and personal correspondence with Fred H. Kiser.
Inventory in the library.

Donnell, John W. D34
Papers, ca. 1948. - 2 v.
In Michigan State University, University Archives and Historical Collections (East Lansing).
Chemical engineer. - Two volumes, entitled A Survey of Chemical Engineering Schools (September 10, 1948).
Gift of John W. Donnell.

Donohue, Jerry, 1920-1985. D35
Papers, ca. 1940-1985. - ca. 25 cu.ft.
In University of Pennsylvania Archives (Philadelphia).
Chemist; University of Pennsylvania faculty member. - Research interests included the structure of the elements, X-ray analysis of crystalline structures, structures of compounds of biological interest, and hydrogen bonding. Donohue was an editor of the Journal of Crystal and Molecular Structure (later known as the Journal of Crystallographic and Spectroscopic Research). He received grants for work at the Laboratory for Research in the Structure of Matter (LRSM). He corresponded with Linus Pauling, Alexander Rich, and others. Collection also contains material relating to research grants for his book The Structure of the Elements and reprints. Bulk of the collection is from the mid-1960's on.

D36 Doty, Wirt P
Business records, 1916-35. 2 ft.
In University of Michigan, Michigan Historical Collections.
Pharmacist, of Detroit. Pharmacy notebook containing recipes for remedies of common ailments; three vols. of records of the Doty Land Company of Detroit; and two vols. of pharmacy accounts.
Gift of Mr. Doty, 1948.

D37 Douglas, Lewis Williams, 1894-
Papers, 1764-1968. 180 ft.
In University of Arizona Library (Tucson)
U. S. Representative from Arizona, business executive, and U. S. Ambassador to Great Britain. Correspondence, diaries, photos, clippings, printed matter, and other papers, dealing with family, friends, education, philanthropy, memberships, and Douglas' professional and public career, such as service in the Arizona State Legislature (1923-25) and in the U. S. Congress (1927-33), U. S. Director of the Budget (1933-34), president of McGill University (1938-39), Ambassador to Great Britain (1947-50), Deputy Administrator of the U. S. War Shipping Administration (1942-44), his mining, oil, and other investment interests, and his association with such organizations as American Cyanamid Company, General Motors, Mutual Life Insurance Company of New York, Rockefeller Foundation, Memorial Hospital, New York City, English-Speaking Union, Southern Arizona Bank & Trust Company, and the University of Arizona's Institute of Atmospheric Physics. Includes papers of Douglas' father, James Stuart Douglas, and of his grandfather, James Douglas, some relating to work in mining technology and associations with various mining companies, chiefly Phelps Dodge Corporation. Correspondence of all three Douglases includes major contemporary figures in politics, mining, and world affairs. James Stuart Douglas' papers include letters (1922-29) from Georges Clemenceau.
Inventory with the collection.
Gift of Mr. Douglas, 1968.

D38 Dover Forge Iron Works, Berkeley Township, Ocean County, N.J.
Records, 1821-50. 39 items.
In Monmouth County Historical Association Library (Freehold, N.J.)
Correspondence, account statements, receipts, bills, orders, narrative survey, broadside, and account book, of a bog iron works. Persons represented include owners Thomas Butcher and Samuel J. Read.
Unpublished finding aid in the repository.
Gift of Mrs. J. Amory Haskell, Red Bank, N.J., 1941 and 1942.

Drabkin, David L., 1899-1968. D39
Papers, 1927-1961. - Ca. 10 linear ft.
In University of Pennsylvania Archives (Philadelphia).
Biochemist. - Collection reflects Drabkin's research interests: water balance; urinary pigments; spectrophotometry of hemoglobin and derivatives; cytochrome C and tissue regeneration; enzymes and metabolism; protein metabolism and biosynthesis in experimental nephrosis; and bile pigment metabolism. Much of the collection is made up of Drabkin's lecture notes. Other material includes correspondence files, reprints, original manuscript materials, instrumentation manuals, and medical and scientific journals.
Folder inventory at the repository.

Drake, Lee D   1882-1957. D40
Papers, 1909-54.
5 ft.
In University of Oregon Library.
Newspaperman and promoter. 1500 letters and 850 photos., the latter of Pendleton, Or., the Pendleton Roundup, and Indians. Mr. Drake was associated with the Astorian budget, the Pendleton East Oregonian, and the Idaho evening times, and active in the Northwestern Publishers Syndicate, of which he was a founder, the Oregon State Chamber of Commerce, the Castilloa Rubber Plantation Co., Chlorine Products Corporation of America, Old Oregon Trails Association, Blue Mountain Baseball League, Pacific Northwest Tourist Association, and the Pendleton Roundup Association.

Drake, Quaesita Cromwell, 1889-1967. D41
Papers, ca. 1955-66. 1 ft.
In University of Delaware Archives (Newark)
In part, transcripts.
Professor of chemistry at the University of Delaware. Letters, notes, maps, photos, and slides, pertaining to Miss Drake's research for a biography of Eben Norton Horsford (1818-1893).
Inventory in the repository.
Gift of Miss Drake's sister, Beata Drake, 1970.

Draper family. D42
Papers, ca. 1835-1908. ca. 3 ft.
In Smithsonian Institution, National Museum of American History, Division of Physical Sciences (Washington, D.C.)
Chiefly scientific publications and reprints, of John William Draper (1811-1882), chemist, historian, Methodist minister, and pioneer in photography, and his sons, John Christopher Draper (1835-1885), physician and chemist, Henry Draper (1837-1882), astronomer and pioneer in astronomical photography, and Daniel Draper (1841-1931), meteorologist and founder of New York Meteorological Observatory; publications of University of the City of New York (New York University) and New York Meteorological Observatory; and letters (ca. 1892-1908) to Daniel Draper acknowledging receipt of publications from New York Meteorological Observatory.
Gift.

D43 Draper, John William, 1811-1882.
Family papers, 1777-1951. ca. 16,100 items.
In Library of Congress, Manuscript Division (Washington, D.C.)
Scientist and historian. Correspondence, family papers, subject files, ms. and printed copies of speeches, articles, and books, financial papers, and miscellany, relating to Draper's scientific work, American Chemical Society, his work as president of the Medical College of New York University (1850-82), and his writings, including his address (1860) before the British Association for the Advancement of Science which prompted the Bishop Wilberforce-Thomas Huxley exchange on Darwin's theory on the origin of species, his Life of Franklin, and his History of the American Civil War (1867-70). Correspondents include Benjamin Alvord Theodorus Bailey, George Bancroft, Frederick Augustus Barnard, George Bell, Schuyler Colfax, George W. Cullum, John Adolphus B. Dahlgren, James Dwight Dana, William Darling, Charles Darwin, John Gibbon, William E. Gladstone, Ulysses S. Grant, William Babcock Hazen, John Frederick William Herschel, Oliver Wendell Holmes, Benson J. Lossing, Valentine Mott, Simon Newcomb, William T. Sherman, Benjamin Silliman, Jr., Ainsworth R. Spofford, Edwin M. Stanton, John Tyndall, Gideon Welles, Josiah Dwight Whitney, and Thomas J. Wood. Includes papers of Draper's sons, John Christopher, Henry, and Daniel, and other members of the Draper, Maury, and Ludlow families. Papers of Daniel Draper relate to his work (1869-1911) with the New York Meteorological Observatory and include correspondence with Cleveland Abbe, Alexander Graham Bell, James McKeen Cattell, Alexandre Gustave Eiffel, Valentine Mott, Charles Piazzi Smyth, and John Tyndall.
Unpublished finding aid in the repository.
Gift of Daniel C. Draper, 1973-74.

D44 Drinker, Cecil K., 1887-1956.
Papers, 1920-22, 1936-49. - 1 box.
In Harvard University Archives (Cambridge, MA).
Professor of Physiology, Dean of Faculty of Public Health - Miscellaneous papers.

D45 Drinker, Henry, 1734-1809.
Papers, 1756-1869.
ca. 25,000 items.
In Historical Society of Pennsylvania collections.
Quaker merchant, ironmaster, and land developer, of Philadelphia. Business records, including correspondence, letter books, estate accounts, cashbooks, receipts, legal documents, deeds, and other business papers; also family papers of some of Drinker's descendants, material relating to public affairs during the Revolution and to Quaker meetings and activities. Data on extensive land operations in several Pennsylvania frontier counties, commodities, prices, road building, maps, bonds, indentures, poll taxes, the settlement of Drinker's and others' estates. Material concerned with settlements on the Susquehanna, Meshoppen, and Sugar Creek, settlers from Connecticut and Delaware; correspondence with Henry Pigon of London; and material on sugar refining. Of Revolutionary interest are letters relating to the Philadelphia Tea Party (1773-78), orders of Council of Safety, William Franklin's proclamation of 1775, a letter (copy) from Benjamin Harrison to George Washington (1775), and other items.
Extracts of letters from Drinker to his business partner, Abel James (James and Drinker, importing firm) published in Pennsylvania magazine of history and biography, v. 14 (1890) p. 41-45.
1. Real property—Pennsylvania. 2. Pennsylvania—Hist.—Revolution—Sources. 3. Sugar—Manufacture and refining—Philadelphia. 4. Friends, Society of. Philadelphia. I. James, Abel. II. James and Drinker, Philadelphia. III. Drinker family. IV. Pennsylvania. Council of Safety.

D46 **Druggists'** records, 1784-1895. 8 v.
In Maryland Historical Society Library (31, 32, 825, 840, 1031, 1052)
Ledgers, daybook, administration account of the estate of George W. Andrews, personal account, record of sales of apothecary supplies, prescriptions, newspaper clippings, and other papers of druggists, partly of Baltimore and Harford Co., Md.
In part, gift of Louis H. Dielman, 1945, L. M. Gurner, and W. Bryant Tyrrell, 1946, and purchase, 1938.

D47 Dryden, George Bascomb, 1869-1959.
Papers, 1901-1959. - 5 linear inches.
In Illinois State Historical Library (Springfield).
Resident of Evanston, Illinois, Chicago businessman and manufacturer, and founder of Dryden Rubber Company, 1901. - Collection contains agreements and audits (for Dryden Hoof Pad Company, Peerless Rubber Horseshoe Company, Charles P. Dryden Company, Peerless Mold and Machine Company and Dryden Rubber Company); correspondence (subjects include a proposed biography of Harvey S. Firestone, the Armed Forces Chemical Association and the publication of "Chemical Warfare Service in World War II", a trip to Europe and Africa in 1929, and the gift of the Dryden Theatre to the George Eastman House in New York); and miscellaneous papers, including "A Historical Survey of the Rubber Industry in Chicago," by H.A. Winkelman, 1946.

D48 Du Bois, Alfred, 1824-1909.
Papers, 1850-1906. 40 items.
In University of Michigan, Bentley Historical Library, Michigan Historical Collections (Ann Arbor)
Asssistant professor of chemistry, University of Michigan. Letters (1863) concerning Du Bois' removal from the university, and miscellaneous university material.

D49 DuBois, J. Harry. History of Plastics Collection.
Papers, ca. 1900-1975. - 3.4 linear ft.
In National Museum of American History Archives Center (Washington, DC).
DuBois was an industrial consultant in petrochemicals and plastics. - Collection includes trade catalogues of the General Bakelite Company, Bakelite Corporation, and competitors in the plastics industry in the 1920's and 1930's, including Dow Chemical Company and General Electric; material relating to the Boonton Rubber Company, including Richard W. Seabury's notebooks, 1911-1923; memoranda of 1916, 1918, 1923, and 1926 concerning the history of Boonton's early use of Bakelite for molded products; catalogues, photographs, and a blueprint of an 1897 hydraulic press from Charles Burroughs Company; research

reports, 1925-1932, of Sigfreid Higgins, on mechanical development research at the Bakelite Corporation; reports on sales development research; and more.
Unpublished finding aid available.

D50 Dudley, William Lofland, 1859-1914.
Papers, 1880-1915. ca. 3 ft. (700 items)
In Joint University Libraries (Nashville, Tenn.)
Professor of chemistry and dean of the Medical School, Vanderbilt University. Correspondence, lecture notes, addresses, articles, invitations, photo, and other papers. Includes material relating to the International Congress of Applied Chemistry, National Collegiate Athletic Association, and Southern Inter-Collegiate Athletic Association; class rosters of Miami Medical College; and blueprints of the Cornell University chemical laboratory.
Unpublished register in the library.
Gift chiefly of the Chemistry Dept., Vanderbilt University; and E. C. Bradley, 1967.

D51 Dugas and LeBlanc, Paincourtville, La.
Account books, ca. 1895-1933. 144 v.
In Louisiana State University, Dept. of Archives and Manuscripts (Baton Rouge)
Account books, daybooks, and ledgers, of a sugar and molasses manufacturing and general merchandising firm; payroll books (1904-33) for Armelise, Magnolia, Westfield, and Whitmel plantations and levee work (1927) in the Fourth Mississippi River District; and miscellaneous letter books, cashbooks, checkbooks and voucher registers, cane and sugar books, journals, ledgers, and daybooks.
Unpublished inventory in the repository.
Information on literary rights available in the repository.

D52 Duhamel du Monceau, Henri Louis, 1700-82.
Papers, 1716-89. - 8 linear ft. (20 boxes and 18 v.).
In American Philosophical Society Library (Philadelphia, PA).
French agronomist and botanist. - Principally on botany and agriculture, this collection includes many manuscripts on trees, shrubs, and plants of different species, copies of botanical essays by others, essays on fruit trees, etc., by Auguste Denis Fougeroux de Bondroy (1732-89), notes and drafts for the latter's revision of Duhamel's Traite des Arbes et Arbustes. Also miscellaneous essays, sketches, and memoranda on bones of birds and animals, electricity, poisons, steam engines, ventilation, temperature and air pressure, mathematics, paleontology ("Observations sur les os d'elephante fossile"), chemistry, metallurgy, entomology, architecture, taxidermy ("Methode pour empailler les oiseaux"); lists of plants; notes on England, Canada, Mexico, China; notes of reading in Pliny, John Evelyn, Alexander Russell, William Derham, and others.
Unpublished finding aid available.

Duke University, Durham, N. C. Library. D53
Notebooks, 1858-1860. - 3 v.
In Duke University Library (Durham, NC).
Notes taken by Robert Galt (Fluvanna County, Va.), on medical and chemistry lectures at the University of Pennsylvania.
Published guide available in the library.

Duke University, *Durham, N. C. Library.* D54
Teachers' papers, 1803-1954. 1544 items and 36 v.
In Duke University Library.
Diaries, journals, record books, school records, lecture notes, and other papers, largely relating to education in the South in the 19th century.
Card index in the library.
Acquired, 1930–

Dumas, Jean Baptiste A., 1800-1884. D55
Unpublished biography, 1924. - 1 volume (235 pp.).
In American Philosophical Society (Philadelphia, PA).
Prominent French chemist who excelled in applied chemical research and was renowned as a teacher. - Mimeographed biography in French by his son, General J.-B. Dumas.

Dun & Bradstreet. D56
Papers, 1760-1955. - 208 volumes.
In Syracuse University Library (Syracuse, NY).
Primarily account and ledger books from the late eighteenth to mid-nineteenth century, mostly from small businesses based in the northeastern United States. The volumes are arranged according to occupation: banking, blacksmithing, butcher, etc. Of particular interest are 5 volumes of the Philadelphia druggist Jeremiah Emlen, covering the years 1810-1826 and including letter books, day books and a ledger. These volumes document Emlen's business dealings in medicines, paints, and related items.
Unpublished finding aid available.

Dunham, Lawrence Boardman, 1882-1959. D57
Papers, 1913-59. 2 ft.
In Rockefeller Foundation Archives (New York City)
Lawyer, business executive, judge, and public official, of New York City. Correspondence, memoranda, reports, speeches, clippings, and scrapbook. Includes material on the Bureau of Social Hygiene, of which Dunham was director, 1928-34,

Clayton Oil Company, criminology, Domestic Relations Court of New York City, of which he was judge, 1934-42, immigration, juvenile delinquency, narcotics, New York City mayoral campaign of 1933, New York City Police Dept., of which he was deputy commissioner, 1913-17, and oil refining. Correspondents include Fiorello LaGuardia, John D. Rockefeller, Jr., and Arthur Woods.
Unpublished inventory in the repository.
Gift of Mrs. Dunham, 1971.

58 **Du Pont Company. Nylon Collection.**
Papers, 1939-1977. - 1.3 linear ft.
In National Museum of American History Archives Center (Washington, DC).
In 1938, the Du Pont Company was issued a patent for a new textile fiber, "nylon." Nylon for many reasons is one of the most important of Du Pont's many developments. It was the result of the chemical industry's first large-scale fundamental research program, and it proved to be the first of a whole family of synthetic fibers for consumer consumption. Collection contains materials pertaining to the research, development, and marketing of nylon.
Unpublished finding aid available.

59 **Du Pont de Nemours (E. I.) and Company.**
Records, 1801-1902. - Ca. 45,000 items.
In Hagley Library (Greenville, DE).
In addition to the initial donation of Du Pont Company records before 1971, an additional donation of records was made and considered to be a continuation of the first group as "Series I". This series includes correspondence, accounts and financial records, inventories and patents, purchase and receiving records, production records, administrative records, and records of mills outside of Delaware.
Described more fully in the Supplement to A Guide to Manuscripts in the Eleutherian Mills Historical Library (1978) pp. 182-87.

60 **Du Pont de Nemours (E. I.) and Company,** *Wilmington, Del.*
Old Stone Office records, 1802-1917.
800 ft. (477,000 items and 668 v.)
In Eleutherian Mills-Hagley Foundation Library (Wilmington, Del.)
Correspondence, journals, ledgers, production and sales books, agents' accounts, purchase, transportation, and banking records, trial balances, legal papers, inventories, receipted bills, and other business records concerning the gunpowder and woolen industries of the Du Pont Co. Includes miscellaneous papers (1913-17) of H. M. Barksdale, a Du Pont Co. executive, covering matters pertaining to personnel, organization, policy, and patent handling; business records of the woolen factory of Victor and Charles Du Pont, the cotton factory of Duplanty, McCall and Co., the tannery of A. Cardon and Co., Grasselli Co., Standard Acid Co., Gunpowder Trade Association, and other companies controlled by or affiliated with the Du Pont Co.; and Du Pont family papers and accounts.
Cataloged in the library.
Open to investigators under library restrictions.
Deposited by the Du Pont Co. and Du Pont family members, 1954-59.

Du Pont de Nemours (E. I.) and Company. D61
Miscellaneous records, 1802-1952. ca. 2600 items.
In Eleutherian Mills Historical Library (Greenville, Del.) (various accessions)
In part, transcripts, photocopies, and microfilm (negative) made in 1955-65 from originals in various repositories and privately owned.
Correspondence, administrative, financial, and production and sales records, records of explosions in the powder mills, and other papers. Most of the records are files of the superintendents of the Repauno Works of the company's Explosives Dept. and consist primarily of inter-company correspondence addressed to Oscar R. Jackson (superintendent, 1884-1901) and T. W. Bacchus (superintendent, 1902-12). Includes records of Burnside Laboratory, Penn's Grove, N. J., Eleutherian Mills and Hagley Yard on the Brandywine River near Wilmington, Del., and the Mooar, Iowa, plant.
Described in A Guide to the Manuscripts in the Eleutherian Mills Historical Library, by John B. Riggs (1970) p. 789-794, 843.

Du Pont de Nemours, (E. I.) and Company. D62
Records, 1902-1972. - Ca. 233,000 items.
In Hagley Library (Greenville, DE).
This entry includes 82 accessions, It brings together in established order a mass of material which reached the Library over a period of years. It is recorded as Series II of the Records of the Du Pont Company in continuation of Series I, which was described on pp. 575-672 of the original Guide. This Series is composed of four parts: (1) Records of Absorbed Companies; (2) Records of E.I. du Pont de Nemours & Co., 1902-72; (3) Records of Affiliated Companies; and (4) Records of Subsidiary Companies.
(1) These records represent activities of 243 companies absorbed by the Du Pont Company. The majority of the items fall within the period from the late nineteenth century to 1915 and total ca. 33,000 items.
(2) These records represent accumulations received from various Du Pont Company offices and are often fragmentary in many instances. There are great quantities of chemical research records in this group. Of particular interest are: Advertising Department records, 1907-69; Experimental Station records, 1908-30; Explosives Department records, 1889-1971; Fabrics and Finishes Department, 1927-57; Foreign Relations Department; Legal Department, 1894-1968; Patents and related papers, 1900-56; Presidential files of T. Coleman du Pont and Walter S. Carpenter, Jr.; Textile Fibers Department, Nylon Division, 1931-45, files of George Preston Hoff, technical director of the division; Vice Presidential files of H.M. Barksdale and J.E. Crane.

(3) The affiliated companies whose records are held are: E.I. du Pont de Nemours & Co. of Pennsylvania; E.I. du Pont de Nemours Export Company; E.I. du Pont de Nemours Powder Company.

(4) The records of the subsidiary companies number slightly over 15,000 items of which those of the Du Pont Rayon Company form the bulk. Other substantial groups are represented by the Du Pont Engineering Co., Du Pont International Powder Co., Du Pont Rayon Co., General Motors Securities Co., Gunpowder Trade Association, and Imperial Chemical Industries, Ltd.

Described more fully in the Supplement to A Guide to Manuscripts in the Eleutherian Mills Historical Library (1978).

D63 Du Pont de Nemours (E.I.) and Company. Chemicals and Pigments Department.
Records, 1896-1942. - 5 boxes.
In Hagley Library (Greenville, DE) (1676).
Miscellaneous records including cost accounting journals for the Niagara Falls facility (1896-1906) and Carney's Point Dye Works (1917); Roessler and Hasslacher Chemical Co. miscellaneous records 1926-50; documents relating to development of a continuous bleaching machine, 1921-1942; typescript mimeographed "History of the Electrochemicals Department", 1941; and a dye formula record book from the Jackson Technical Laboratory, 1918-1931.
Unpublished inventory available.

D64 Du Pont (E.I.) Company. Du Pont, Washington.
Papers, 1912-77. - 4 vols.
In Du Pont Historical Museum (Du Pont, WA).
Explosives plant. - Collection includes: (a) one volume of papers reporting on the celebration of the fiftieth anniversary of the plant in 1959, an employee list, and historical notes about the company; (b) a one volume scrapbook (1912-1977) including clippings, correspondence, programs, and memorabilia about the company plant; (c) a one volume scrapbook of photographs depicting the fiftieth anniversary of the plant; (d) one volume of photographs taken during the last four days of dynamite manufacturing at the Du Pont plant in 1976.
The general reference file of the Museum also contains miscellaneous information about the plant.
Card catalog in the Museum.

D65 Du Pont de Nemours (E.I.) and Company. Employee Relations Department and Predecessors.
Records, 1911-1972. - 14 boxes.
In Hagley Library (Greenville, DE) (1615).
Records of predecessor divisions and departments: 1) Safety Division, Engineering Department (1911-1919) - records documenting safety work, patriotic campaigns, accident statistics during World War I. (2) Service Department (1919-1950) - records of the Personal Relations Section (wage rates), Industrial Relations Section (study sheets on company labor - management policies, 1902-1924), and Safety & Fire Protection Division (annual reports, studies, historical records on development of safety work at Du Pont, newsletters, safety contests, and company insurance reserves).
Records of the Employee Relations Division (1950 - present): Annual reports to Executive Committee, 1950-1972, covering departmental work in labor and industrial relations, management training programs, recruitment of technical personnel, personnel research, and safety and fire protection.
Unpublished finding aid available.
Restricted.

D66 Du Pont de Nemours (E.I.) and Company. Experimental Station, Chemical Department.
Organization charts, 1928-37. - 1 item:26 pp.
In Hagley Library (Greenville, DE) (1753).
Charts of the Chemical Department technical staff at the Experimental Station, 1928-1937. These are photocopies of selected records borrowed for copying at Hagley.

D6 Du Pont de Nemours (E.I.) and Company. Office of the President.
Papers, 1909-1952. - 39 linear feet.
In Hagley Library (Greenville, DE) (1662).
Chiefly from the term of Lammot du Pont as President (1926-40) and as Chairman of the Board (1940-48). The remainder is from the Presidencies of his predecessors Pierre S. du Pont and Irenee du Pont 1914-26 and from the period 1948-52, overlapping the Board Chairmanship of Walter S. Carpenter, Jr. and the early years of the Presidency of Crawford H. Greenewalt.
Unpublished finding aid available.
Restricted.

D68 Du Pont de Nemours (E.I.) and Company. Textile Fibers Department.
Records, 1920-1970.
In Hagley Library (Greenville, DE).
Typescript drafts of histories of various aspects of du Pont fiber research and development including rayon, Qiana, terylene, and also "A Village Study of Old Hickory, Tennessee" dated 1926. Associated with the typescripts are correspondence, reports, clippings, and photographs.
Unpublished finding aid available.

D69 Du Pont de Nemours (E.I.) and Company. William P. Allen, Executive.
Papers, 1928-30. - 5 inches.
In Hagley Library (Greenville, DE) (1832).
Records from Allen's tenure as Vice President and member of the Board of Directors. Includes correspondence, subject reference files, minutes of meetings, and other documents. Subjects include employees relations, pigment research and production, and merger of Grasselli Chemical Company and the Krebs Pigment & Color Corporation with the du Pont Company.
Unpublished inventory available.

D70 Du Pont de Nemours (E.I.) and Company. Willis F. Harrington, Executive.
Papers, 1929 - 1942. - 18 linear feet.
In Hagley Library (Greenville, DE). (1813).
These records trace the career of Willis Harrington (1882-1960) as a Vice President and member of the Executive Committee of the Du Pont Company from 1929 to 1942. The papers are organized by subject within each year and cover the following major areas of Harrington's duties: employee relations, research and development, and purchasing. His papers also contain files which document important aspects of the firm's industrial diversification and general operations.
Unpublished finding aid available.

D71 Du Pont, Alfred Victor, 1798-1856.
Papers, 1847-56. 340 items.
In Eleutherian Mills Historical Library (Greenville, Del.) (512)
Partner in E. I. du Pont de Nemours and Company, manufacturers of gunpowder. Business and personal correspondence, financial papers, printed matter, and other papers, relating to real estate, loans, powder sales, C. I. du Pont and Company (woolen manufacturers at Louviers, Del.), properties at Louviers and Rokeby near Wilmington, Del., sale of Rockland Manufacturing Company in New Castle Co., Del., a paper mill at Louisville, Ky., other business matters, and personal affairs.

Includes mss. on gunpowder in French and English dating from the early 19th century. Correspondents include Alexis Irénée du Pont, Alfred Victor du Pont, Jr., Charles Irénée du Pont, Charles Irénée du Pont, Jr., Henry du Pont, and Charles Warner.
Gift, 1956.

D72 Du Pont de Nemours, Eleuthère Irénée, 1771-1834.
Miscellaneous papers, ca. 1787-1884. 200 items.
In Eleutherian Mills Historical Library (Greenville, Del.) (various accessions)
In part, microfilm (negative) made in 1963 from originals in the Library of Congress and photocopies made 1955-62 from originals owned by the American Philosophical Society, Historical Society of Pennsylvania, Massachusetts Historical Society, and Winterthur Museum.
Gunpowder manufacturer. Correspondence; reports and notes on gunpowder, powder mills, and saltpetre; biographical notes concerning Du Pont de Nemours; reports (1861) from captains of the Delaware Home Guards; receipts; personal and household accounts; records of bonds, mortgages, and real estate; and other papers of Du Pont de Nemours and his descendants. Persons represented include Victorine (du Pont) Bauduy (1792-1861), Alexis I. du Pont (1816-1857), Alfred Victor du Pont (1798-1856), and Henry du Pont (1812-1889).
Described in A Guide to the Manuscripts in the Eleutherian Mills Historical Library, by John B. Riggs (1970) p. 743-745.
Gifts and purchases, 1954-63.

D73 Du Pont de Nemours, Pierre Samuel, 1739-1817.
Miscellaneous papers, 1764-1816. 480 items.
In Eleutherian Mills Historical Library (Greenville, Del.) (various accessions)
In part, transcripts (typewritten), photocopies, and microfilm (negative) made in 1957-65 from originals in various repositories.
French economist, statesman, and diplomat. Correspondence, memoirs, essays, and other literary mss., financial accounts, and other papers, of Du Pont de Nemours and his second wife, Françoise (Robin) du Pont de Nemours (1748-1841). Subjects covered by the correspondence include political and economic matters, reports on literary and cultural interests from Paris, affairs of the Institut national de France, emigration of the Du Ponts from France to the U. S. and plans for creating a French colony there, business affairs of Madame de Staël pertaining to the firm of Du Pont de Nemours, père et fils & cie., E. I. du Pont as a powder manufacturer in the U. S., and the negotiation preceding the Louisiana Purchase in 1803. Principal correspondents include Carl Friedrich, Margrave of Baden, Philippe Nicolas Harmand, Thomas Jefferson, and Marie Anne Pierrette (Paulze) Lavoisier.

Described in A Guide to the Manuscripts in the Eleutherian Mills Historical Library, by John B. Riggs (1970) p. 779-789.
Many of the letters have been published in Carl Friedrichs von Baden, Brieflicher Verkehr mit Mirabeau und Du Pont, edited by Carl Knies (1892); Politische Correspondenz Karl Friedrichs von Baden, 1783-1806, edited by Bernard Erdmannsdorffer and Karl Obser (1888-1901); and The Correspondence of Jefferson and Du Pont de Nemours, edited by Gilbert Chinard (1931).
Gifts and purchases, 1954-65.

D74 Du Pont, Ernest, 1880-1944.
Miscellany, 1905-08; n.d. - Ca. 60 items.
In Hagley Library (Greenville, DE) (acc. #1438).
Du Pont Company chemist. - Notebook with drawings, measurements, and cost estimates for guncotton stock house at Carney's Point works, 1905; weekly reports on Stabillite, continuous process for nitroglycerin, submitted by Ernest du Pont, 1907-08, 49 items; Hudson Maxim reports, "Advantages of Stabillite: reasons why it has come to stay" and "Philosophy of erosion"; annual report on Stabillite for year ending 1907; Experimental Station, orders to H. E. Kaighn, Fin Sparre, and Ernest du Pont for tests on Stabillite, 1908, 5 items; Schutte & Koerting Co., Philadelphia, advertising brochure regarding their moist ventilator, with memorandum on prices; eulogy on Francis G. du Pont (1850-1904) addressed to the Church Club of Delaware.
Described more fully in the Supplement to A Guide to Manuscripts in the Eleutherian Mills Historical Library (1978).

D75 Du Pont, Eugene, 1840-1902; and Eugene du Pont, Jr., 1873-1954.
Personal and business papers, 1827-1928. - 3,175 items.
In Hagley Library (Greenville, DE) (acc. #1503).
President of Du Pont Company, 1889-1902 (Eugene, Sr.). - Eugene, Sr.'s papers number about 1,650 items, and include correspondence, and drafts of letters and of an article concerning coal, explosives, and government powder mills. The correspondence is alphabetically arranged and covers a wide range of topics relating to Du Pont Company business and to other powder companies. Other papers of Eugene, Sr. include family records and estates, personal correspondence and material relating to personal investments.
Papers of Eugene, Jr. (1,500 items) include business and personal correspondence, and bills and receipts.

Described more fully in the Supplement to A Guide to Manuscripts in the Eleutherian Mills Historical Library (1978).

D76 Du Pont, Eugene, Jr. 1873-1954.
Papers, 1835-1956. - Ca. 10,000 items.
In Hagley Library (Greenville, DE) (Acc. # 1599).
Great-grandson of du Pont company's founder; Assistant Director of Sales 1902-1917; a Director of the du Pont company 1917-1954. - Collection composed of 3 groups of papers: (1) Du Pont company correspondence between Vice President Alfred I. du Pont and General Manager Frank L. Connable, 1906-09. - 300 items (1 box); (2) Personal papers of Eugene du Pont, including items from family members and Frank Connable, 1835-1956. - ca. 4000 items (13 boxes); (3) Files of the Kinlock Club near Georgetown, SC of which Eugene du Pont was founder and officer, 1906-1935. - ca. 5700 items (24 boxes).
Unpublished finding aid available.

D77 **Du Pont family.**
The Eleuthera Bradford du Pont collection, 1799-1834.
1835 items.
In Eleutherian Mills Historical Library (Wilmington, Del.)
Contains early records of E. I. du Pont de Nemours and Co., including the firm's original charter of organization, articles of agreement (1801); papers concerning the Talleyrand loan to the du Ponts, company lawsuits, early administrative problems, procurement of capital and raw materials; and records of contracts, including powder contracts with the U. S. Government.
Acquired since 1954.

D78 **Du Pont family.**
The Longwood manuscripts, 1780-1954.
ca. 530,500 items.
In Longwood Library (Kennett Square, Pa.)
Personal and business papers of various members of the family, and of family business firms. The bulk of the material (covering 1893-1954) relates to the affairs of Pierre Samuel du Pont, manufacturer, financier, and philanthropist, and includes records and papers of the E. I. du Pont de Nemours and Co. Earlier family papers (1780-1906) concern Pierre Samuel du Pont de Nemours, French economist, statesman, and author; his second wife, Marie Françoise Robin du Pont de Nemours; Victor du Pont de Nemours, French diplomat and American merchant and manufacturer; his wife, Gabrielle Joséphine du Pont de Nemours; Eleuthère Irénée du Pont de Nemours, powder manufacturer and founder of E. I. du Pont de Nemours and Co.; his wife, Sophie Madelaine Dalmas du Pont de Nemours; accounts of the firm (1800-1902), with business and legal papers; material pertaining to other family members, business associates, and enterprises, including Alfred Victor du Pont, Lammot du Pont, Charles Dalmas, Peter Bauduy, Nicholas Van Dyke; textile firms (Du Pont, Bauduy & Co.; Duplanty, McCall & Co., and Rockland Manufacturing Co.); and drawings, charts, maps, and land records relating to family properties, chiefly along the Brandywine, near Wilmington, Del.
Selected items have been published.
In part, closed to investigators.

D79 Du Pont family.
Papers, 1793–1902.
30 ft. (33,000 items)
In Eleutherian Mills-Hagley Foundation Library (Wilmington, Del.)
Family correspondence, business papers, and material relating to social, cultural, and religious matters. Some of the papers were collected by Allan J. Henry for the writing of biographies of Francis Gurney Du Pont and Alexis Irénée Du Pont. Includes letters, receipts, articles of agreement, notices of sales, settlements of account, and other papers relating to Alfred Victor Du Pont, and personal papers, family, school and business correspondence, maps, plant and machine drawings, and other technical papers of Lammot Du Pont, relating to the manufacture of explosives.
Cataloged in part; finding guides in the library.
Open to investigators under library restrictions.
Deposited by the E. I. Du Pont de Nemours Co. and members of the Du Pont family, 1954-59.

D80 Du Pont, Francis Gurney, 1850-1904.
Papers, 1861-1904. - Ca. 5000 items.
In Hagley Library (Greenville, DE) (Acc. # 504, 1402, 1462, 1490, 1600).
Chemist; Vice-President and General Manager of Du Pont Corporation. - Personal and professional papers including correspondence with firms doing business with the Du Pont Company, plant specifications, records of production, technical notes and papers, and accidental explosion studies and reports.
Described more fully in the Supplement to A Guide to Manuscripts in the Eleutherian Mills Historical Library (1978) pp. 49-52, and in the original Guide (1970) pp. 746-751.

D81 Du Pont, Francis Victor, 1894-1962.
Papers, 1804-1962. 2500 items.
In Eleutherian Mills Historical Library (Greenville, Del.) (351)
Engineer and political figure, of Delaware and Maryland. Correspondence (1922-62), financial records, testimonials, notes on genealogy and biography, scrapbooks, clippings, and other personal and business papers. Includes ca. 100 papers (1914-17) of Du Pont's father, T. Coleman du Pont (1863-1930), president of E. I. du Pont de Nemours and Company and U. S. Senator from Delaware, relating to his business career. Much of the correspondence relates to Delaware and national politics, especially 1949-60, covering Du Pont's term as State finance chairman for the Republican Party in Delaware. Other papers relate to the Delaware Memorial Bridge, Delaware State Highway Dept., Executive Committee of the Highway Research Board of the National Academy of Sciences and National Research Council, Bureau of Public Roads of the U. S. Dept. of Commerce, American Road Builders' Association, and Coleman du Pont Road, Inc. Correspondents include Irénée du Pont, Lammot du Pont, and Pierre S. du Pont.
Open to investigators under restrictions accepted by the library.
Described in A Guide to the Manuscripts in the Eleutherian Mills Historical Library, by John B. Riggs (1970) p. 751.
Gift, 1962.

Du Pont, Henry, 1812-1889. D82
Miscellaneous papers, 1845-92. 344 items.
In Eleutherian Mills Historical Library (Greenville, Del.) (various accessions)
In part, photocopies made in 1959-60 from originals in the Historical Society of Delaware, the University of Rochester, and privately owned, and microfilm (negative) made in 1963 from originals in the Library of Congress.
Senior partner in E. I. du Pont de Nemours and Company. Correspondence of Du Pont and his wife, Louisa (Gerhard) du Pont; and correspondence and papers concerning Du Pont's service with the Delaware militia and Delaware political affairs. Includes letters from James Irénée Bidermann (1817-1890), of Paris, containing references to the attitude of the French government towards the Civil War in the U. S.
Described in A Guide to the Manuscripts in the Eleutherian Mills Historical Library, by John B. Riggs (1970) p. 751-752.
Gifts and purchases, 1956-63.

Du Pont, Lammot, 1880-1952. D83
Miscellaneous papers, 1801-1954. 308 items.
In Eleutherian Mills Historical Library (Greenville, Del.) (362, 384, 408)
In part, transcripts.
President of E. I. du Pont de Nemours and Company. Correspondence, personal citations and awards, and annual reports, financial statements, and related correspondence (1907-25) concerning E. I. du Pont de Nemours and Company. Includes correspondence of Thomas S. Grasselli and others relating to the role of Lammot du Pont (1831-1884) in supplying gunpowder for the Crimean War and other early letters and papers pertaining to Du Pont history; and correspondence of other members of the Du Pont family, especially Alfred Irénée du Pont (1864-1935), Bessie (Gardner) du Pont (1864-1949), Irénée du Pont (1876-1963), and Pierre S. du Pont (1870-1954). Other correspondents include C. M. Bailey, Albert Blum, Philip H. Chase, Crawford H. Greenewalt, Thomas Greenwood, and H. E. Passavant.
Described in A Guide to the Manuscripts in the Eleutherian Mills Historical Library, by John B. Riggs (1970) p. 773-774.
Gifts, 1962-65.

Dushman, Saul, 1883-1954. D84
Papers, 1924-54. ca. 7 ft.
In Smithsonian Institution, National Museum of American History, Division of Electricity and Modern Physics (Washington, D.C.)
Physical chemist. Correspondence (1936-54) with publishers, editors, scientists, academic institutions, and corporations; lectures (1926-52); books, articles, and reviews (1939-52); and research notes, technical data, and reprints (1924-54). Includes material relating to quantum mechanics, electromotive force, atomic structure, electron emission, unimolecular force, high vacuum, and other research

interests, and to Dushman's work as assistant director of General Electric Company Research Laboratory, Schenectady, N.Y.
 Unpublished finding aid in the repository.
 Gift.

D85 Du Vigneaud, Vincent, b. 1901.
 Papers, 1927-75. - 57 ft.
 In New York Hospital Cornell Medical Center, Medical Archives (New York, NY).
 Professor of Biochemistry, Cornell University Medical College, New York, N.Y.; Nobel Prize winner in Chemistry in 1955 for synthesis of polypeptide hormones. - Correspondence, memoranda, speeches, reports, scientific papers, research notebooks, office files, notations regarding laboratory equipment, requests from colleagues for samples of chemical compounds, awards, and slides.

A naphtha-fractionating tower at Exxon's Baytown, Texas, petroleum refinery, 1949. Courtesy Exxon Corporation.

East Tennessee Medicine Company (Johnson City)
Records, 1891-1925. - 2 ft.
In East Tennessee State University Library, Archives of Appalachia (Johnson City)
Producer of medicines, ca. 1885-1935. - Correspondence (1891-1925), financial records (1911-1924), form letters, place cards, advertising material, and list of company agents. Includes correspondence (1916-1921), financial records (1915-1919), legal documents (1916-1919), and minutes of stockholders meetings (1916, 1917) of Ferguson Drug Company, commercial drugstore also in Johnson City; and correspondence (1907-1922) and financial records (1905-1923) of Ferguson Drug Company's secretary-treasurer, Joseph W. Cass.
Gift of Alvin Gerhardt, executive director, Rocky Mount State Museum, Piney Flats, Tenn., 1980.
Unpublished finding aid in the repository.

East Tennessee and Western North Carolina Railroad Company.
Records, 1868-1921. - ca. 3 ft.
In East Tennessee State University Library, Archives of Appalachia (Johnson City)
Railroad transporting products of Cranberry Iron and Coal Company, Cranberry, N.C., to Johnson City, Tenn. - Correspondence, ledgers, and minute books, of the railroad and iron mine and processing plant.
Gift, 1980.
Unpublished finding aid in the repository.

Ebaugh, William Clarence, b. 1877.
Records, 1899-1914. - 7 ft.
In Case Western Reserve University Libraries, Department of Special Collections (Cleveland, OH).
Chemist, Professor of Chemistry, Denison University, Granville, Ohio. - The papers include student notebooks, correspondence and notes regarding Ebaugh's professional activity as a chemist, faculty member, Secretary of Denison Scientific Association, and editor of Journal of the Scientific Laboratories. Correspondents include paleontologists August F. Foerste and Raymond C. Moore.
Gift of Denison University.
Unpublished finding aid available.

Edgeworth, Lovell.
Student notebook, ca. 1796. - 1 v. (of 2). 989 pages Holograph.
In University of California - Los Angeles Library, Department of Special Collections.
Student of Joseph Black. - Volume contains lecture notes of Black's chemistry course taken by Edgeworth while a student at Edinburgh University.
Catalog card entry.
Library purchase, 1980.

Edison, Thomas Alva, 1847-1931.   E5
Sloane collection, 1850-1929. - Ca. 50 items.
In Edison Birthplace Museum (Milan, OH).
Inventor. - Chiefly photos, drawings, newspapers, other printed matter, and memorabilia; together with a few letters by Edison (mostly photocopies), notebook (1929) kept at Fort Myers, Florida, concerning Edison's experiments with making rubber from plants growing in the U.S., and 4 p. from notebooks (1880-86) relating to the Edison effect lamp.
Private loan collection of Mrs. John E. Sloane, Edison's daughter.
Access restricted.

Edsall, John Tileston, b. 1902.   E6
Videocassette recording, 1972.
In American Philosophical Society (Philadelphia, PA).
Biochemist. - Lecture given to the Biochemistry Department of Harvard University in April 1972 entitled " A Fifty Year Historical Perspective of Protein Chemistry."

Edsall, John Tileston, b. 1902.   E7
Papers, 1931-1976. - 35 ft.
In Harvard University Archives (Cambridge, MA).
Teacher of biochemical sciences and biological chemistry at Harvard University. - Professional correspondence, laboratory notebooks and experimental notes of Edsall, his undergraduate and graduate students, and research fellows in biology, and other papers.
Access restricted.
Information on literary rights available in the repository.
Acquired from Professor Edsall, 1966, 1969, 1974, 1977, 1979.

Edwards, John Haldane.   E8
Papers, 1890-1912. 1 ft.
In University of Michigan, Bentley Historical Library, Michigan Historical Collections (Ann Arbor)
Superintendent and secretary-treasurer of Hancock Chemical Company, manufacturers of dynamite and nitric acid in Dollar Bay, Mich. Business correspondence, financial records, tax records, inventories, contracts, account books, miscellaneous material concerning the manufacturing of blasting powder, and notes relating to the strike of company workers in 1905.

E9 **Effinger Brewing Company, Baraboo, Wisconsin.**
Records, 1902-1968. - 3 ft.
In State Historical Society of Wisconsin (Madison).
Family-owned and managed brewery, founded by Ferdinand Effinger, Sr.; existed 1885-1966. - Administrative, financial, legal, and other records, including correspondence with federal agencies (1933-1966), minute books (1911-1945), audit reports (1933-1965), tax returns (1926-1965), brewery blueprints, plans, and specifications (1913, 1962), and material documenting the decline of the family business as it lost its regional market to larger corporations.
Unpublished finding aid available.

E10 **Egleston, Thomas,** 1832-1900.
Papers, ca. 1870-1900. ca. 200 items.
In Columbia University Libraries.
Founder of the Columbia University School of Mines. Correspondence and technical reports relating to mining engineering and metallurgy. Correspondents include Seth Low.

E11 **Ehrlich, Paul, 1854-1915.**
Papers, ca. 1877-1981. - Ca. 37 items.
In Leo Baeck Institute (New York, NY).
Chemist; Nobel Prize winner, 1908; known for his discovery of a cure for syphilis. - Collection includes materials documenting his life and work: university degree (Breslau); photographs of Ehrlich and others; reprints; letters; articles by and about Ehrlich; magazine clippings; photo albums; plaques; press releases; portfolio of material commemorating the 100th anniversary of Ehrlich's birth; invitations, menus, programs relating to other commemorative events in Ehrlich's life; and other material. In German.
Unpublished finding aid available.

E12 **Eisenschiml, Otto, 1880-1963.**
Papers, 1936-1963. - 4.58 linear ft.
In Illinois State Historical Library (Springfield).
Chemist, author, and Civil War/Lincolniana expert. - Correspondence, research materials, notes, and manuscripts of articles, stories, and speeches concerning the Civil War, Abraham Lincoln's assassination, the Orpet-Lambert poisoning case, chemistry, and publication of Eisenschiml's books and articles. Correspondents include Lincoln research colleagues E. B. (Pete) Long, Ralph Newman, David R. Barbee, and Margaret Bearden; editors Harrison and Platt and T. O'Conor Sloan III; and businessman Gordon Beaham III.
5 linear inches of photographs were transferred out of the collection.
Unpublished finding aid available.

E13 **Eliot, Charles William, 1834-1926.**
Papers, 1869-1924. - 199 feet.
In Harvard University Archives (Cambridge, Massachusetts).
Chemist and President of Harvard College. -Personal and official correspondence, copies of speeches, photos, and other items relating to Harvard and its presidency.
Information on literary rights available in the repository.
Shelf list in the repository and indexes in correspondence.

E14 **Emergency Committee of Atomic Scientists.**
Records, 1946-51. 12 ft. (ca. 12,000 items)
In University of Chicago Library.
Correspondence; official papers; financial records; memoranda; minutes; papers of various scientific meetings; material relating to fundraising and educational campaigns; phonograph recordings of discussions at meetings; wire recordings of a conference held at Princeton, N. J. (Nov., 1947); and a film entitled, The atomic bomb: peace or disaster? Includes letters of R. F. Bacher, Hans A. Bethe, Edward U. Condon, Albert Einstein, Harold C. Urey, and V. F. Weisskopf.
Unpublished guide in the library.
Gift of the committee, 1952.

E15 **Englis, Duane T., 1891-1974.**
Papers, 1912-1967. - 7.1 ft.
In University of Illinois Archives (Urbana).
Professor of Chemistry at the University of Illinois. - Correspondence, research notebooks, published articles, abstracts of theses prepared under Englis' supervision (1933-1952), adviser's material (1929-1958), laboratory work and test records (1912-1957), student notebooks, retirement banquet material (1959), subject files on analytical chemistry, artichoke research, quantitative methods, corn, food technology, A. E. Staley Manufacturing Company, starch sugars, and water treatment, and other papers. Includes material relating to analysis of carbohydrates, ion exchanger, spectrophotometric analysis, color of sugar products, levulose from artichokes, water analysis, professional activities and consulting, and courses on chemistry, water chemistry, and food technology. Correspondents include Justin J. Alikonis, Harold A. Fiess, Jack M. Gillette, Gordon O. Guerrant, Donald J. Hanahan, and Russell J. Kiers.
Unpublished finding aid in the repository.

E16 Ephrata collection, 1723-1951. 2 v.
In Pennsylvania Historical and Museum Commission collections (Harrisburg)
Scrapbooks of items relating to astrological, alchemical, theosophical, literary, devotional, folklore, medical, magical, and historical subjects, of interest to a religious community of German Seventh Day Baptists in Lancaster Co., Pa., and later at Snow Hill, Franklin Co., Pa. Persons represented include Jacob Martin (1725-1790).
Unpublished listing in the repository.
Formerly part of the library of Samuel W. Pennypacker.

E17 Eppes, Richard, 1824-1896.
Student notebook, 1842-1843. - 226 p., 8 1/4 x 6 1/2 in. holograph.
In Virginia Historical Society (Richmond).
Bound volume. - Concerns lectures at the University of Virginia of Robert Empie Rogers (concerning chemistry) and George Tucker (concerning economics).
In the Eppes family (of Appomattox Manor, Hopewell, Va.) muniments, 1722-1948.

E18 Erdmann, Otto, 1804-1869.
Papers, 1829-69. - ca. 25 items.
In University of Pennsylvania Libraries, E.F. Smith Collection (Philadelphia).
German chemist.
Cataloged by item on cards in the repository.

E19 Evans, William Ernest, d. 1930.
Papers, 1846-1930. - 431 items.
In East Carolina University Library, East Carolina Manuscripts Collection (Greenville, NC).
North Carolina physician. — Papers include notes (1884) from chemistry course at Hampden-Sydney College, Virginia.
Unpublished finding aid in the library. Also described in "East Carolina Manuscripts Collection, Bulletin No. 8."
Gift of Mrs. Elizabeth Savage, Greenville, NC, 1978, 1981.

E20 Everest, David Clark, 1883-1955.
Papers, 1891-1957. 135 ft.
In State Historical Society of Wisconsin collections (Madison)
Paper manufacturer, of Wausau, Wis. Correspondence, speeches, articles, subject files, and other papers, relating to Everest's business, professional, civic, philanthropic, political, and social concerns. Includes material relating to paper mills with which he was associated; correspondence, reports, minutes, and news releases received while consultant to U.S. Office of Production Management (1940's); personal financial papers (1929-34, 1949-50); material relating to Western Paper Manufacturers Association, concerned with counteracting development of unions; papers from Everest's participation in 1928 and 1936 Republican Presidential campaigns; reports (1950) by National Planning Association on Marathon Corporation, of which he was president and chairman of the board, as case study of industrial peace; and family genealogy and other personal papers.
Unpublished finding aid in the repository.
Gift of D.C. Everest Foundation, Wausau, Wis., 1958-60.

E21 Explosives.
Papers, 1912-1918, n.d.
In Hagley Library (Greenville, DE) (Acc. # 1435).
Miscellaneous papers on the manufacture of high explosives and tetranitroanile (TNA). (1) Patent to B. J. Flurscheim for manufacture of nitroderivative of aniline, 1912; John F. Heffernan, deposition concerning manufacture of high explosives for the government at the Noblestown, Pa., and the services of Oscar Byron and Louis G. Teetsell, 1918; Christian G. Storm, deposition of like character, 1918; comparative Trauzl lead block power tests performed for the Imperial Russian Government at the Noblestown plant of Aetna Explosives Co. 1916; notes by Flurscheim and Byron, undated; Flurscheim's instructions for manufacturing TNA, undated.

E22 Eyring, Henry, 1901-1981.
Papers, ca. 1920-1981. - 90 boxes.
In University of Utah Libraries, Special Collections Department (Salt Lake City).
Professor of Chemistry and Metallurgy; administrator. - Correspondence, notebooks, diaries, date books, unpublished and published scientific papers and manuscripts, photographs, audio and video tapes, films, newspaper clippings, and books. Collection is wide-ranging and fully covers Eyring's career, from undergraduate days at the University of Arizona to administrative positions at the University of Utah. Eyring was a major force in theoretical chemistry, and made significant contributions to quantum chemistry, chemical kinetics, liquid theory, the theory of optical activity, the nature of natural and synthetic fibers, and the physical understanding of biological processes. He also developed a first-rate graduate program in chemistry at the University of Utah, and served as Dean of the Graduate School and Distinguished Professor in the Chemistry Department for many years. He was also President of both the American Chemical Society and the American Association for the Advancement of Science.
Unpublished finding aid available.

F1 **Fairfield, N. Y. Seminary.**
Records, 1810–1922.
4 ft. (2700 items)
In Syracuse University Library.
Papers of the Fairfield, N. Y. Academy (1803–ca. 1895) and its sister institution, Fairfield Medical College (1812–40) Includes lecture notes (ca. 1828–68) in chemistry and physics of William Mather, the personal papers of Alonzo C. Mather together with the personal papers of Messrs. Buell, Chassell, Mather, Watkins, and other faculty members and their families.
Gift, 1954.

F2 **Fajans, Kasimir, 1887–1975.**
Papers, 1912–75. – 13 ft.
In University of Michigan, Bentley Historical Library, Michigan Historical Collection (Ann Arbor).
Physical chemist, Director of Institute of Physical Chemistry, University of Munich, and Professor of Chemistry, University of Michigan. – Correspondence, lecture notes, and other papers, largely concerning Fajan's professional interests, especially relating to his teaching and publications; and material pertaining to his concern for Jewish scholars in Germany before and during World War II.
Unpublished finding aid in the repository.

F3 **Fall River Iron Works,** *Fall River, Mass.*
Records, 1821–1909.
54 ft.
In Harvard Business School, Baker Library.
Letters, general accounts, and production and sales records, relating to the manufacturing of hoop and bar iron and nails. Includes records of Annawan Manufactory (producers of printed cloth), 1825–99; letters and general and production accounts of Metacomet Mills, 1846–1903; records of Fall River Gas Works, 1847–79; records of Bay State Steamboat Co., 1850–70; and accounts for individual ships and general accounts of a steamboat line operating between Fall River and Providence, 1829–81.
Inventory in the library. Also described in List of business manuscripts in Baker Library, compiled by Robert W. Lovett (2d ed., 1951)
Deposited by American Print. Co., 1930.
I. Annawan Manufactory, Fall River, Mass. II. Metacomet Mills, Fall River, Mass. III. Fall River Gas Works, Fall River, Mass. IV. Bay State Steamboat Company, Fall River, Mass.

F4 **Falls City Brewing Company, Louisville, Kentucky.**
Records, 1906–1978. Microfilm: 4 reels.
In University of Louisville Archives, (KY).
Founded in 1905; last surviving local beer producer and distributor in Louisville. – Collection includes Board of Directors and Annual Stockholders Minutes; Annual Financial Audit Reports; "An Appraisal of the Falls City Ice & Beverage Company, May 23, 1931," and a "Descriptive Manual of Cost and General Accounting System, August 1, 1936."
Researchers must secure written permission from the President, Falls City Industries, Inc., or a designated company representative for access to the Board of Director's Minutes, 1970 to 1978.
Finding aid available.

F5 **Fankuchen, Isidor, 1904–1964.**
Papers, 1935–64. 11 ft.
In American Institute of Physics, Center for History and Philosophy of Physics collections (New York City)
Physicist, chemist, educator, and American editor of Acta crystallographica. Personal, scientific, and other correspondence, MSS. of writings, reprints, photos, doctoral thesis (Cornell University), data books, samples of specimens, and lecture and research notebooks. The bulk of the collection is post-1947 correspondence. Subjects include Fankuchen's work in the fields of x-ray diffraction and crystallography, and special topics such as academic freedom. Correspondents include John Desmond Bernal, George Lindenberg Clark, Norman Fordyce McKerron Henry, Richard Fiske Jarrell, Raymond Pepinsky, Aaron Sidney Posner, Dorothy Wrinch, the American Crystallographic Association, the American Society for X-Ray and Electron Diffraction, the International Union of Crystallography, the National Bureau of Standards, National Research Council, National Science Foundation, the U. S. National Committee on Crystallography for the International Union of Crystallography, and representatives of various academic institutions, journals, publishing companies, technical and humanitarian societies, industry, and Government agencies.
Unpublished guide in the repository.
Information on literary rights available in the repository.
Gift of Dina D. Fankuchen, 1965.

F6 **Farmer, Moses Gerrish, 1820–1893.**
Papers, ca. 1813–1904. – 5 boxes.
In University of California - Los Angeles Library, Department of Special Collections.
Inventor and pioneer electrician, known for his patents and applications of electricity, including first electric fire-alarm system in the U.S., a miniature electric railroad, the means for duplex and quadruplex telegraph, electrotyping (depositing aluminum electrolytically), an incandescent electric lamp, and a "self-exciting" dynamo. – Papers include Farmer's personal correspondence, much of it dealing with patents; biographical material, including an autobiography and other Farmer writings; and family correspondence, including legal papers, diaries, and photographs.
Purchased by the Library in July 1962.
Unpublished finding aid available.

F7 **Fay family.**
Papers, 1850–68.
45 items.
In California State Library.
Correspondence and business papers, of Logan, David, Patrick, and John Fay, brothers who came to California in 1849 to work in the mines. Includes 35 letters written by the Fays to their family in New York and papers relating to the soap business which they established in San Francisco.
Card index in the library.
Gift of John W. Scott, 1911–13.

Federal Glass Company, Columbus, Ohio.
Records, 1879-1979. 10 ft.
In Ohio Historical Society collections (Columbus) (MSS 665)
Administrative files, appraisals, efficiency studies, financial records, fuel studies, histories, legal documents, sales and marketing material, union records, contract negotiations and revisions, and other records; together with merger report and record of 1958 merger with Federal Paper Board, Inc., and material relating to American Glassware Association and National Association of Manufacturers of Pressed and Blown Glassware.
Unpublished inventory in the repository.
Gift of Federal Paper Board Company, Inc., 1979.

Federation of American Scientists.
Records, 1946-58. ca. 10 ft.
In University of Chicago Library.
Correspondence, administrative files, minutes of meetings, financial records, subject files on topics dealing with atomic energy, newsletters, press releases, and miscellaneous documents.
Unpublished guide in the library.
Gift of the federation, 1960.

Ferry, John D., b. 1912.
Papers, 1951-1963. - 1 box.
In University of Wisconsin Archives (Madison).
Professor of Chemistry. - Administrative and financial reports covering contracts with the following agencies: Hercules Powder Company, prime contractor for Navy Ordnance, 1951-56; Office of Naval Research, 1951-63; and Picatinny Arsenal, 1953-62.
Received in December, 1977.
Unpublished finding aid available.

Ferry, Ronald M., 1891-1970.
Papers, 1930s-1970. - 1 box.
In Harvard University Archives (Cambridge, MA).
Professor of Biochemistry. - Miscellaneous papers.

Fessenden, Reginald A., 1866-1932.
Papers, 1887-1935. - 60.5 ft.
In North Carolina State Office of Archives and History Collections (Raleigh).
Physicist and inventor. - Correspondence, with other scientists and inventors, reflects his teaching career at Purdue and Western University of Pennsylvania, as well as his employment with the U.S. Weather Bureau and his inventing career, particularly the continuous wave principle of wireless transmission. Subjects include electricity, magnetism, the nature of molecules, "ether," and the sun; scientific terminology, standards, and measures; and technology of metallurgy and alloys, insulations, X-rays, microphones, microfilm, and wireless telegraphy. Other papers reflect his later career as an electrical engineer and include notebooks, instructions, reports on experiments and tests, and photographs. In addition, there are page proofs of a book by Fessenden, articles, an incomplete autobiography, clippings, and miscellaneous personal papers.

Fieser, Louis P., 1899-1977. F13
Papers, ca. 1950-1965. - 5.5 linear ft.
In Harvard University Archives (Cambridge, MA).
Professor of organic chemistry. - Collection contains Fieser's miscellaneous correspondence; card files relating to students in Chemistry 20, including student course cards, evaluations by graduate teaching fellows, and letters of recommendation written by Professor Fieser; reprints; and Chemistry in Three Dimensions, 1963, published by author.
Transferred from Chemistry Department, December 4, 1978.
Unpublished finding aid available.
Access restricted. Contact archivist for further information.

Findlay, Alexander. F14
Diary, 1924-1925. - 1 folder.
In Stanford University Archives (Stanford, CA).
Professor of Physical Chemistry, University of Aberdeen. - Diary kept during a visiting professorship at Stanford University. Also includes news clippings.

Finke, Almore H F15
Papers, 1941-52.
40 items.
In State Historical Society of Wisconsin collections.
Mainly correspondence of Dr. Finke, director of dental health in Sheboygan, Wis., pertaining to fluoridation of Sheboygan's water supply.

F16 Fischer, Emil, 1852-1919.
Papers, 1876-1919. – 37 boxes, 12 cartons, and oversize material.
In University of California, Berkeley, Bancroft Library.
Forms part of the repository's History of Science collection.
German biochemist.– Correspondence, writings, autobiography, subject files relating to Fischer's research, to work during World War I, and to professional activities; laboratory notebooks (his own and his students'); reprints, clippings, photos, and material from scientific societies. Includes correspondence and papers of his son, Hermann.
Unpublished finding aid in the repository.

F17 Fisher family.
Papers, 1855-1899. – 35 items.
In Oregon Historical Society Library (Portland).
Ezra T. T. Fisher, surveyor, Willamette Valley. – Collection includes correspondence of Ezra T. T. Fisher (b. 1835) and family, 1858-1924; survey daybook, 1855-1858, with accounts for surveys, farm supplies, loans for claims, medicinal recipes (1 v.), documents, including indentures, order for mineral surveys; miscellaneous materials relating to mines in Oregon.
Gift of Miss L. Fisher, 1970.
Unpublished finding aid available.

F18 Fisk, Charles Frederick.
Papers, ca. 1861-1965. 13 cartons.
In University of California, Berkeley, Bancroft Library.
Business correspondence, reports (ca. 1930's and 40's), deeds for mining and farm property in Alpine, Amador, Calaveras, and Tuolumne Counties, Calif., mining stock certificates and mine assessment receipts, accounts, family papers (1861-1916), and other papers. Includes letters to Frank W. Fisk and material relating to the Penn Chemical Company and the New Penn Mine (ca. 1948-65), to bee culture, goat raising, and other ranching operations in Kings Co., Calif., to real estate operations in Oakland, Calif., the Corcoran Land and Farm Company, and other investments and business enterprises.

F19 Fletcher, Frank Ward, 1853-1922.
Papers, 1871-1922. ca. 1 ft. and 4 v.
In University of Michigan, Michigan Historical Collections.
Lumberman of Alpena, Mich., and regent of the University of Michigan. Correspondence of Fletcher, mainly with his father, George N. Fletcher, relating to the lumbering and paper manufacturing business in Alpena and Detroit, Mich.; personal and financial papers; and miscellaneous business papers. Includes letter press book (1885-98) of Fletcher and his father; letter press book (1871) of Fletcher's father; letters on State politics and University of Michigan affairs; and two account books of Frank Fletcher.
Gift of Philip K. Fletcher, 1936.

F20 Flory, Paul J., 1910-1985.
Papers, ca. 1935-1985. – Ca. 65 linear feet.
In University of Pennsylvania, E. F. Smith Collection (Philadelphia).
Professor of Polymer Chemistry, Stanford University; Nobel laureate in Chemistry, 1974. – Collection includes correspondence, notes, and drafts for most of Flory's more than 350 publications; research notes; reprints collected by Flory; a complete set of Flory reprints; and files containing drafts and notes of the many lectures delivered by Flory around the world, including drafts of his 1974 Nobel address.
Gift of Mrs. Emily Flory, December 1985.
Finding aid available.

F21 Forbes, George Shannon, 1882-1979.
Papers, 1889-1971. – 14 ft.
In Harvard University Archives (Cambridge, MA).
Teacher of chemistry at Harvard University. – Personal and professional correspondence, reports, laboratory work papers and notebooks, lecture notes and other material relating to teaching, personal cashbooks, and biographical and bibliographical material.
Access restricted.
Information on literary rights available in the repository.
Acquired from Professor Forbes, 1972.

F22 Forbes, Robert Humphrey, 1867-1968.
Papers, 1870-1967. – 56 linear ft.
In Arizona Historical Society (Tucson).
Professor of Chemistry, Director of Territorial and State Department of Agriculture, agronomist, state legislator. – Correspondence, research notes and papers, research and legislative proposals dealing with chemistry, agriculture, legislation, water and environmental problems and conservation in Arizona and abroad. Forbes was head of the Arizona Department of Agriculture from 1901-1918 and worked as an agronomist for the State Department, 1916-1931, before becoming a state legislator, 1938-1954.
Unpublished finding aid in the repository.

F23 Fordyce, George, 1736-1802.
Lectures, ca. 1770. – 931 p.; 21 cm.
In National Library of Medicine (Bethesda, MD).
Lectures on chemistry given at London, ca. 1770.

F24 Foreman, Edward R., 1808-1885.
Papers, ca. 1868-1879. — 0.33 linear ft.
In Smithsonian Institution Archives (Washington, DC).
Chemist; professor. — Foreman was an assistant in the Smithsonian Institution from 1848 to 1852, when he became Chief Examiner in the Patent Office. The papers include lectures and notes on chemistry, ethnology, and natural history, especially botany; a sketch of the geology of Arkansas; and various lists of specimens.

F25 Foskett & Bishop Company, New Haven, Conn.
Records, 1893-1947. 855 items.
In New Haven Colony Historical Society collections (Conn.)
Ledgers and account books, work contracts, tax records, insurance policies and records, building plans and architectural drawings, and other records of an engineering and contracting firm, which often worked with local architects Douglas Orr and Westcott & Mapes. Clients include Yale University, New Haven railroad, Armstrong Rubber, A. C. Gilbert (toy manufacturers), Sargent's (hardware firm), and New Haven Clock Company.

F26 Fox, Denis Llewellyn, b. 1901.
Manuscript (photocopy). — 28 cm.
In University of California, Berkeley, Bancroft Library.
Emeritus Professor of Marine Biochemistry, Scripps Institution of Oceanography. — Autobiography written between the years 1969 and 1974, primarily for his family and friends. Covers personal reminiscences of his childhood in and around Napa, California; undergraduate student days at the University of California, 1921-25; and professional work. Includes copies of his letters and poems.
Description in repository.

F27 Franck, James, 1882-1964.
Papers, 1905-1966. — 13 linear ft.
In University of Chicago Library.
Professor, Department of Chemistry (1938-47); Manhattan Project (1942-45); one of the founders of Atomic Scientists of Chicago. — Correspondence (late 1910's to 1964) with leading physicists and chemists, including among the latter: Fritz Haber, Otto Hahn, Robert L. Platzman, Weldon Brown, Norman Nachtrieb, Nathan Sugarman, Robert S. Mulliken, and Zay Jeffries; manuscripts of articles relating to photosynthesis (1935-64); and reports, memoranda, and correspondence concerning the postwar atomic scientists' movement (1944-53), including the papers of the Jeffries Committee (1944) and of the Franck Committee (1945).
Published finding aid available.

F28 Frank, Adolf, 1834-1916.
Papers, ca. 1857-1957. — 4.6 ft.
In Leo Baeck Institute (New York, NY).
German industrial chemist. — Business and scientific correspondence of Adolf Frank, concerning fixation of atmospheric nitrogen, affairs of the Charlottenburg Gas Works, scandals surrounding Professor Paul Wagner, and the German chemical industry during the First World War, including a letter from Fritz Haber. Laboratory notebooks of Frank, his son Albert, and their associates; research papers, lectures and professional opinions of Frank; clippings on the chemical industry, and material relating to patents. Also business records of chemical concerns, including the Cyanid Gesellschaft, the Vereinigte Chemische Werke, Brandenburg, and the Bayerische Stickstoff Werke, containing its correspondence with I. G. Farben. Obituaries, photo albums, and clippings on Frank and on German politics and culture. In German, English, and French.
Gift of Robert Frank, 1968.
Unpublished finding aid available.

F29 Frank, Joseph Otto, 1885-1949.
Papers, 1926-1949. — 1 box.
In State Historical Society of Wisconsin (Madison).
Professor of Chemistry at the State Normal School at Oshkosh, Wisconsin. — Primarily correspondence concerning research Frank did for the Calumet Baking Powder Company and the General Foods Company.
Gift of Mrs. Glenn Sharratt, Madison, Wisconsin.
Unpublished finding aid available.

F30 Frear, William, 1860-1922.
Papers, 1850-1922. – 15.75 ft (5 ft. processed).
In Pennsylvania State University Libraries (University Park).
Professor of Agricultural Chemistry, Pennsylvania State College. – Correspondence, memoranda, press releases, reports, test analysis results, notebooks, pamphlets, and architectural plans, relating to Frear's professorial duties, his role in drafting state and federal pure food laws, and work in the chemistry of foods and chemistry of soils and fertilizers, particularly: service as chairman of Federal-State Joint Committee on Food Definitions and Standards; chemist for Pennsylvania Bureau of Foods and Pennsylvania Department of Agriculture; President of Association of Official Agricultural Chemists; and Chairman of National Pure Food and Drug Congress. Correspondents include Alva Agee, N. B. Critchfield, Thomas Edge, Edwin E. Sparks, and H. W. Wiley. Family correspondence and financial records are also included.
Gift of Donald E. H. Frear and Svend Pedersen, 1964, transfer from Pennsylvania State University College of Agriculture, 1966, and gift of Dr. Mary Frear Keeler, 1980.
Unpublished finding aid available.

F31 Freeland, Emile C.
Papers, 1919-1975. – 5 linear ft.
In Louisiana State University, Department of Archives (Baton Rouge).
Chemical engineer and LSU alumnus (1918). – Papers consist of reports (1926 - 62) on Latin American industries selected as being of interest to workers in tropical industries. Additional papers consist of reports (1960's) on Pakistani and Taiwanese industries, prepared for the United States Agency for International Development; technical articles, pamphlets and notes by Freeland; letters relative to sugar cane bagasse and the by-products that are processed from it; and papers relative to Freeland's memoirs, Tales of a Sugar Tramp. Also included are papers and other materials which pertain to sugar culture and processing.

F32 French, Charles C., b. 1910.
Papers, 1914-71. – 3 lin. ft.
In Washington State University Libraries, Manuscripts Division (Pullman).
Professor of Chemistry and administrator at Randolph - Macon College and President of Washington State University. – Correspondence, photographs, certificates, memorabilia and other papers.
Container list at repository.

Fries, Amos Alfred, 1873–   F33
Papers, 1903-52. ca. 4 ft.
In University of Oregon Library (Eugene)
Army officer. Correspondence and other papers relating to Fries' positions as chief of gas service of the American Expeditionary Forces, and chief of Chemical Warfare Service (1920-29), to his activities with the Friends of the Public Schools of America and with patriotic societies, and to the planning and construction of The Dalles-Celilo Canal. Correspondents include Gretta Deffenbaugh, Elizabeth (Kirkpatrick) Dilling, and George W. Robnett.

Frisch, John G., 1899-1952.   F34
Papers, 1921-1952. – 9 boxes and 4 v.
In State Historical Society of Wisconsin (Madison).
Dentist. – Materials relating to Dr. Frisch's leadership of the campaign for fluoridation of public water supplies. Comprises 5 boxes of his correspondence (1931-52), including many letters from Dr. Frederick S. McKay of Colorado Springs, Colorado, another leader in the field of fluoridation; articles and speeches by Frisch; statistical summaries of fluoridation experiments in Wisconsin and other states; a pre-fluoridation survey made in Madison (2 v., 1947), and a post-fluoridation survey (1 v., 1951), together with lecture notes (2 v., 1921-22) taken by Frisch at the Marquette University dental school.

Frondel, Clifford, b. 1907.   F35
Papers, 1960-64. – 1 folder, 2 boxes.
In Harvard University Archives (Cambridge, MA).
Professor of Mineralogy. – Scientific and miscellaneous correspondence; reprints and pamphlets.

Fulmer, Elton, 1864-1916.   F36
Letterbooks, 1894-1901. – 2 v.
In Washington State University Library (Pullman).
Professor of Chemistry. – Personal and professional outgoing correspondence.

Fulton, Gardiner, 1782-1861.   F37
Papers, 1818-78. 72 items.
In Eleutherian Mills Historical Library (Greenville, Del.) (513, 969)
Agent of E. I. du Pont de Nemours & Company, at Frankford, Pa. Business and personal papers, consisting chiefly of letters from E. I. duPont de Nemours & Company concerning orders, shipments,

credit, and payments; powder account book (1845-49); receipts (1818-54); and personal accounts and letters concerning the Fulton family.

In part, described in A Guide to the Manuscripts in the Eleutherian Mills Historical Library, by John B. Riggs (1970) p. 842.

Gifts, 1956 and 1967.

F38 Funk, Casimir, b. 1884.
Papers, 1944-1963. - 6 items.
In Library of Congress, Manuscript Division (Washington, DC).
Author and biochemist. - Correspondence (1954-1963), clipping, photograph, and typescript of autobiography: "Life of C. F.: A Retrospect at the Age of 60."
Gift of Casimir Funk, 1963.

F39 Fuson, Reynold Clayton, b. 1895.
Papers, 1924-1964. - 0.6 ft.
In University of Illinois Archives (Urbana).
Professor of Chemistry at the University of Illinois. - Correspondence and other papers relating to appointments, Alpha Chi Sigma, the American Chemical Society, book publication, the Center for Advanced Study, the Class of 1930 of the University of Illinois, committees, contacts with Italian and German chemists, Fuson's research and university work, former students' graduate work and employment, the Fuson Fund, John R. Kuebler Award, speaking engagements, sabbatical leaves, and travel. Correspondents include Roger Adams, Wallace H. Carothers, Elmer P. Kohler, and A. P. Tanberg.
Received 1965.

The Bridesburg plant of Rohm and Haas Company, Philadelphia, 1917. Courtesy Rohm and Haas Company.

G1 **Galland-Burke Brewing and Malting Company, Spokane, Washington.**
Records, 1891-1913. - 6 ft.
In Washington State University Library (Pullman).
Accounts, sales records, production records, letterbooks, minutes and other papers.
Unpublished finding aid available.

G2 **Garford, Arthur Lovett, 1858-1933.**
Papers, 1877-1933.
72 ft. (ca. 37,000 items)
In Ohio Historical Society Library.
Industrialist, politician, and philanthropist. Correspondence, speeches, articles, balance sheets, lists, account books, 2 scrapbooks (1912-16), a list of chairmen and officials of the Roosevelt Memorial Association, printed material, other papers, and photos, concerning the industrialization of Elyria, Ohio, financing the Republican Party in Ohio, the participation of businessmen in politics, the Republican-Progressive split in Ohio, 1912-16, (Garford was the Progressive candidate for governor in 1912, and for senator in 1914), the foreign born and Negro voter, labor relations and labor in politics, temperance, prohibition, woman suffrage, the tariff, conservation, Garford's business career, his philanthropic activities, mining in southwestern United States, and the Martin bomber in World War I. Includes records (2 boxes, 1921-28) of the Cleveland Automatic Machine Co., records (2 boxes, 1919-24) of the Electro Alloys Co., daybooks and account books (8 v., 1891-1902) of the Garford Manufacturing Co., and a ledger (1904-06) of the Worthington Ball Co. Correspondents include John Berry, Walter F. Brown (530 items, 1906-33), Theodore Burton, Meyers Y. Cooper (223 items, 1916-33), James B. Cox, Harry M. Daugherty, Charles Dick, Simeon Fess, James R. Garfield (120 items, 1910-33), John H. Hammond, Warren G. Harding, Myron T. Herrick (85 items, 1904-22), Harold Ickes, Jay Ford Laning, James V. Martin (ca. 100 items, 1917-18), George W. Perkins (73 items, 1912-19), Gifford Pinchot, William C. Procter (53 items, 1916-23), Theodore Roosevelt, William G. Sharp, William H. Taft, Leonard Wood, and Nathaniel Curwin Wright.
Described in Inventory of the Arthur Lovett Garford papers, 1961.
Gift of Katharine Garford Thomas, 1948-61.

G3 **Garrett, Alfred Benjamin, b. 1906.**
Papers, 1954-1963. - 1.5 linear ft.
In Ohio State University Archives (Columbus).
Professor of Chemistry. - Collection concerns the involvement of Garrett in university research programs: Advisory Research Council, 1954-1961; Institute of Research in Vision, 1958-1963; and Natural Resources Institute.
Gift of Professor Ernest W. Bowerman, 1974.

G4 **Garvan, Francis P., 1875-1937.**
Papers, 1917-1956. - 297 ft.
In University of Wyoming, American Heritage Center (Laramie).
President, The Chemical Foundation, 1919-1937. - Financial records, minutes, licenses, patent records and applications, scrapbooks, speeches and extensive corporate records including correspondence, legal documents and printed material, all pertaining to The Chemical Foundation, founded in 1919 by Francis P. Garvan, President Woodrow Wilson, A. Mitchell Palmer, and Joseph H. Choate, Jr. Includes manuscripts: "Chemical Foundation History," and "The American Chemical Industry History," by Edward J. Muhs. As U.S. Alien Property Custodian during World War I, Garvan was able to secure important chemical patents held exclusively by Germany for the United States. The Chemical Foundation was founded to buy these patents for use by private industry. In 1922, the United States and Germany disputed the sale of same and the collection reflects the ensuing court battles. Also includes records of scholarships and grants for scientific research, which Garvan believed would stimulate the U.S. chemical industry.
Gift of Mrs. Francis P. Garvan and Anthony N.B. Garvan, 1973.
Finding aid available.

G5 **Gaskill, James R   M   1820-1894.**
Papers, 1854-1926. 185 items.
In Illinois State Historical Library.
Physician. Personal correspondence, Civil War material, and miscellaneous papers of a surgeon who served with the 45th Illinois Infantry. The correspondence is with Gaskill's wife, Clara, and with friends and relatives in Greenville and Columbia, Ill., and in Marine (later Marine Mills), Washington Co., Minn., and concerns Gaskill's store in Marine, and his Army service. The Civil War papers consist of official correspondence and medical reports for the 45th Illinois Infantry (1863-65) including field orders and circulars; records concerning recruits examined; disability discharges; reports and invoices on hospital and medical property; and reports on sick and wounded. Includes a circular (1854) giving prices and products of a wholesale drug and chemical company, and a notebook (1879) giving formulae for prescriptions.
Unpublished guide in the library.
Gift of Philip D. Sang, 1962.

G6 **Gasser, Herbert Spencer, 1888-1963.**
Papers, 1935-61. 20 ft.
In Rockefeller University Archives, Rockefeller Archive Center (Pocantico Hills, North Tarrytown, N.Y.)
Physiologist and director of Rockefeller Institute for Medical Research. Administrative, professional, and personal correspondence, mss. of writings and addresses, bibliography, collected reprints, biographical material, memorabilia, laboratory notebooks, microscopic slides, lantern slides, electromicrographs, specimens, and photos; together with records relating to Gasser's work on behalf of the John J. Carty Medal and Award for the Advancement of Science (1946-52); as official investigator for National Defense Research Committee and Office of Scientific Research and Development (1940-49); on an advisory committee

with Vannevar Bush on synthesis of penicillin (1944-46); his activities for International Physiological Congresses (1937-50) and consideration of an International Union of Physiology; antivivisection campaigns for National Society for Medical Research; and subject files dealing with administration of the Rockefeller Institute. Correspondents include George Bishop, Detlev W. Bronk, James Conant, Sir Henry Dale, J.C. Eccles, Joseph Erlanger, Wallace Fenn, John F. Fulton, A.V. Hill, Joseph Hinsey, Alan Hodgkin, and Walter J. Meek.

Unpublished finding aid in the repository and in Rockefeller University Archives in New York City.

Gift of Rockefeller University.

G7 General Economy, *Northampton Co., Pa.*
Records, 1750-71.
80 ft.
In Moravian Archives (Bethlehem, Pa.)
Accounts of the farms, forests, dairy, orchards, stage coach, ferry, store, taverns, and schools, together with business records of the following occupations: apothecary and surgeon, baker, blacksmith, brick and tile maker, cloth weaver, cobbler, cooper, distiller and dyer, foundry operator, fuller, gristmiller, gunstock maker, hatter, locksmith, mason, millwright, oil miller, pewterer, potter, nailsmith, saddler, saddletree maker, sawmill operator, silversmith, skinner, soap maker, stocking weaver, tailor, tanner, tawer, turner, wagon builder, and wheelwright. Some of the records extend beyond the cessation of the Economy in 1762. Material is related to the repository's records of the Moravian congregation in Bethlehem, Pa.
Open to investigators under restrictions of the repository.
Information on literary rights available in the repository.

G8 General Tire and Rubber Company, Akron, Ohio.
Records, 1940-1973. - 100 cu. ft. and 300 microfilm reels.
In University of Akron, American History Research Center (Akron, OH).
Records of Akron tire and rubber company. - Collection contains court records, exhibits, technical reports, manufacturers' literature, and other material, both original and microfilm, relating to legal proceedings, especially concerning oil-extended rubber patents. Includes material on the German rubber industry; other major American rubber manufacturers; major American oil companies; and government agencies.
Unpublished finding aid available.

G9 Gershenfeld, Louis, 1895-
Bertha and Louis Gershenfeld papers, 1924-76.
1 ft. (104 items)
In Philadelphia Jewish Archives Center (Pa.)
Bacteriologist, of Philadelphia. Papers, publications, photos, and miscellany, of Gershenfeld and his wife, Bertha. Includes bacteriology text and reprinted articles by Gershenfeld, and papers and books of B'nai B'rith, Dropsie College, Galen Pharmaceutical Society, Mikveh Israel Congregation, and National Council of Jewish Women.
Gift of the Gershenfelds, 1976.

G10 Giauque, William Francis, 1895-1982.
Papers, ca. 1920-1981. - 30 cartons.
In University of California, Berkeley, Bancroft Library.
Nobel Prize winner for chemistry, 1949. - Correspondence, writings, reports, student theses, notebooks, drafts of publications, grant applications for research and the building of low temperature labs, and departmental papers from UC Berkeley. Includes manuscripts on Giauque's wartime research involving the liquification of hydrogen and oxygen and other contract work for the government.
Partially processed collection; folder list available in repository.

G11 Gibbes, Lewis Reeves, 1810-1894.
Papers, 1793-1894. ca. 5700 items.
In Library of Congress, Manuscript Division (Washington, D.C.)
Mathematician and naturalist. Principally correspondence, together with some printed material, clippings, specimen lists, resolutions, and miscellany, chiefly 1838-94, relating mainly to Gibbes' career as professor of mathematics, physics, and astronomy at the College of Charleston (S.C.). Includes correspondence with other scientists relating to astronomy, physics, geology, meteorology, chemistry, botany, zoology, American Association for the Advancement of Science, and Smithsonian Institution. An addendum to the collection contains correspondence of James McBride (1784-1817), botanist, of South Carolina, including letters from Thomas Smith Grimké and John C. Calhoun, the latter discussing matters relating to the War of 1812. Gibbes' correspondents include Stephen Alexander, Alexander D. Bache, Jacob Whitman Bailey, E.R. Beadle, Amos Binney, Langdon Cheves, Thomas Cooper, James Dwight Dana, James D.B. DeBow, Henry William DeSaussure, Charles Ellet, Elizabeth Fries Lummis Ellet, James Espy, Alex M. Forster, Benjamin Apthorp Gould, Asa Gray, Samuel S. Haldeman, Joseph Henry, Edward C. Herrick, John Lawrence LeConte, Joseph LeConte, Elias Loomis, Joseph Lovering, G.E. Manigault, Francis Markoe, Matthew F. Maury, C.G. Memminger, Robert Treat Paine, James L. Petigru, C.C. Pinckney, William C. Redfield, E.S. Ritchie, John D. Runkle, Jared Sparks, William Stimpson, David Humphreys Storer, William Henry Trescot, Michael Tuomey, Joseph Winlock, and William Wurdemann.
Unpublished finding aid in the repository.
Gift of Miss S.P. Gibbes, 1932; purchase, 1916.

G12 Gibbs, Wolcott, 1822-1908.
Papers, 1854-1910.
In Harvard University Archives (Cambridge, MA).
Prominent inorganic and analytical chemist and research scientist; Professor of the Application of Science to the Useful Arts at Harvard. - Small collection including printed material by and about Gibbs; a few letters by him to R.T. Jackson, 1885-98; a wastebook; lectures, memorials, and testimonials about him by colleagues.

G13 Gibbs, Wolcott, 1822–1908.
  Letters, 1860–1902.
  300 items.
  In Franklin Institute Library (Philadelphia)
  University professor. Letters to Ogden Nicholas Rood relating to Rood's work in inorganic chemistry and Gibb's work in physics.
  Unpublished guide in the library.
  Open to investigators under library restrictions.
  Gift of the Rood family, 1904.

G14 Gilbert family.
  Papers, 1900–73. 25 ft.
  In University of Maine, Raymond H. Fogler Library (Orono)
  Papers of a Bangor, Me., family and records relating to the Great Northern Paper Company.
  Access restricted.
  Information on literary rights available in the library.
  Gift of Charles Gilbert, 1972.

G15 Gilchrist family.
  Papers, 1867–1945. – 7 boxes.
  In University of Michigan, Bentley Historical Library, Michigan Historical Collections (Ann Arbor).
  Alpena, Michigan family. – Collection includes correspondence, letterpress books, financial papers, and other materials largely relating to the family's business enterprises in lumbering, sugar manufacturing, ferry and excursion lines, mining, and banking; also, records of business affairs in Alpena and other areas of northern Michigan, Minnesota, Ohio, Wisconsin, Missouri, Oregon, and Mississippi; family members represented in the collection include Frank Gilchrist and two of his sons, Frank R. and Ralph Gilchrist, as well as members of related Fletcher and Potter families.
  Unpublished finding aid available.

G16 Gilchrist, Harry L., 1870–1943.
  Papers, 1925–29. – 0.1 lin. ft.
  In Howard Dittrich Museum of Historical Medicine, Allen Memorial Library (Cleveland, OH).
  Physician; member of Chemical Warfare Department. – Chiefly correspondence with C. J. Couseno concerning Couseno's gas machine.
  Unpublished guide available.

G17 Gilchrist, Huntington, 1891–1975.
  Papers, 1913–73. ca. 15,000 items.
  In Library of Congress, Manuscript Division (Washington, D.C.)
  Public official and business executive. Correspondence, travel diaries, personal and official documents, reports, speeches, research materials, newspaper clippings, and other printed material, chiefly 1919–55, documenting Gilchrist's interest in international affairs, especially his activities on behalf of the League of Nations, United Nations, and Institute of Pacific Relations. Topics associated with his career include the mandates system and the Permanent Mandates Commission of the League of Nations, administration of Danzig and the Saar region, development of the city in China, site selection for United Nations headquarters, and, to a lesser extent, plague control and the drug aureomycin, which were connected with his work at American Cyanamid. Correspondents include Charles C. Bauer, Nicholas Murray Butler, Richard S. Childs, John W. Davis, Norman H. Davis, Sir Eric Drummond, Allen Dulles, Raymond B. Fosdick, Hugh Gibson, Leland Harrison, Edward M. House, Manley O. Hudson, Philip Jessup, James G. McDonald, Denys P. Myers, James T. Shotwell, and Arthur Sweetser.
  Unpublished finding aid in the repository.
  Deposited by Mr. Gilchrist, 1974.

G18 Gilchrist, Peter Spence, 1861–1947.
  Correspondence, 1901–11. 10 ft.
  In University of North Carolina Library, Southern Historical Collection (Chapel Hill) (3393)
  Chemical engineer, of Charlotte, N.C. Business letters (chiefly 1904–10) received by Gilchrist relating to the building of fertilizer plants and the installation of works for phosphate processes along the eastern seaboard.
  Gift of Cecil W. Gilchrist, 1958.

G19 Gilder, William Henry, 1838–1900.
  Papers, 1878–94. ca. 400 items.
  In Dartmouth College Library (Hanover, N.H.)
  Forms part of the repository's Stefansson collection on the polar regions.
  Arctic explorer and journalist. Correspondence (1887–92), writings, diary (1887) of round-trip between Winnipeg and York Factory, Man., and newspaper clippings. Includes material on Jeannette relief expeditions, other expeditions of the period, and development of new types of military gunpowder and firearms. Correspondents include Edward William Bok, Angelo Heilprin, Thomas C. Mendenhall, Albert Operti, and Frederick Schwatka.
  Unpublished finding aid in the repository.

G20 Gilpin, Joshua, 1765–1840.
  Papers, 1770–1868. 131 items and 64 v.
  In Eleutherian Mills Historical Library (Greenville, Del.) (various accessions)
  In part, transcripts (typewritten), photocopies, and negative or positive microfilm made from originals in private hands or owned by the Matthew Bolton Collection, Assay Office, Birmingham, Eng., the New-York Historical Society, the Historical Society of Pennsylvania, the Pennsylvania Historical and Musuem Commission, and the University of Virginia Library.

Philadelphia businessman, manufacturer, canal company executive, and poet. Correspondence; journals and notebooks (1790-1801, 1830-33), recording travels in the U. S., Great Britain, and Europe, with special attention to industrial development, internal improvements, mechanical devices, and public institutions; notes; memoranda; and other papers, relating to the bleaching of pulp in the manufacture of paper, instructions for dyeing, fabrics, colors, and papermaking machinery. Includes correspondence of Gilpin's brother, Thomas Gilpin (1776-1853), and of their uncle, Miers Fisher (1748-1819).

In part, described in A Guide to the Manuscripts in the Eleutherian Mills Historical Library, by John B. Riggs (1970) p. 843-844, and in Joshua Gilpin: An American Manufacturer in England and Wales, 1795-1801, by Harold B. Hancock and Norman Wilkinson (Transactions of the Newcomen Society for the Study of the History of Engineering and Technology, v. 39, 1960-61, p. 15-28, 57-66).

Purchases and gifts, 1955-70.

G21 Gittings family.
Papers, 1815-96. 31 items and 2 v.
In Maryland Historical Society Library (Baltimore) (1667)
Correspondence of David Sterett Gittings (1797-1887), Mary Sterett (Gittings) Bryarly (d. 1957), and other family members; accounts (1815-30) of Richard Gittings (1764-1830) at the Bank of Baltimore; cookbook (1822-80) of Mrs. Rachael Garrett, of Elkridge Landing, Howard Co., Md., containing recipes for food and dyes and a list (1822-59) of slaves belonging to David S. Gittings; and will (1819) of Deborah Ridgely Sterett (1747-1819). Includes letters (1818-19) relating to the medical education received by David S. Gittings at the University of Edinburgh, with descriptions of the cities of Edinburgh and London. Correspondents include Richard Gittings, Mary (Sterett) Gittings, Caroline Howard, Charles W. Howard, Rev. John R. Keech, and Robert Y. Welford.
Gift of Miss Victoria Gittings, 1962.

G22 Glenn family.
Papers, 1842-1919.
ca. 100 items.
In Minnesota Historical Society collections.
Letters to Andrew W. Glenn from his sons, Harry W. and Horace H.; a volume (1842-70) kept by members of the McMillan family containing recipes, minutes of a literary society in Iowa, farm accounts, plat of an orchard, texts of sermons, essays, minutes (1842-43) of the Franklin Institute (a literary society) and fragments of poems and essays; and two letters (1868-69) written by Aggie Wallace from a school in Nashville, Tenn., describing southern attitudes towards northern teachers and the activities of the Ku Klux Klan. The letters (1898, 1901) from Horace H. Glenn relate to his activities with a surveying crew in South Dakota and with a logging crew near Two Harbors and Marcy, Minn., and describe the life and character of lumberjacks. The letters (1917-19) from Harry W. Glenn, a private in Co. B, First Gas Regiment during World War I, give information on training camps, gas attacks, Red Cross services in the U. S. and France, and French civilian life during the war.
Descriptive inventory in the repository.
Gift of Mrs. W. A. Mortenson, 1956.

Glenz, Adolph H. G23
Prescription book, early twentieth century. - 1 v.
In State Historical Society of Wisconsin (Madison).
Madison pharmacist. - Early twentieth century prescription book belonging to Glenz, containing recipes for medicines, ointments, hand lotions, hair tonics, furniture polish, floor wax, and other substances.
Gift of Mrs. Joseph Dragotto, Glendale, Wisconsin, November 8, 1974.
Unpublished description available.

Glyol Chemical Company, Baltimore. G24
Minute book, 1901. - 1 v.
In Maryland Historical Society Library (Baltimore).
Minutes of Baltimore corporation formed to manufacture and distribute glyol, a preparation invented by Dr. Wirt A. Duvall. Covers first four months only, and includes constitution and agreement covering use of formula.

Goddard, Robert Hutchings, 1882-1945. G25
Papers of Robert H. Goddard and Esther Christine (Kisk) Goddard, 1901-1979. - 70 ft. (168 boxes).
In Clark University Library (Worcester, MA).
Rocket and space pioneer. - Personal and professional papers of Goddard and his wife Esther C. Goddard (1901-1982), who was his constant companion in rocket research, interpreter of his work, and collector and editor of his papers. Includes correspondence, 48 pocket diaries, autobiographical and biographical data, writings, speeches, lecture notebooks, reports, theses, contracts, memoranda, rocket test data, patents, honors, awards, memorials, philatelic material, sound recordings, motion picture film, and photos (processed and available for research; register supplement for photos in preparation). Scrapbooks and other memorabilia added after Esther Goddard's death.
Collection includes material on Werner Brugel, Harry F. Guggenheim, Robert W. E. Lademann, Willy Ley, Charles A. Lindbergh, William Joffe Numeroff, Hermann Oberth, Robert Esnault Pelterie, G. Edward Pendray, Army Air Corps, Chemical Warfare Service, Signal Corps, Clark University, Curtiss-Wright Corporation, Daniel and Florence Guggenheim Foundation, NASA, Navy Bureau of Aeronautics and Bureau of Ordnance, Smithsonian Institution, and The Papers of Robert H. Goddard (1970).
Gift of Esther C. Goddard, 1964, and trustee of Goddard estate, 1978.
Unpublished register available.

G26 Godfrey, Almon T., 1880(?)-1923.
Papers, 1906-1979. – 19 items.
In Hope College Archives (Holland, MI).
Professor of Chemistry at Hope College, 1904-1923. – Clippings, photographs, correspondence. Also includes manuscripts: "Dr. Almon Tanner Godfrey," a student paper by Suzanne M. Gould, 1973.

G27 Goetz, Alexander, 1897-1970.
Papers, 1919-1976. – 17 boxes.
In California Institute of Technology Library (Pasadena).
Professor of physics, California Institute of Technology. – Correspondence, mss. of writings, patent specifications, reports, technical films, and personal papers. Includes material relating to Goetz's work at Caltech's Cryogenic Laboratory and Rare Metals Institute and to his research on industrial applications of silver, particularly in desalination of water (during World War II), development of membrane filters for U.S. Chemical Corps for detection of microbiological agents, and studies of aerosols and atmospheric pollution. Some of the early papers are in German. Bulk of the collection is dated after 1930. Correspondents include E.T. Bell, P.W. Bridgman, L.A. DuBridge, E.U. Condon, A. Dember, P.S. Epstein, A. Fassler, A.B. Focke, J.A. Hedvall, R.C. Hergenrother, S. Kyropoulos, M. von Laue, Giovanna Mayr, and R.A. Millikan.
Finding aid in the repository.

G28 Gold and Silver Extraction Company.
Records, 1895-1904. – 500 items.
In Colorado Historical Society (Denver).
The Gold and Silver Extraction Company of America, Ltd., was incorporated in May, 1890, under the laws of Colorado. H.A.W. Tabor was president and Dr. M. Werner of Saguache was general manager. – Collection contains correspondence, business papers, and annual reports.

G29 Gold, Harry, 1899-1972.
Papers, ca. 1923-1967. – Ca. 50 ft.
In New York Hospital-Cornell Medical Center, Medical Archives (New York, NY).
Professor of Clinical Pharmacology. – Correspondence, notes (especially concerning his research into digitalis), reports, scientific papers (mostly in reprint form), lectures and slides.
Gift of Muriel Gold Morris, Naomi Gold Steinberger, and Stanley Gold.
Unprocessed.
Unpublished finding aid available.

G30 Goostray, Stella, 1886-1969.
Papers, 1899-1969. – 23 boxes and 1 package.
In Boston University, Mugar Memorial Library, Special Collections, Nursing Archive.
Nurse, educator, administrator, and author of articles and books on nursing; in 1977 she was elected to the American Nurses' Association Hall of Fame. – Collection includes professional correspondence, addresses and articles, tapes, photos, citations, memorabilia, an untitled manuscript for a biochemistry textbook, and correspondence and other material pertaining to Goostray's published memoirs. Also included are tributes and biographies she wrote and material she collected concerning such nurses as Florence Nightingale, Mary Adelaide Nutting, Sophia French Palmer, Linda Anne Judson Richards, Mary May Roberts, Isabel Maitland Stewart, Julia Catherine Stimson, and Effie Jane Taylor.

G31 Gordon, George, ca. 1819-1869.
Letters, 1866-67. ca. 60 items.
In California State Library (Sacramento)
Letters to John T. Doyle concerning sugar refining and operations of San Francisco and Pacific Sugar Refinery, which Gordon founded in 1857.
Unpublished finding aid in the repository.
Gift of Guy C. Miller, Palo Alto, Calif., 1943.

G32 Gordon, Louis, 1914-1966.
Papers, 1934-1965. – 3 ft.
In Case Western Reserve University Archives (Cleveland, OH).
Professor of Chemistry. – Correspondence, manuscript articles, including Gordon's doctoral dissertation, student notebooks, lecture and laboratory notes, photos, reprints, and reference files collected by Gordon. Subjects include analytical chemistry and chemical precipitation. Also includes course material related to Gordon's appointment as Centennial Professor of Physical Sciences at the University of Kentucky, 1965.
Unpublished finding aid available.

G33 Gorham, John, 1783-1829.
Papers, 1824-27. 1 box.
In Harvard University Archives.
Professor of chemistry and materia medica at Harvard University. Correspondence and other papers.
Information on literary rights available in the repository.

34 Gourdin-Young family papers, ca. 1727-1889.
ca. 3 ft., 76 items, and 6 v.
In South Carolina Historical Society collections (Charleston) (11-170, 24-267/273)
Residents of Charleston, S.C. - Family records (1727-1852) of the Hamilton, Lewis, and Miller families; family correspondence (1809-1852) of Eliza S. Courtney; papers of attorney Thomas S. Grimké (1786-1834) relating to the estate of Jonathan Ashe; professional correspondence, accounts, receipts, and other legal papers (1847-1859), of Theodore S. Gourdin and Col. Tandy Walker (d. 1853), relating to the purchase, operation, lease, sale, and collection of debts owed the Kennesaw Paper Mill, Marietta, Ga., managed and then leased by George P. Ensign before being sold through Gen. A.J. Hansell; correspondence relating to Gourdin's land transactions in Macon, Ga., through John Rutherford; bonds made to Gourdin as sheriff of Charleston (1853-1855); correspondence (1866-1889), relating to collections, bonds, pleas, petitions, writs, and summonses, issuing from Charleston attorneys Rutledge & Young, together with papers relating to the cases of Eliza S. Lee vs. Henry Gourdin & C.J. Durham, Arthur Middleton vs. Bickly Mazyck Desel, and Clarke vs. Port Royal Railroad; genealogical notes and family Bible; telegraphic code books (ca. 1880) for commission merchants Gourdin, Matthiessen & Company, Charleston, S.C., and Gourdin, Young & Frost, Savannah, Ga.; and vestry accounts (1828-1849), vestry daybook (1823-1825), and papers relating to St. John's Episcopal Church, Berkeley County, S.C., and Strawberry Chapel, St. John's Parish, S.C.
Gift of Mr. and Mrs. Joseph R. Young, 1981.

35 Graham, Helen (Tredway), 1890-1971.
Papers, 1898-1971. - 11 ft.
In Washington University School of Medicine Archives (St. Louis, MO).
Professor of Pharmacology at Washington University School of Medicine. - Correspondence, card files, newspaper clippings, laboratory and lecture notes, relating to Dr. Graham's research on the physiology and pharmacology of the peripheral nerve and the role of histamine, and to her interest in civil liberties, air pollution control, educational reform, and other civic affairs.
Gift of Dr. Graham's son, Evarts A. Graham, Jr., 1971.
Information on literary rights available in the repository.
Unpublished finding aid available. The repository also has 10 reels of microfilm (negative) available for interlibrary loan.

G36 Grahame, Israel J
Business records, 1852-97. 19 v.
In Historical Society of Pennsylvania collections.
Pharmacist, of Philadelphia. Receipt books, daybooks, ledgers, cashbooks, stock book, and prescription books.
Gift of Mr. and Mrs. E. Perot Walker, 1965.

Grain Belt Breweries, Inc., Minneapolis. G37
Records, 1890-1976. - 23 cu. ft.
In Minnesota Historical Society (St. Paul).
Minneapolis brewery owned by G. Heileman Brewing Company. - Correspondence, minutes, financial records, trademark information, annual reports, scrapbooks, in-house publications, and miscellaneous papers.
Gift of Irwin L. Jacobs, October-November 1981.
Unpublished finding aid available.

Granite-BiMetallic Consolidated Mining Company. G38
Records, 1881-1922. 21 ft.
In Montana Historical Society collections (Helena)
Correspondence and financial, production, and legal records, including payrolls, lease agreements, inventories, assays, and production reports, of the company and two predecessor firms, Granite Mountain Mining Company and Bi-Metallic Mining Company, relating to mining operations near Philipsburg, Mont. Includes correspondence between company officers in St. Louis, Mo., and management of mines near Philipsburg; weekly reports sent to St. Louis; and correspondence with other companies and individuals about equipment ordered and services provided.
Unpublished finding aid in the repository.
Gifts of Peter Antonioli, Butte, Mont., 1954, and Barry Engrav, Missoula, Mont., 1973.

Grasselli family. G39
Papers, 1778-1967. - 6.25 linear ft.
In Case Western Reserve University Archives (Cleveland, OH).
Immigrant family prominent in chemical manufacturing. - E. R. Grasselli (1810-1882) came to the United States from Alsace in 1836. He married, established a family, and eventually settled in Ohio. The papers primarily document the growth of the Grasselli Chemical Corporation and its subsidiaries. Collection consists of the personal and business papers of Eugene R., Caesar A., and Thomas S. Grasselli, the directors of the company. The business papers include agreements between the Grasselli Chemical Corporation and the Standard Oil Company, and some financial reports and minutes of the Grasselli subsidiaries. The business correspondence documents the exchange of information on new chemical processes between the Grassellis and other firms in the U.S. and Europe. Personal papers include wills; deeds; certificates of marriage; correspondence, some of it translated from French; diaries; ledgers; patents; letter press books; and various financial records.
Donated by Caesar A. Grasselli II to the Western Reserve Historical Society in 1967, and following years.
Available to the serious scholar.
Unpublished finding aid available.

G40 Gray, William.
Patents, 1898–1911. ca. 300 items.
In Nebraska State Historical Society collections.
Vice president of the Eureka Manufacturing Company, Lincoln, Neb. U. S. and foreign patents relating to Gray's inventions of an improved pressure cooker and grain, ore and mineral separators.
Gift of Viola Gray.

G41 Greenewalt, Crawford, b. 1902.
Diary, 1943-45. - 1 vol.
In Hagley Library (Greenville, DE) (1889).
Chemical engineer; President and Chairman of the Board of Du Pont Company. - A photocopy of Greenewalt's Manhattan Project diary which documents the Chicago atomic energy experiments that led to the first chain reaction and the author's involvement in that work. The original remains with the Department of Energy and a few passages in the photocopy have been censored by the Federal Government.
Restricted.

G42 Griffin, Edward Lawrence, b. 1899.
Oral history interview, 1973. - 2 reels; 2 hours.
In National Library of Medicine (Bethesda, MD).
Chemist. - Interview conducted by Adelynne H. Whitaker in Lawrence, Kansas, January 31, 1973. Interview concerns the enforcement of the Insecticide Act of 1910 and the Federal Insecticide, Fungicide, and Rodenticide Act of 1947. Dr. Griffin served as a chemist in the Department of Agriculture and the Food, Drug, and Insecticide Administration, 1913-38. From 1938 to his retirement in 1955, he was an administrator with the Insecticide Division, Department of Agriculture.
Transcript with index (90 p.).
Information on restrictions available in the Library.

G43 Griffin, Lucille Mouton, ca. 1900-1980.
Papers, 1798-1967. - 7 linear ft.
In University of Southwestern Louisiana, Southwestern Archives and Manuscripts Collection (Lafayette).
Collection contains the papers of Alexander Mouton (1953-1938), sugar, railroad, and steamboat engineer, Director of the mint at Mexico City, and Director of the U.S. Mint at New Orleans. The volume of the Alexander Mouton series is ca. 2 linear ft., and the date span is 1872-1967. The papers cover topics concerning Mouton's life, Louisiana history, Mouton family genealogy, sugar and syrup manufacture, and the mints.
Gift of Lucille Meredith Mouton Griffin.
Information on literary rights available in the repository.
Finding aid available.

G44 Grimes, J. Bryan, 1884-1923.
Papers, 1760-(1880) - 1935. - Ca. 5,700 items.
In East Carolina University Library, East Carolina Manuscripts Collection (Greenville, NC).
Planter and politician from prominent eastern North Carolina family. - Papers include notes (1894) from chemistry course at the University of North Carolina.
Unpublished finding aids in the library. Also described in "East Carolina Manuscripts Collection, Bulletin No. 2."
Gift of Mrs. Ida W. Grimes, Washington, NC, 1968.

G45 Griswold, John Augustus, 1818-1872.
Papers, ca. 1837-95. 8 ft. (3450 items)
In New York State Library, Manuscripts and Special Collections (Albany)
Steel manufacturer, mayor, and U.S. Representative, of Troy, N.Y. Correspondence (mainly incoming), memoranda, bills, receipts, printed circulars and business and Government reports, and other papers, relating chiefly to John A. Griswold Company, and related iron and steel companies and railroads. Includes political correspondence, particularly letters from Roscoe Conkling; detailed account sheets for the ironclad gunboat Patapsco; and 33 patent applications for barbed wire, bale ties, and fenceposts, and canceled checks of J. Wool Griswold.
Unpublished finding aid in the repository.
Gift of Mrs. S. G. Tenney, Williamstown, Mass., 1962.

G46 Group Against Smog and Pollution, Missoula, Mont.
Records, 1968-70. ca. 2 ft.
In University of Montana Library (Missoula)
General correspondence, minutes (1968), news releases, and miscellaneous papers.

G47 Grubb family.
Papers, 1834-69.
1000 items.
In Historical Society of Pennsylvania collections.
Papers relating to the St. Charles Furnace, Henry Clay Furnace, Manada Furnace, and the Chestnut Hill ore bank. Includes Union Canal boat permits (1849-50) and the correspondence of Clement B. and Edward B. Grubb.
Presented by Mrs. William S. Morris, 1949.

Grubb, Peter.
   Forge and furnace account books, 1765-1880.
   89 v.
   In Historical Society of Pennsylvania collections.
   Business accounts of Peter and Henry Bates Grubb, early ironmasters of Pennsylvania, containing material on the development and commercial activities of some leading furnaces and forges. Includes records of Hopewell Forge, Mount Vernon Furnace, Mount Hope Furnace, Codorus Forge, Codorus Ore Bank, Manada Furnace, Columbia and St. Charles Furnaces, Chestnut Hill Ore Bank, Cornwall Ore Bank, with related gristmill and provision accounts, including those of Clement Grubb.
   Described in Guide to the manuscript collections of the Historical Society of Pennsylvania, 2d ed., no. 253.
   Gift of Daisy E. R. Grubb, M. Lilly Beall, Mrs. William M. Thornton, Jr., Mrs. William T. Morris, and Mrs. I. Wistar Morris.

Gustafson, Ben G., 1903-1982.
   Papers, 1924-1977. - 12 ft.
   In University of North Dakota, Chester Fritz Library, Orin G. Libby Manuscript Collection, (Grand Forks, ND).
   Professor of Chemistry, administrator, North Dakota State Legislator. - Correspondence, minutes, reports, lectures, class notes, and other materials relating to Gustafson's career as a Professor of Chemistry and as Dean of Continuing Education at the University of North Dakota. Papers relating to Gustafson's post-retirement activities in senior citizens organizations and North Dakota state politics are also included.
   Deposited by Ben G. Gustafson in 1978.
   Unpublished inventory available.

Guthrie, Boyd, 1916-1964.
   Papers, 1916-1964. - 0.33 ft.
   In University of Utah Libraries, Special Collections Department (Salt Lake City).
   Director of Anvil Points Project, 1949-1956. - Located near Rifle, Colorado, the Anvil Points Project was a demonstration plant operated by the U.S. Bureau of Mines to learn more about mining and refining of oil shale. Included are scientific articles and reports on oil shale development, 1916-1943; historical sketches of the project, accompanied by photos; limited correspondence; and a map.
   Gift, ca. 1971.
   Unpublished finding aid available.

Gutleben, Dan, 1878-1969.
   Manuscript: Gutleben's Sugar Thesaurus. 1958-1965. - 2 v.
   In University of California, Berkeley, Bancroft Library.
   Volume 1: Copies of his work on the sugar beet industry in Ohio and Indiana and on the history of monosodium glutamate. With related correspondence.
   Volume 2: Compilation of materials documenting the history of the California and Hawaiian Sugar Refining Corporation in Crockett, California, including correspondence, notes, photographs and blueprints of the plant.
   Description in repository.

Gutleben, Dan, *collector*. G52
   Gutleben collection: sugar research materials, 1917-36.
   1 box.
   In New York Public Library.
   Papers relating to the sugar industry in the U. S., particularly Arbuckle Bros., Brooklyn, N. Y., Belcher Sugar Refining Company, St. Louis, Mo., Pennsylvania Sugar Company, Philadelphia, and the Sugar Institute.
   Gift of Mr. Gutleben, 1967.

Gutleben, Dan. G53
   Papers, 1945-59. 7 folders.
   In Michigan State University, Archives and Historical Collections (East Lansing)
   Reports written by Gutleben as chief engineer for Pennsylvania Sugar Company; history of the sugar industry in the U.S. and its possessions, especially Hawaii and Puerto Rico; and monographs on the first successful American beet sugar factory and on monosodium glutamate.
   Gift of Mr. Gutleben.

Guymon, James Fuqua, 1911-1978. G54
   Papers, 1938-1978. - 27 cartons.
   In University of California, Davis, Shields Library, Special Collections.
   Professor of Oenology, UCD. - Collection contains papers concerning wine and brandy technology, distillation, refrigeration, biochemistry of alcoholic fermentation, and aging phenomena.

Gwathmey, Allan Talbott, 1903-1963. G55
   Papers, 1942-1966. - 98 items.
   In Virginia Historical Society (Richmond).
   Chemistry professor. - Includes correspondence of Allan Talbott Gwathmey while Professor of Chemistry at the University of Virginia and President of the Board of Trustees of the Virginia Institute for Scientific Research, Richmond.

H1 Haber, Fritz, 1868-1934.
Haber-Willstaedter collection, 1906-64.
ca. 120 items.
In Leo Baeck Institute collections (New York City)
Transcripts (handwritten) and photocopies (positive).
German chemist. Chiefly photocopies of correspondence (1911-33) between Haber and the chemist Richard Willstaedter (1872-1942), relating to personal matters and research; 10 letters (1906-28) by Haber; newspaper clippings of articles on Haber and his work; photocopies of family tree; and other papers. In German and English.
Gift of Mrs. Ernest Bruch, Willstaedter's daughter, 1961; gifts and purchases from Karl Weigert, Philippsburg, Pa., 1959-63.

H2 Hague, James Duncan, 1836-1909.
Papers, 1824-1936. - Ca. 24,000 items.
In The Huntington Library (San Marino, CA).
Mining engineer. - Family and business affairs of James D. Hague, mainly mining. In particular, the Clarence Rivers King sub-collection. King (1842-1901), geologist, mining engineer, and writer, became head of the U.S. Geological Survey in 1878. Hague had been his assistant on the U.S. Geological Exploration of the Fortieth Parallel a decade earlier. The sub-collection contains King's scientific papers (including those concerning a reconnaissance in Arizona in 1865 and 1866, and Mexican mining companies), 43 scientific notebooks (including material on the California Geological Survey [1864-66], and the United States Geological Exploration of the Fortieth Parallel [1867-72]), some correspondence, manuscripts, and photographs. Also included is material concerning James D. Hague's publication of the Clarence King Memoirs.
Gift of Eleanor, Marion, and James Hague, 1947 and 1971.
Unpublished finding aid available.

H3 Haldeman family.
Papers, 1801-85. ca. 7600 items.
In Eleutherian Mills Historical Library (Greenville, Del.) (840)
Correspondence and other business and personal papers, chiefly of Jacob M. Haldeman (1781-1857), operator of the Cumberland Forge, New Cumberland, Pa., and his son, Richard Jacobs Haldeman (1831-1886), editor of the Patriot and Union, Harrisburg, Pa., and U. S. Representative from Pennsylvania. Jacob Haldeman's papers include correspondence, accounts, invoices, orders, receipts, and other business records relating to his forge, flour mill, general store for workmen, and farms. Papers of Richard J. Haldeman include correspondence (1859-85) with politicians, real estate dealers, and friends, with much emphasis on Pennsylvania politics during the Civil War; diaries of travel in Europe (1852-55) while a student at Heidelberg and Berlin and as an attaché in the legations at Paris, St. Petersburg, and Vienna; speeches; household bills; canceled checks; and bankbooks. Other papers include letters and estate papers (1816-19) of Samuel Jacobs, of Colebrook Furnace, Lebanon, Pa.; shipping account (1801-02) of John Haldeman (1753-1822), of Locust Grove, Lancaster Co., Pa.; and papers (1849-51) concerning the management of farm property of Jacob M. Haldeman and Thomas Elder.
Described in A Guide to the Manuscripts in the Eleutherian Mills Historical Library, by John B. Riggs (1970) p. 847.

H4 Halford, Ralph Stanley, 1914-1978.
Papers, 1940-1959. - 2 boxes.
In Columbia University, Rare Book and Manuscripts Library (New York City).
Professor of Chemistry; Dean of the Graduate Faculties, Columbia. - Includes correspondence, manuscripts, typescripts, conference papers, scientific drawings, photographs, and printed material. Chiefly Halford's manuscripts and typescripts of his writings for scientific journals and papers presented at various conferences and symposia, with related correspondence of colleagues. Also files on teaching of chemistry, photographs of his spectrometers, a copy of his patent for recording spectrometers, and printed material (Halford's writings).
Gift of Mrs. Ralph S. Halford (via Chemistry Library), 1980.
Unpublished finding aid available.

H5 Hall, Charles Martin, 1863-1914.
Papers, 1882-1909. - Ca. 70 items.
In Case Western Reserve University Libraries, Department of Special Collections (Cleveland, OH).
Photocopies (positive).
Inventor of the electrolytic process for the manufacture of aluminum and President of the Aluminum Company of America. - Letters from Hall to his father, the Reverend H. B. Hall, and his sister, Julia B. Hall, relating to family matters, Hall's scientific experiments, patent suits, problems in the manufacture of aluminum and company growth; together with a notebook (1888-89) describing experiments, "History of C. M. Hall's aluminum invention, by his sister Julia B. Hall," newspaper clippings describing processes Hall developed, copies of patent claims, and obituary notices.
Unpublished finding aid available.

Halsband, Ruth Alice, 1901-1971.
   Papers, 1895-1971. – 3 boxes.
   In Columbia University, Rare Book and Manuscripts Library (New York City).
   Research chemist. – Halsband received her A.B. from Smith College in 1922 and her Ph.D. from Columbia in 1946. Her husband Robert was also on the Columbia faculty. Her papers include correspondence, her chemistry dissertation, articles on chemistry, Columbia University and Smith College memorabilia, photographs, and drawings. There are many letters from Robert Halsband concerning the progress of his books. Letters to and from her family and friends have been added.
   Gift of Professor Robert Halsband, 1980; gift of the children of Ruth Alice (Norman) Halsband, 1981.

Hamilton, Cliff Struthers, 1889-1975.
   Papers, 1924-1970. – 3 ft.
   In University of Nebraska Libraries, Archives/Special Collections Department (Lincoln).
   Professor of Organic Chemistry. – Bulk of the collection is student theses done under Dr. Hamilton's direction. In addition, the collection contains correspondence relating to chemistry, notebooks, clippings, and reprints of articles.
   Unpublished inventory available.

Hamilton Manufacturing Company, *Hamilton, N. Y. (Albany Co.)*
   Records, 1796-1804.
   1 box.
   In New York Historical Society collections.
   Minutes, account books, and miscellaneous papers of a glass manufacturing company.

Hamilton, T. M. ("Ted"), b. 1905.
   Papers, 1922-1984. – 317 folders.
   In University of Missouri-Columbia Library, Western Historical Manuscripts Collection and State Historical Society of Missouri Manuscripts.
   Saline County, Missouri farmer and amateur archaeologist. – Collection contains mostly correspondence on archaeological and historical topics, as well as personal, business, and local civic affairs; particularly correspondence with Federated Metals Division of American Smelting and Refining Company, Somersville, New Jersey (1974-1978), concerning identification of archaeological artifacts by metallurgical analysis; and correspondence with John X. Sopkovick (1958-1961), concerning colonial metallurgical processes, particularly in respect to firearms manufacture.
   Gift of T. M. Hamilton, April 17, 1981, and May 5, 1984.
   Unpublished finding aid available.

Hammett, Louis Plack, 1894- H10
   Papers, ca. 1965-70. ca. 750 items.
   In Columbia University Libraries (New York, N.Y.)
   Professor of chemistry, Columbia University.
   Notes and writings relating to Hammett's book, Physical Organic Chemistry (1970).
   Gift of Mr. Hammett, 1978.

Hammond family. H11
   Business records, 1835-1916. 16 ft.
   In Cornell University Library, Collection of Regional History and University Archives (2139, 2140)
   Correspondence, financial statements, orders, invoices, vouchers, board and freight bills, boat operation receipts, memoranda notebooks, and other papers from Crown Point and Ironville (formerly Irondale), Essex Co., N. Y. Many of the records concern the operations of the Crown Point Iron Company and its predecessors, Penfield & Taft, Penfield & Harwood, and Penfield & Hammond, and include journals, cashbooks, payroll records and board bills for furnace, furnace company store, and company railroad employees, and canal and lake steamer shipping accounts. Includes miscellaneous accounts of Hammond & Company, Charles Franklin Hammond, the Sugar Hill Company; grocery and general merchandise daybooks, cashbooks, customers' ledgers, inventories, order books, and other accounts of John and Thomas Hammond, the Hammond Store, Barker & Wyman, and others.
   Unpublished guide in the library.
   Gift of Eugene Barker, 1961.

Hanford, Franklin, 1844-1928. H12
   Papers, 1860's - 1927. – 4 boxes.
   In New York Public Library.
   Naval officer. – Correspondence, notebook on chemistry and physics as applied to military science with emphasis on explosives, fuses, and electricity and other papers. Correspondents includes W. H. Samson, the Journal of American History, and the Naval Academy Graduates Association.
   Acquired, 1929.

Hardgrove, Timothy A   1874-1953. H13
   Papers, 1930-53.
   1 v. and 9 boxes.
   In State Historical Society of Wisconsin collections.
   Dentist, of Fond du Lac, Wis. Papers relating to the campaign for fluoridation of public water supplies, including correspondence with other dentists, among them John Frisch of Madison, Wis., and Frederick McKay of Colorado Springs, Colo., and with officials of the American Dental Association; minutes of meetings of the State fluoridation study committee, of which Hardgrove was chairman; and materials relating to his research on the relationship of dental abnormality to tic douloureux, a muscular twitching of the face.

H14 Hare, Robert, 1781-1858.
Papers, 1764-1859. - Ca. 1200 items.
In American Philosophical Society Library (Philadelphia, PA).
Chemist. - Letters on personal and business matters; drafts of letters to editors of journals on such varied topics as fish guano, slaughterhouses, paper money, and the meaning of the term "Yankee annexations." There were originally over 300 scrolls, since disbound, which contained drafts of letters, essays, and lectures, which Hare composed on ordinary sheets of paper, then pasted end to end, and rolled up. The essay and lecture topics include: chemistry, storms, slavery, currency, fire-fighting, capital punishment, railroads, Smithsonian Institution, Michael Faraday, religion and spiritualism, riots in Philadelphia, epidemics, underwater blasting, and Ralph W. Emerson; there is some verse. Correspondents include: Alexander D. Bache, Franklin Bache, William Ellery Channing, Robley Dunglison, Michael Faraday, John Fisher, Richard Fisher, John K. Kane, Thomas S. Kirkbride, Joseph Henry, Matthew F. Maury, Charles Partridge, Benjamin Silliman, Benjamin Silliman, Jr., Petty Vaughan, William Vaughan.
Accessions, 1961, 1975, 1980.
Unpublished finding aid available.

H15 Hare, Robert, 1781-1858.
Lecture, 1848. 1 item.
In Smithsonian Institution, National Museum of American History, Division of Physical Sciences (Washington, D.C.)
Professor of chemistry at the University of Pennsylvania, Philadelphia. Ms. of a lecture prepared as an introduction to a chemistry course.
Gift.

H16 Hare, Robert, 1781-1858.
Papers, 1811-1852. - Ca. 80 items.
In University of Pennsylvania Libraries, E. F. Smith Collection (Philadelphia).
Professor of Chemistry at the University of Pennsylvania. - Letters and miscellaneous invitations, certificates and acknowledgements including extensive correspondence with Benjamin Silliman at Yale.
Cataloged separately on cards at the repository.

H17 Hare-Willing family.
Papers, ca. 1744-1905. - Ca. 1300 letters. 53 v.
In American Philosophical Society Library (Philadelphia, PA).
Collection of letters, letterbooks, account books, diaries, scrapbooks, etc., concerning the families of Robert Hare and Thomas Willing (1731-1821). Robert Hare's son, also named Robert (1781-1858) was the noted chemist, whose mother was Margaret Willing. The letters and other documents include early family material, as well as documents written by numerous family relations, and some obviously only collected by them.
The Willing family letters (1744-1863, ca. 460) are diverse, concerning family matters, business, society, and comments on the Civil War. There are numerous letters from Thomas Willing, many concerning his banking career, as President of the Bank of North America and later at the First Bank of the U.S.
The Hare family letters (1781-1890, ca. 800) are more extensive and diverse, including much on travel in the U.S. and elsewhere. There is a letter from Robert Hare Jr. concerning steam engines, and letters from Horace Binney Hare concerning his education at Harvard, 1860, his trip to San Francisco and the West, 1862, and numerous letters written while a soldier in the Civil War. There are many letters from Horace Binney (1780-1875) to his daughter Esther, who was married to John Innes Clark Hare (1816-1905), concerning family travel and court cases. There are also letters from outside the family, such as those from Dorothea L. Dix.
The bound volumes include, among others: Robert Hare letterbooks (1824-25, 1841-57), estate records, and laboratory expense accounts (1818-1860); G.H. Hare's journal or log of cruises aboard the U.S. United States(1841) and U.S. Flint(1845); Horace Binney Hare's 1812 journal of his trip to San Francisco. There are account books and accounts (1754-95) kept by Thomas Willing; accounts of the controversy over the estate of John Innes Clark; and records of the First Colored Wesley Methodist Church of Philadelphia (receipt book, 1820-48; minute book, 1927-44). re are also Philadelphia court records, and minutes of the Common Council of the city, 1832.
Unpublished finding aid available.

H18 Harmon, Robert Rogers, 1900-1967.
Papers, 1934-65. 8 ft. (ca. 6000 items)
In University of Virginia Library (Charlottesville) (9657)
Chemical engineer. Business correspondence and patents of the Southern Welding and Machine Company, of Charlottesville, Va., of which Harmon was president, together with records of engineering problems and experiments, especially concerning gas scrubbing, and blueprints and specifications for products. Includes a few personal letters regarding Sigma Chi fraternity.
Unpublished list of contents with the collection.
Gift, 1971.

H19 Harper, Henry Winston, 1859-1943.
Papers, 1834-1943. 4 ft.
In University of Texas Library, Texas Archives (Austin)
Scientist and teacher, of Texas. Correspondence, reports, notes, memoranda, bibliographies, monographs, articles, poems, travel sketch, biographical sketches, speeches, lectures, student papers, genealogies, theses abstracts, scientific reports, technical and scientific bulletins, minutes of meetings, school and university budgets, receipts, bank records, invoices, scrapbooks, newspapers, magazines, and clippings, relating to Harper's career as physician, chemist, pharmacist, metallurgist, and teacher, especially to his work as a metallurgist in Mexico; as professor of chemistry and dean of the Graduate School of the University of Texas, as a scientist in civic and public service, particularly in connection with establishment of the University of Texas Mineral Survey and with the conduct of sanitary investigation of the El Paso city water supply, to his membership and leadership in scientific, learned, and honorary societies, to his contributions to scientific journals, and to family history and genealogy.
Described in The University of Texas Archives; a Guide to the Historical Manuscripts Collections (1968) p. 157-158.

H20 Harris, James Courtland, d. 1968.
Papers, 1909-1956. - 5 mss. v.
In Louisiana State University, Department of Archives (Baton Rouge).
Distinguished sugar chemist and 1911 LSU graduate. Harris held managerial posts in several Latin American countries, primarily Cuba.

H21 Harris, William P., 1859-1942.
Diaries, 1914-1937. - 44 v.
In University of Michigan, Bentley Historical Library, Michigan Historical Collections (Ann Arbor).
Manager of the Huron Portland Cement Company mill and the Michigan Alkali Company quarry at Alpena, Michigan. - Daily entries deal with Harris's activities in the cement industry; also memorandum books.
Unpublished finding aid available.

H22 Harrison Brothers & Company.
Records, 1828-90. ca. 350 items.
In Eleutherian Mills Historical Library (Greenville, Del.) (840, 861)
Correspondence (ca. 1830-90), journal (1828-34), ledger (1868-75), petty cash ledgers (4 vols., 1873-83), order book (1850-74) with experiments and notes, receipt book (1832-50), bills, checks, agreements, notes on paint manufacturing processes, drawings of machinery, and papers of a paint manufacturing plant mainly located in Philadelphia. Persons represented include John Harrison (1773-1833), his sons, Thomas and Michael Leib Harrison, and his grandsons, George Leib Harrison, Jr. and Thomas Skelton Harrison (1837-1919).
In part, described in A Guide to the Manuscripts in the Eleutherian Mills Historical Library, by John B. Riggs (1970) p. 848.
Purchase, 1965.

H23 Harston, C B 1911-
Papers, 1939-74. 4 ft. (ca. 1600 items)
In Washington State University Libraries (Pullman)
Extension soils specialist at Washington State University. Correspondence, financial records, reports, minutes, research data, photos, and other papers, concerning soil analysis and conservation methods. Includes records of a Tennessee Valley Authority-Washington State University cooperative fertilizer project (1939-64) and material on Whitman County Solid Waste Disposal Subcommittee (1969-74).
Unpublished finding aid in the repository.
Gift of Mr. Harston, 1977.

H24 Harte, Robert A., b. 1911.
Papers, 1945-52. - 2 boxes.
In National Library of Medicine (Bethesda, MD).
Nutritional biochemist. - Correspondence, memoranda, and reports pertaining to the revision of the U.S. Pharmacopia and standards for the preparation of amino acids. Among the correspondents are E. Fullerton Cook.

H25 Harteck, Paul, b. 1902.
Papers, 1927-1979. - 4.5 linear ft.
In Rensselaer Polytechnic Institute Archives and Dept. of Special Collections (Troy, NY).
Professor of Physical Chemistry. - Correspondence, research notes, legal papers, speeches, lectures, drafts of articles and proposals, newspaper clippings and photographs comprise this collection which documents Harteck's career. The most substantial series of correspondence is between Harteck and Karl Frederich Bonhoeffer and spans several decades.
Gift of Paul Harteck, 1982.
Unpublished finding aid available.

H26 **Hartford residents' correspondence**, 1807-1947. ca. 150 items.
In Trinity College Library (Hartford, Conn.)
In part, photocopies (negative) and transcript (typewritten).
Correspondence (1873-1914) between C. H. Clark, W. D. Howells, and Mark Twain; letters (1927-47) to Henrietta Gardiner, three of which are from Laura E. Richards; letters (1890-1904) to Frederic Penfield, received while serving as consul in Egypt; letters (1807-40) from Daniel Wadsworth to Benjamin Silliman; letters (1857) from Thomas H. Gallaudet concerning work with the deaf; letters (1886-87) from H. Carrington Bolton to Charles K. Wead, concerning a chemistry periodical list; and a typed copy of Homer Worthington Brainard's ms., "Thomas Hooker, His Life and Writings" (ca. 1914).
Information on literary rights available in the library.
A composite collection created from miscellaneous collections of letters, plus the Homer W. Brainard ms., which is stored separately.

H27 Hartman, Frank A 1893-
Papers, 1918-70. ca. 3 ft. (ca. 300 items)
In Holy Family College Archives (Philadelphia, Pa.)
In part, photocopies of originals owned by Mr. Hartman.
Radium specialist and consultant, and owner of Radium Services, Philadelphia, Pa., 1919-56. Correspondence (1927-70) with academic institutions, officials of government and cancer societies, and Philadelphia radiologists, relating to Hartman's services, including gifts of radium and exhibits of radioactive minerals; reports, catalogs, and price lists (1918-56) of radium products; photos; and other papers relating to Hartman's life and career.
Unpublished finding aid in the repository.
Access restricted in part until Mr. Hartman's death.
Gifts of Mr. Hartman, 1971 and 1975.

** NOTE: The following records of Harvard University and associated organizations in the Archives which are less than fifty years old require special permission for use. Contact the archivist for further information. **

H28 [Harvard University] Association of Harvard Chemists (est. 1911).
Papers, 1911-ca. 1951. - 10 v.
In Harvard University Archives (Cambridge, MA).
Publications and reference material; correspondence, minutes, pledge book, financial records, ballots, and other miscellaneous records.

H29 [Harvard University]. Boylston Chemical Club (est. 1887).
Records, 1885-1927. - 4 v.
In Harvard University Archives (Cambridge, MA).
Minutes, accounts, notes for talks by T. W. Richards.

H30 [Harvard University]. Chemical Engineering Society (est. 1906).
Papers. - 1 folder.
In Harvard University Archives (Cambridge, MA).
General materials folder.

H31 Harvard University. Chemical Laboratories.
Records, 1858-ca. 1950. - 201 v.
In Harvard University Archives (Cambridge, MA).
Mainly correspondence, but also routine records, including: accounts, apparatus inventories, bills, chemical inventories, class and professor accounts, course lists, exams, forms inventory records, office memoranda, opening and closing notices, orders records, original information memoranda dealing with administration, petty cash books, printing samles, student records, quotations and orders, scrapbook, shipment receipts, student accounts; also papers of C. W. Eliot, including correspondence and breakage accounts, records relating to new buildings, and reports.

H32 [Harvard University] Davy Club (est. 1834).
Minutes, 1838-1843. - 1 v.
In Harvard University Archives (Cambridge, MA).
In addition to the minutes, collection includes some notes on geology and natural history lectures.

H33 Harvard University. Department of Biochemistry and Molecular Biology.
Records, 1927-1967. - 27 v.
In Harvard University Archives (Cambridge, MA).
Correspondence and other papers, examinationa, student folders, and other records relating to students.

H34 Harvard University. Department of Chemistry.
Records, 1870-1969. - 61 v.
In Harvard University Archives (Cambridge, MA).
Mainly twentieth century. Correspondence, minutes, letters of recommendation, library general correspondence and interlibrary loan and periodicals records, committee reports; also routine records including accrediting for membership in the American Chemical Society, applications for positions, concentration and enrollment cards, examinations, forms, fund records, records relating to W. Gibbs Laboratory, recommendations for appointments, research in the department, student evaluation cards, catalogues of apparatus, records relating to draft deferments (1940-1946), and opinions of returning graduate students.

H35 Harvard University. *Graduate School of Business Administration. Baker Library.*
Account books and manuscripts of economic material in German and relating to German enterprises, 1621–1770.
3 v.
In Harvard Business School, Baker Library.
Journal (1720–47) of Leo Bruno, commission merchant of Nuremberg; an account (1769–70) of the smelting of silver and copper, written by B. V. Ephraim and J. H. Muntz; and papers (1621–1746) of the Stadion family of Arnegg, including agreements between the town of Ulm and a German nobleman about the right to lease part of a stone pit in Arnegg. The material is related to the Kaiserlich Deutsche Gesandschaft für China Collection.
Gift of Charles L. Bernheimer, 1940; purchased from E. Baer, 1937.

H36 Harvard University. *Graduate School of Business Administration. Baker Library.*
Account books of dyers, Massachusetts and Vermont, 1813–46.
5 v.
Accounts of Brainerd Stebbins, Barre, Vt., 1813–24; Samuel H. Whitney and Otis Dickinson, Granville, Mass., 1823–32; Malden Dye House, Hingham and Boston, Mass., 1830–46; and Levi Fiske, Shelbourne, Mass., 1834–38.
Described in List of business manuscripts in Baker Library, compiled by Robert W. Lovett (2d ed., 1951)
Gifts of Charles E. Tuttle and Norman S. B. Gras; purchased from John E. Bowman and Mrs. G. E. Cash. The Whitney-Dickinson accounts acquired 1935; the Fiske accounts acquired 1938.

H37 Harvard University. Graduate School of Engineering.
Records, 1897-1964. - 12 v.
In Harvard University Archives (Cambridge, MA).
Minutes of meetings (including Lawrence Scientific School), correspondence and invoices of the Gordon McKay Library, student records, correspondence and other papers concerning the building of Pierce Hall, and other records.

H38 Harvard University. Graduate School of Engineering.
Records, 1908-1955. - 293 v.
In Harvard University Archives (Cambridge, MA).
Minutes of faculty meetings; minutes of professors; notebooks on budgets, corporation records and votes; correspondence of the dean's office and other correspondence; processed material, minutes, forms, correspondence, reports and other records of committees and subdivisions such as the Administrative Board, Qualifying Committee in Civil Engineering, Committee on Degrees, Library; routine records grouped alphabetically including or relating to applicants for admission, ballots for commencements marshall, course materials including examinations, day pages, final grades, forms and form letters, grade cards and sheets, processed material for distribution among faculty, record cards of deceased former students, student folders, folders of deceased former students; also records relating to special activities or considerations such as students' petitions regarding M.C.E. 1910, Civilian Pilot Training Program 1939-1942, War Training courses 1942, Proposed Union of Faculties of Engineering and Arts and Sciences 1949; and other records.

H39 [Harvard University]. Harvard Chemical Club (est. 1879).
Records, 1879-1900. - 1 v.
In Harvard University Archives (Cambridge, MA).
Minutes of the Harvard Chemical Club, 1879-1900.

H40 [Harvard University]. Harvard-Radcliffe Chemical Society (est. ca. 1947).
Papers. - 1 folder.
In Harvard University Archives (Cambridge, MA).
General materials folder.

H41 [Harvard University]. Harvard-Tech Chemical Club (est. 1922).
Papers. - 1 v.
In Harvard University Archives (Cambridge, MA).
Publications and reference material.

H42 Harvard University. Office of the President: James B. Conant, 1933-1953.
Papers, 1933-1953. - 572 containers.
In Harvard University Archives (Cambridge, MA).
Professor of Organic Chemistry; President of Harvard University. - A collection of papers, including those of the Corporation as well as those of the President, deposited by the President's office. Includes correspondence, reports, speeches, calendars of appointments, and other records.
Partially indexed, but well organized alphabetically within groups of years.

H43 [Harvard University]. Rumford Chemical Society (est. 1833).
Papers, 1854-1863. - 8 v.
In Harvard University Archives (Cambridge, MA).
Publications and reference material; reports of meetings, constitution and lists of members.

H44 Harvey, Edmund Newton, 1887-1959.
Papers, 1923-59. – Ca. 7,000 items, 19 notebooks (7 linear ft.).
In American Philosophical Society Library (Philadelphia, PA).
Physiologist; Consisting of correspondence, notes, memoranda, extracts from publications, reprints, drafts of essays, poems, magazine and newspaper articles, comic strips, popular songs, etc, on bioluminescence, collected principally 1945-59 for his History of Luminescence from the Earliest Times until 1900 (APS Memoirs 44, 1957). Other topics are chemistry, military medicine, natural history, professional associations, etc. There are 24 letters to Talbot Howe Waterman about the chapter "Light Production," which Harvey wrote for Waterman's Physiology of Crustacea (New York, 1960-61). Other correspondence is with persons in Princeton and other universities, the National Institutes of Health, and the United States Department of Agriculture. The notebooks contain reports to the United States Office of Scientific Research and Development on wound ballistics, bubble formation, decompression sickness, etc. (19 v.).
Presented by Princeton University, 1964, and Talbot Howe Waterman, 1961.

H45 Hassid, William Zev, 1897-1974.
Papers, ca. 1815-1974. – 5 boxes.
In University of California, Berkeley, Bancroft Library.
Professor, College of Agriculture and Department of Biochemistry, UC, Berkeley. – Letters written to Hassid and copies of letters by him; speeches and papers; and reprints of articles documenting his professorial career and his research on carbohydrate chemistry.
Photographs removed to Pictorial Collection.
Description in repository.

H46 Hastings, Albert Baird, b. 1895.
Papers, 1915-76. – 54 boxes, 13 v., and case items.
In National Library of Medicine (Bethesda, MD).
Biochemist. – Autobiographical data, correspondence (from individuals as well as relating to conferences and organizations), lectures and talks, photos, research data, certificates and diplomas, printed matter and reprints. Correspondents include C. Sidney Burwell, Vannevar Bush, James B. Conant, and others.
Hastings was with the U.S. Public Health Service, 1917-21; the Rockefeller Institute, 1921-26; the University of Chicago, 1926-35; was Head of the Department of Biochemistry, Harvard University, 1935-59; and was with the Scripps Clinic and Research Foundation, 1959-66. Collection contains reports, memoranda, and other data pertaining to the Scripps Clinic and Research Foundation.
Hastings material also includes an oral history interview, conducted by Peter D. Olch in La Jolla, CA, December 1967, and February and May 1968, consisting of 26 reels (39 hours). Interview is an autobiographical memoir which includes descriptions of the institutions, research, and individuals important in Hastings' career. Transcript, with index and appendix (2 v.; 706 p.).
Gift of Albert Baird Hastings, 1976.
Finding aid available.
Information on restrictions available in the Library.

Hastings, John, b. 1787. H47
Papers, 1810-50. 376 items.
In Dickinson College Library (Carlisle, Pa.)
New England artist and ink manufacturer. Family and business papers, centered in Massachusetts and eastern New York. Includes 4 sketches in oil and pencil.
Card catalog in the library.
Purchase, 1956.

H48 Hauser, Samuel T., 1833-1914.
Papers, 1832-(1864-1914)-1941. – 29.1 linear ft.
In Montana Historical Society (Helena).
Early Montana pioneer, businessman, banker, and politician. Interests included mining, smelting, banking, townsite development, real estate, irrigation, cattle ranching, coal mining and coke roasting, branch railroads, and hydroelectric power. Materials reflect the entire range of Hauser's activities. More than 50 percent is incoming correspondence (1864-1914) and outgoing correspondence (especially complete for the years 1881-1911). Financial records, legal documents, organizational records, and reports support the correspondence series. 32 subgroups, divided into 4 sections (banking, mining, railroads, and miscellaneous) make up the remainder of the material.
Unpublished finding aid available.

Hawkins family. H49
Papers, 1740-1898. 20 ft.
In University of North Carolina Library, Southern Historical Collection (322)
Family correspondence and business papers of members of the Hawkins family, of North Carolina and of the allied families of several States. Persons represented include John Davis Hawkins (1781-1858), William J. Hawkins (1819-1894), Philemon Benjamin Hawkins (1823-1891), and Colin M. Hawkins, who were planters, manufacturers of machinery and phosphates, railroad builders and execu-

tives, bankers, and commission merchants. Includes account books, letter books, and order and shipping records of Hawkins, Williamson & Company, of Baltimore, Md., the North Carolina Phosphate Company, of Raleigh, and Castle Hayne, N. C., the Pioneer Manufacturing Company, of Raleigh, and the Raleigh Gas Light Company.
Unpublished description in the library.
Gifts and purchases from the Andrews family, before 1940 and in 1952, and 1956.

H50 **Haynes, Elwood,** 1857-1925.
Papers, 1862-1932.
ca. 300 items.
In Howard County Historical Museum (Kokomo, Ind.)
Correspondence, legal documents, drawings, clippings, newspapers, magazine articles, family photos., memorabilia, and other papers relating to the automobile invented by Haynes in 1893-94. Includes descriptive and advertising material of the Haynes car, bills of lading of metal shipments received, material on Haynes Stellite, and the MS. of a paper given before the 8th International Congress of Applied Chemistry on "Alloys of cobalt with chrome, iron, and other metals."
Open to investigators under museum restrictions.
Information on literary rights available in the repository.
Gift of March Haynes and Bernice Haynes Hillis, 1958-59.

H51 **Hazen, Allen,** 1869-1930.
Papers, 1884-1929. - 7.5 cu. ft.
In Massachusetts Institute of Technology, Institute Archives and Special Collections (Cambridge, MA).
Hydraulic engineer. - Collection includes business correspondence, material on water filtration and analysis, sanitary engineering, student notes from M.I.T., notebooks on water analysis from Lawrence Experimental Station, and photographs. Additional material expected.
Restricted; consult Institute Archivist for further information.
Box list available in the Institute Archives.

H52 **Hedrick, Benjamin Sherwood,** 1827-1886.
Papers, 1848-93.
6033 items and 4 v.
In Duke University Library.
Professor of chemistry and U. S. Patent Office official. Chiefly letters to Hedrick. The early correspondence is between Hedrick and Mary Ellen Thompson, his future wife. Other correspondence concerns life at the University of North Carolina, Hedrick's dismissal from the University in 1856 for his Republican and anti-slavery opinions, and his life in the North during the Civil War period. Many of the post-1861 papers relate to Hedrick's position as chemical examiner at the Patent Office. Other topics include Reconstruction, the economic plight of the South, and politics, including Hedrick's attempt to win political office in North Carolina (1868). Correspondents include Kemp P. Battle, Daniel R. Goodloe, Horace Greeley, Hinton Rowan Helper, David L. Swain, John Torrey, and Jonathan Worth.
Card index in the library.
Acquired, 1959.

H53 **Hedrick, Benjamin Sherwood,** 1827-1886.
Papers, 1856-1900. 212 items.
In University of North Carolina Library, Southern Historical Collection (325)
Professor of agricultural chemistry and Assistant Commissioner of Patents in the U. S. Patent Office. Papers, chiefly 1873-80, consisting of family, scientific, and business correspondence. Includes a letter relating to Hedrick's dismissal from the University of North Carolina because of his political views in 1856; and several MSS. on chemistry.
Unpublished description in the library.
Gift of the Misses Hedrick, before 1940, and in 1943.

H54 **Hegler family.**
Papers, 1809-1919. ca. 2 ft.
In Ohio Historical Society collections (Columbus) (196)
In part, transcripts (typewritten), translated from German to English.
Family correspondence, account books, indentures, land titles, and other legal documents, Ohio military pollbooks and militia papers (1813-32), receipts, statements, surveys, and plats; daybook (1830), of Abraham Hegler (1789-1865), cattle importer, railroad promoter, and State legislator, of Ross Co., Ohio; and journals and account books (1881-95) of a fertilizer manufacturing company owned by Almer Hegler (1854-1920) and Elijah Hopkins, Washington Court House, Ohio.
Gift of Almer and M. F. Hegler, Washington Court House, Ohio, 1916 and 1921.

H55 **Heidelberger, Michael,** 1888-
Papers, 1911-67. 6 boxes.
In National Library of Medicine (Bethesda, Md.)
Physician associated with the Rockefeller Institute for Medical Research, College of Physicians and Surgeons of Columbia University, New York City Public Health Research Institute, and New York University School of Medicine. Personal and general correspondence, and material relating to associations, conferences, congresses, and institutes. Organizations represented include American Association of Immunologists, American Chemical Society Conference on Immunochemistry, Institute of Microbiology, New York City, International Congress for Microbiology, National Academy of Sciences, National Science Foundation, New York Academy of Sciences, New York City Public Health Research Institute, and World Health Organization.
The repository also has the Heidelberger oral history memoir transcript, v. 2 of which is a catalog of papers.

H56 **Helena Mining and Reduction Company.**
Records, 1883-1888. - 1.5 linear ft.
In Montana Historical Society (Helena).
Gold and silver mining operation located in Wickes-Corbin, Montana Territory. Organized in 1883. Major properties owned by the company were the Alta Mine and Mill, the Comet Mine and Mill, and the Rumley Mine. The company operated three mines, two furnaces, and six charcoal kilns. - Collection includes small quantity of interoffice correspondence for the Helena Mining and Reduction Company; daily and monthly expense reports; daily furnace reports for the Alta and Comet mills; and mine production reports for the Alta, Comet, and Rumley mines.
This collection was separated from the Helena Banking Collection.
Unpublished finding aid available.

H57 Hellerman, Leslie, b. 1896.
Papers. – Ca. 20 linear ft.
In Johns Hopkins University, Alan Mason Chesney Medical Archives (Baltimore, MD).
Physiological chemist. – Hellerman served in the Department of Physiological Chemistry at Hopkins under William Mansfield Clark and Albert L. Lehninger. Hellerman's principal interest was in enzymes as mediators of biological reactions, and he carried on his work for parts of four decades. His papers include correspondence files, research notes, and unpublished manuscripts.
Unprocessed collection; not open for research.
No finding aid available.

H58 Hench, Atcheson Laughlin, 1891-1974.
Papers, 1811-1974. 27 ft. (ca. 24,000 items)
In University of Virginia Library (Charlottesville) (10000)
Professor of English, University of Virginia. Correspondence, diaries, lecture notes, articles, speeches, memoirs, legal documents, poetry, short stories, autograph albums, clippings, photos, and other family, personal, and professional papers. Includes papers of the Hench and Michie families documenting the family tanning business; genealogical material on the Drangold, Ickes, and Showalter families; Presbyterian Bible lessons; papers (1935-50) relating to the work of Virginia Bedinger (Michie) Hench (1898-1971) with Albemarle County Child Welfare Association, Community Chest, and Red Cross; diaries and photos (1917-19) from Hench's World War I duty in France at U.S. Army Base Hospital No. 7; clippings on his brother, Philip Showalter Hench (1896-1965); and material on Pittsburgh, Pa. (1885-1923), travel in Europe (1887) and the West (1915), cancer, and life at University of Virginia (1930's and 40's). Hench's professional papers pertain to American Dialect Society, American Name Society, American Speech, Britannica Book of the Year, College English Association, English Institute, Modern Language Association, Virginia Quarterly Review, University of Virginia Dept. of English, Raven Society, his alma mater, Lafayette College (Easton, Pa.), his year in Austria as Fulbright professor, and his dictionary advisory work. Correspondents include Edmund Wilson.
Unpublished finding aid in the repository.
Gifts, 1975 and 1976.

H59 Hench, Philip Showalter, 1896-1965.
Papers, 1903-64. ca. 5 ft.
In University of Virginia Library (6230, 6230-a)
Physician, of Rochester, Minn., who shared in the 1950 Nobel Prize for medicine and physiology. Correspondence, articles, speeches, notes, printed material, photos, memoranda, and other papers relating to Hench's career and his family, to his travels, palindromic rheumatism and its naming, the development of cortisone, the Nobel Prizes, especially for 1950, Walter Reed and the conquering of yellow fever, Hench's collecting of papers of the members of the Yellow Fever Commission, his service in the U. S. Army Medical Corps during World War II, his appearance for the Merck Company before Estes Kefauver's Senate Sub-committee on Anti-trust and Monopoly, music, and Sherlock Holmes.
Information on literary rights available in the library.
Deposits, 1960, 1964.

H60 Henco Brewery, Spokane, Washington.
Records, 1899-1909. – 26 items.
In Washington State University Library (Pullman).
Accounts, letters, ledgers, receipts, records and other papers.
Unpublished finding aid available.

H61 **Henderson, Lawrence Joseph,** 1878-1942.
Papers, 1905-42. 3 ft.
In Harvard University Archives.
Professor of biological chemistry at Harvard. Correspondence, lectures, notebooks, and other papers.
Open to investigators only upon prior application to the repository.
Information on literary rights available in the repository.

H62 **Hendrickson family.**
Papers, 1691-1838. 2 boxes.
In Rutgers University Library (1764)
Correspondence and other papers (1753-87) of Daniel Hendrickson, a merchant; documents and accounts (1791-1801) relating to distilling activities of Hendrick Hendrickson; ledgers and other account books (1747-1816) of other members of the Hendrickson family of Middletown, N. J.; and deeds and accounts relating to the activities of the sloop, Catharine.

H63 Henry Bower Chemical Manufacturing Co., Philadelphia, Pennsylvania.
Records and related items, 1854-1957. – 1,725 items.
In Hagley Library (Greenville, DE) (acc. #1032, 1070).
Firm was founded in 1855 and produced crude drugs and chemicals. Collection includes accounts (journals, vouchers, ledgers, cash books, bank deposit books; posting sheets, etc.); production records (expense records, cost of materials registers, employee time sheets, etc.); inventories (raw materials, inventory cards, and memorandum book); and miscellaneous records (minutes of directors' and stockholders' meetings, inspectors' weekly reports, minute books, dues account books, miscellaneous correspondence, etc.).
Described more fully in the Supplement to A Guide to Manuscripts in the Eleutherian Mills Historical Library (1978).

Herrick, John Peirce, 1868-1961, collector.
Oil history collection, 1797-1952. 9 ft.
In Pennsylvania State University Library (University Park)
In part, transcripts (typewritten) and photocopies.
Correspondence, articles, agreements, leases, deeds, rights-of-way, indentures, stock shares, certificates of incorporation, newspaper and magazine clippings, pamphlets, and other papers, relating to oil history, gas, early wells, drilling, flooding, refining, pipelines, railroads, and transportation. Areas covered are mostly in New York State, particularly around Bolivar and Olean, and areas of Pennsylvania concerned with the drilling of oil. Includes material on the wreck of the B. B. & K. Railroad in 1844 and the completion of the first oil pipelines in the 1880's. Most of the papers are dated in the 1930's and 1940's.
Information on literary rights available in the library.
Gift of Mr. Herrick's son, John L. Herrick, 1961.

Herrington, Barbour Lawson, 1904–
Papers, 1931-64. ca. 3 ft.
In Cornell University Libraries, Collection of Regional History and University Archives (Ithaca, N. Y.) (21/20/742,/tr.118)
Professor of dairy chemistry at Cornell University. Correspondence, minutes, reports, memoranda, and printed matter. Includes illustrated reports of experiments performed by Herrington's students, Herrington's Fulbright lectures in Finland and Denmark, a tape recording concerning the early history of the Dept. of Dairy Industry, and material on the College of Agriculture at Cornell, the dairy industry, summer research at Sheffield Farms and at the Dry Milk Laboratories, Bainbridge, N. Y., the American Dry Milk Institute, the Dairy Institute Association, the Dairy Institute conferences at Cornell, the Society of Photographic Scientists and Engineers and its Ithaca chapter, and the Ithaca Photographic Society.
Gift of Mr. Herrington, 1964.

Herschel family.
Papers, 1809-71. ca. 4000 items.
In University of Texas at Austin, Humanities Research Center Library.
Correspondence, diaries (1820-71), memorandum books (1825-53), account and housekeeping books (1822-53), 12 vols. of travel journals (1809-50), and scientific papers, of Sir William Herschel (1738-1822), British astronomer, his sister, Caroline Lucretia Herschel (1750-1848), who assisted him in his work, and his son, Sir John Frederick William Herschel (1792-1871), astronomer and scientist. Includes 12 star charts; mss. of writings on astronomy, chemistry, mathematics, photography, physics, and the actinometer and barometer, and material on the Greenwich Observatory, terrestrial magnetism survey, the Standards Commission, the University Commission, and Sir John's position as Master of the Mint. Correspondents include George Biddell Airy, Charles Babbage, Mary Baldwin, William Henry Bateson, H. D. Harnap, Lady Mary Herschel, Benedetto Pistruci, Sir Edward Sabine, James Calder Stewart, William Samuel Stratford, and Sir Charles Edward Trevelyan.
Described in John Herschel and Victorian Science; Exhibition of Manuscripts, Photographs, Drawings, and Books from the History of Science Collection (1966).
Purchase.

Herter, Christian A., 1865-1910. H67
Papers. - Ca. 2 linear ft.
In Johns Hopkins University, Alan Mason Chesney Medical Archives (Baltimore, MD).
Biochemist. - Herter was interested in chemistry as a service to scientific medicine. He helped to finance the Journal of Biological Chemistry and he was instrumental in the early organization of the Rockefeller Institute for Medical Research. He donated $25,000 to Johns Hopkins School of Medicine in 1902 for a memorial lectureship. Collection includes reprints, unpublished manuscripts, and a correspondence file containing several letters from Paul Ehrlich, as well as other correspondence.
Unprocessed collection; not open for research.

Herty, Charles Holmes, 1867-1938. H68
Papers, 1884-1938. - 64 linear ft.
In Emory University, Robert M. Woodruff Library (Atlanta, GA).
Chemist, publicist for chemistry and the chemical industry, forest researcher. - The papers document almost every interest and activity of Herty's long career: forest and pulp and paper research; the early development of the American chemical industry; medical and health research; national defense and chemical warfare. Types of material include personal and professional correspondence; financial and legal records; manuscripts, notes, photographs, clippings, and copies of articles and speeches dealing with Herty's research and interests; records of his work with professional organizations; and family photographs and memorabilia.
Transferred to Emory University by the American Chemical Society in 1957.
Published finding aid available.

H69 Higginson, Henry Lee, 1834–1919.
Business records, 1799–1919.
30 ft.
In Harvard Business School, Baker Library.
Boston businessman and philanthropist. Correspondence and other papers relating mainly to investments, including a ledger and journal of James J. Higginson (1856–65), an account of Salisbury & Higginson (1799), and material on Ecuador Coal Co., Gage Co., Ecuador Trading Co., Cherry River Land Association, Reynolds Chocolate Co., New Metals Process Associates, Jones Step Process Trust, John T. Jones Holding Co., Burn-Boston Battery, Submarine Signal Co., Standard Lessee Corp., Standard Alcohol Co., St. Louis, Cable & Western Railroad, and St. Louis & Suburban Railroad. Correspondents include J. Alexander Agassiz, Charles Fairchild, John M. Forbes, Henry Lee, Charles E. Perkins, Samuel G. Ward, and Charles W. Wetmore.
Inventory in the library.
Gift of Seth T. Gano, 1946.

H70 Hilby, Francis Martin, 1860–1918.
Papers, 1880–1923. ca. 2600 items.
In California State Library (Sacramento)
Druggist, of Monterey, Calif. Cashbooks (1884–91) and other business papers (1880–1918) and material relating to organizations with which Hilby was associated, together with correspondence, clippings, pamphlets, and photos.
Unpublished finding aid in the repository.
Gift.

H71 Hildebrand, Joel Henry, 1881-1983.
Papers. - 6 cartons and 1 box.
In University of California, Berkeley, Bancroft Library.
Professor of Chemistry, UC Berkeley. - Correspondence, articles, bibliography, lecture outlines, biographical materials, book reviews, subject and research files on various chemical compounds and processes, National Science Foundation and other grant files, other varied writings relating to Hildebrand and his career. Some materials regarding academic appointments and awards for Hildebrand's colleagues.
Partially processed collection; folder list available in repository.

H72 Hill, Crawford, 1862-1922.
Papers of Crawford and Louise S. Hill, 1899-1955. ca. 30,000 items.
In State Historical Society of Colorado collections (Denver)
Businessman, of Denver, Colo. Personal and business correspondence and other papers, relating to the family's business interests. Includes records of the estates of Cora Cowan, Crawford Hill, and Mary B. Neely; and business records relating to Boston and Colorado Smelting Company, Denargo Land Company, Denver, Dolly Varden Mining Company, Hill Land and Investment Company, Denver, Hill Securities, Land and Development Company, Denver, Inland Oil and Refining Company, Florence, Colo., and Republican Publishing Company, Denver.
Unpublished guide in the repository.

H73 Hill, Walter Nickerson, 1846-1884.
Papers, 1860-1905. ca. 1200 items.
In Brown University Library (Providence, R.I.)
Chemist at the U.S. Naval Torpedo Station, Newport, R.I. Chiefly correspondence (1870-84) between Hill and ordnance specialists, scientists, political, military, and social figures, business associates, and his wife, Katharine L. (Smith) Hill; together with lectures, drafts of articles, scrapbooks, memorabilia, printed matter, and photos; and Hill family correspondence and genealogies. Includes correspondence relating to Hill's application for appointment as professor of mathematics in the Navy. Correspondents include Gen. Henry L. Abbot, Lammot duPont, Levi Parsons Morton, Gen. John T. Wilder, and Harvard professors Oliver Wolcott Gibbs and John Trowbridge.
Unpublished finding aid in the repository.
Gift of Catharine (Hill) Ostrander, 1969.

H74 Hinckley, Robert H., b. 1891.
Papers, 1891-1977. - 46 ft.
In University of Utah Libraries, Special Collections Department (Salt Lake City).
Businessman; government official. - Although the collection primarily features material from the federal agencies with which Hinckley was associated and from his work with the American Broadcasting Company, there is some correspondence with the officers of Amalgamated Sugar Company about production and processing techniques.
Gift, 1974.
Unpublished finding aid available.

H75 Hitchcock, Ethan Allen, 1798–1870.
Papers, 1810–73. 6 ft. (3000 items)
In Library of Congress, Manuscript Division.
Army officer and author. Correspondence, reference notes, MSS. of articles and books, military papers, and printed matter. The reference notes are those of Hitchcock, of his brother, Samuel Hitchcock and of various friends, and are supplemented by various vols. annotated by Hitchcock. The correspondence relates to his long military career at West Point, in Florida, in Mexico, in California and in other posts; his U. S. Dept. of War experience during and after the Civil War; and his interest in metaphysics, philosophy, alchemy, and hermetic philosophy. Correspondents include John Bell, W. W. Bliss, George Washington Cullum, William Greenleaf Eliot, Dr. C. M. Hitchcock, Ethan Allen Hitchcock (his nephew), Henry Hitchcock, Samuel Hitchcock, Charles H. Larnard, Francis Lieber, Mary Mann, James Miller, Theodore Parker, Winfield Scott, and William Jenkins Worth.
Unpublished finding aid in the Library.
Gift and purchase from Mrs. Hitchcock, Mrs. W. A. Croffut, and Mrs. Anita V. Nicholls, 1905-38.

H76 Hitchins, Clayton S.
Student notebook, 1932-1933. - 1 v.
In Cornell University Libraries, Department of Manuscripts and University Archives (Ithaca, NY).
Bound volume of Clayton Hitchins' typed notes for Chemistry 375, taught by Dr. Ralph Colton Tallman, 1932-1933.

H77 Hogness, Thorfin, 1894-1976.
Papers, 1945-1949. - 0.5 linear ft.
In University of Chicago Library.
Professor, Department of Chemistry (1930-59); Director, Chemistry Division, Metallurgical Laboratory (1944-55). Material relating to postwar scientists' movement, including Senate and House bills and amendments as well as other print and near-print material.
Unpublished finding aid available.

H78 Holden, Edward Fuller, 1901-1925.
Papers, 1922-25. 50 items.
In University of Michigan, Bentley Historical Library, Michigan Historical Collections (Ann Arbor)
Instructor in mineralogy and member of Mineralogical Laboratory, University of Michigan. Correspondence, notes, and articles, relating to Holden's research.

H79 Holley, Alexander Lyman, 1832-1882.
Papers, 1864-83. ca. 20 ft. and 4 reels of microfilm.
In Smithsonian Institution, National Museum of American History, Division of Mechanical and Civil Engineering (Washington, D.C.)
In part, microfilm of letters whose location is unknown.
Mechanical engineer and metallurgist. Chiefly papers relating to Holley's career, especially his design of Bessemer steelmaking plants and other steel production components, including drawings and tracings (1864-82) by Holley or his staff or copied by Holley on European trips, relating to steel production, steam engines, and railway engineering; together with microfilm of letters of Holley and his assistants concerning design and construction of steel production facilities, and correspondence of Holley's wife, Mary Holley.
Unpublished finding aid in the repository.
Gift.

H80 Holter, Anton M., 1831-1921.
Papers, 1861-1968. - ca. 83 ft.
In Montana Historical Society collections (Helena)
Merchant and businessman, of Helena, Mont. Correspondence, financial records, organizational material, legal documents, minutes, and other papers, relating to Holter family business interests, managed in subsequent generations by Norman B. Holter (1868-1959) and Norman J. "Jeff" Holter (b. 1914). Principal family businesses represented include A.M. Holter and Brother (general merchandise, 1867-1886), A.M. Holter Hardware Company (1886-1958), Holter Company (investments, 1917-1963), Holter Realty Company (1917-1968), and Holter Research Foundation, a non-profit research Laboratory (beginning 1947), all in Helena. Other persons and business represented include Samuel T. Hauser (1833-1914), Aubrey M. Holter (1883-1945), Edwin O. Holter (1871-196-?), and Blackfoot Cattle Company, Flesher, Mont., Boston and Seven Devils Copper Company (Idaho), Helena and Frisco Mining Company (Idaho), Helena and Victor Mining Company (Montana), Humbird and Sand Point Lumber Company (Idaho), Judith Farms Company (Montana), Little Ben Mining Company, Landusky, Mont., Maginnis Mining Company, Maiden, Mont., Montana Flour Mills, Great Falls, Mont., Pacific Coast and Norway Packing Company (Washington and Alaska), Parrot Silver and Copper Company, Butte, Mont., and United Missouri River Power Company, Helena, Mont.
Gift of Norman J. Holter, 1978.
Unpublished finding aid in the repository.

H81 Hommell, Philemon E., 1862-1935.
Papers, 1860-1932. - 2 boxes.
In Rutgers University Library, Special Collections Department (New Brunswick, NJ).
Physician. - Hommell was thirty-sixth President of the New Jersey Pharmaceutical Association; an organizer of the Rutgers College of Pharmacy; and its first professor of botany, physiology, and materia medica. The collection includes letters received, 1911-1932; medical student notebooks, 1880-81; college lecture notes in botany, physiology, and other subjects; business records from the Hommel Pharmacy in Jersey City 1884-1898; prescription guide books, 1860; and other items relating to the pharmacy business.
Unpublished finding aid available.

H82 Hooker, Albert H., Jr.
Papers, 1909-59. - 3.5 cu. ft.
In Washington State Historical Society (Tacoma).
Scrapbooks, diaries, correspondence, maps, and photographs detailing Hooker's World War I and World War II experiences, along with records pertaining to the Hooker Petrochemical Company.
These papers are reproductions only.

H83 **Hooper (Samuel) and Company,** *Boston.*
Records, 1837-89.
30 ft.
In Harvard Business School, Baker Library.
Correspondence, account books, and other papers of Samuel Hooper and Co. (before 1859 known as William Appleton and Co.), a firm trading with China and other Far Eastern countries, South America, and (before the Mexican War) California. Includes letter books and diaries of William Appleton, other members of the Appleton family, and various members of the Lawrence family, papers of the captains of many of the ships employed by the company, and records relating to Benzinger Coal and Iron Co., St. Mary's, Pa. (1847-81), Boston Oil

Co. (1876-77), Boston Sugar Refinery (1837-43), Erie Basin Dock Co., South Brooklyn, N.Y. (1864-80), and other manufacturing and railroad interests of the Appleton and Lawrence families.
Inventory in the library. Also described in List of business manuscripts in Baker Library, compiled by Robert W. Lovett (2d ed., 1951)
Gifts of Gordon Dexter and W. Appleton Lawrence, 1927-50.

H84 Hoover, Herbert Clark, Pres. U.S., 1874-1964.
Special collections, 1657, 1750-1964. 625 ft.
In Herbert Hoover Presidential Library (West Branch, Iowa)
Correspondence, drafts, galley and page proofs, research notes, collateral translations, and other papers (1908-61), relating to the translation of Agricola's De Re Metallica by Hoover and Lou (Henry) Hoover (1912); appointments calendar (1901-64); articles, addresses, and public statements (1892-1964) with correspondence, research material, drafts, galley proofs, and clippings relating to them; clipping file (1908-64) relating to Hoover's relief activities, service as U.S. Secretary of Commerce and President, and issues of interest to him; correspondence, research material, and printed material, relating to Hoover genealogy, particularly the book Genealogy of the Herbert Hoover Family, by Hulda Hoover McLean; and misrepresentations file (1908-61) containing correspondence, affidavits, printed material, and other papers, assembled in response to misrepresentations of Hoover's statements and actions. Includes 2 Latin imprints (1657) of Agricola's De Re Metallica.
Unpublished finding aid in the repository.
Access restricted, in part.
Gift of Mr. Hoover, 1960.

H85 Hope College, Holland, Michigan. Department of Chemistry.
Records, 1960-date. - 2 linear ft.
In Hope College Archives (Holland, MI).
Minutes, 1968-date; clippings, papers, photographs, ratings brochures, annual reports: 1964, 1969, 1982, 1983, 1984.

H86 Hopkins, Arthur John, 1864-1939.
Papers, 1893-1939. 4 ft.
In Amherst College Library (Mass.)
Chemist. Chiefly lectures, notes, and articles, on chemistry and alchemy, together with Hopkins' correspondence relating to his research on alchemy.
Unpublished finding aid in the repository.
Gift.

H87 Hopkins, B. Smith, 1873-1952.
Papers, 1917-1963. - 0.6 ft.
In University of Illinois Archives (Urbana).
Professor of Inorganic Chemistry at the University of Illinois. - Correspondence relating to research work on the rare earths, fractionation of rare earth solutions, element 61 or "illinium," shipment and use of rare earth salts, ionic migration method of separating rare earths, X-ray spectrographic analysis of rare earths, Hopkins' faculty appointment (1919), work of graduate students, textbooks and revisions, X-ray equipment, laboratory notebooks (1925-1931), chemical manufacturing, and Hopkins' genealogy; and a posthumous list of Hopkins' publications, consisting of 12 books and 129 articles. Correspondents include Gerald Druce, of England; J. Allen Harris, of Canada; G. Hevesy, of Copenhagen; Charles James; David Kinley; R. J. Meyer, of Berlin; William A. Noyes; Luigi Rolla, of Florence; Manne Siegbahn, of Uppsala; S. Urbain, of Paris; and Leonard F. Yntema.

H88 Horsford, Eben Norton, 1818-1893.
Papers, 1755-1944. 14 ft.
In Rensselaer Polytechnic Institute Archives (Troy, N.Y.)
Chemist and educator. Correspondence, notes, bills, and receipts, relating chiefly to Horsford's scientific and business activities and to his teaching at Lawrence Scientific School of Harvard University; mss. and proofs of poems by his first wife, Mary L'Hommedieu (Gardiner) Horsford (1824-1855); and papers of the L'Hommedieu and Gardiner families. Includes much correspondence relating to the Rumford Chemical Works (R.I.) and the monument to Norumbega, the alleged Norse city in New England, in Watertown, Mass.
Unpublished inventory in the repository.
Gift of Mrs. Augustus H. Fiske, 1966.

H89 Houdry, Eugene Jules, 1892-1962.
Papers, 1931-1980. - 36 items.
In Library of Congress, Manuscript Division (Washington, DC).
French-American chemist, chemical engineer, and inventor. - Notes, sketches, blueprints, mathematical calculations, formulas, charts, near-print and printed matter, speeches, and lectures relating to petroleum chemistry, air pollution and public health.
Gift of Jacques Houdry, 1965; and of Jeffrey K. Stine, 1981.
Finding aid available.

H90 Howe, Henry Marion, 1848-1922.
Papers, 1884-1916. ca. 350 items.
In Columbia University Libraries.
Professor of metallurgy at Columbia University. Correspondence, lecture notes, scrapbooks of metallurgical photos, blueprints, mss. of writings, and other papers relating to departmental affairs at Columbia and Howe's work as a metallurgical consultant to various mining and metal companies.

Howe Sound Mining Company, Holden Village, Washington.
Records, 1930-60. - 87 lin. ft.
In University of Washington Libraries, Archives and Manuscripts Division (Seattle).
Correspondence (1930-1954), financial records, reports, and payroll records of Washington's largest copper, gold and zinc mine. The bulk of the records are between the years 1940-1957.
Unpublished inventory available.

Howell, Edward Vernon, 1872-1931.
Papers, 1725-1929.
ca. 1600 items.
In University of North Carolina Library, Southern Historical Collection (1000)
In part, transcripts.
Dean of the School of Pharmacy, University of North Carolina and collector. Personal and professional correspondence (1900-29), together with historical materials largely connected with North Carolina (1725-1871) Includes items relating to Richard Henderson (1735-85) and the Transylvania Co.; the Revolutionary War in North Carolina; the Civil War; the wholesale and retail drug firm of Williams and Haywood, located in Raleigh, N. C. (1858-69); and members of the Boylan, Burton, Lewis, Moore, and Williams families of eastern North Carolina in the 18th and 19th centuries. Also includes copies of ballads collected in Avery County, N.C. (1917-18); account books of physicians and merchants; photos. of Confederate leaders; and copies of papers written by, and articles concerning Henri Harrisse (1820-1910), the French scholar who taught at the University of North Carolina from 1853 to 1857.
Unpublished finding aid in the library.
Gift of James K. Kyser and Emily Royster Howell Kyser, 1946.

Huffman, Eugene Harvey, b. 1905.
Papers, ca. 1945-1967. - 2 cartons and 1 box.
In University of California, Berkeley, Bancroft Library.
Chemist at Lawrence Radiation Laboratory, UC Berkeley. - Correspondence, reports, reprints, research notes regarding Huffman's work in analytical chemistry, including materials on ion exchange and plans for an analytical chemistry laboratory.
Partially processed collection; folder list available in repository.

Hull, Alfred.
Day books and account book, 1828-1839. - 4 v.
In Old Sturbridge Village Research Library, (Sturbridge, MA).
Tanner and shoemaker, Derby, Connecticut. - Day book, 1821-1828; 1828-1835; 1835-1838; account book, 1823-1824.

Hummel, Ray Orvin, 1880-   H95
Papers, 1900-38. 1 ft. (ca. 500 items)
In Nebraska State Historical Society collections (Lincoln)
Physician, of Lincoln, Neb. Chiefly lecture and laboratory notes and laboratory drawings made in chemistry and zoology classes while an undergraduate at the University of Nebraska; together with notes and drawings (1900-04) made during classes at Northwestern University Medical School, and topical reference files with notes on common diseases and disabilities.
Gift of Dr. Hummel's son and daughter-in-law, Mr. and Mrs. Paul A. Hummel, Los Angeles, Calif., 1974.

Hunt, Reid, 1870-1948.   H96
Papers, 1904-1936. - 1 box.
In Harvard University Archives (Cambridge, MA).
Professor of Pharmacology. - Correspondence and papers.

Huntingdon Co., Pa., iron and forge industry, 1820-89. 107 H97 items.
In University of Pittsburgh Libraries, Archives of Industrial Society (Pa.)
Business correspondence, memoranda, financial records, and legal documents, of H. P. Dorsey; together with correspondence of members of the Porter family, especially Alfred Porter, D. N. Porter, and J. Porter. The two collections concern the making and forge operations of the Colerain Forge, Huntingdon Furnace, and the Pennsylvania Furnace, Huntingdon Co., Pa.
Gift of Richard W. Simon, 1965.

Hussey, Robert E.   H98
Papers, 1928-1937. - 3 v.
In Virginia Polytechnic Institute Library (Blacksburg).
Professor of Organic Chemistry. - Three research notebooks (ca. 600 pp.), containing notes on chemistry experiments.

11 I. G. Farben Trust, Propanganda Division.
Mimeographed bulletins, 1932-1942. - 1 folder.
In Hoover Institute on War, Revolution, and Peace, Stanford University (Stanford, CA).
Relates to conditions in the chemical industry in Germany during World War II. In German.

12 Indian Head Mills, inc.
Records, 1839-1947.
45 ft.
In Harvard Business School, Baker Library.
Correspondence, minutes, stockholders' records, general ledgers, journals, cashbooks, trial balances, accounts, and production records of a cotton textile manufacturing firm under its former name Naumkeag Steam Cotton Co. Includes letters of Edmund Smith, David Pingree, Edmund Dwight, Henry D. Sullivan, Frederick Dexter, and N. G. Simonds, treasurers; Josiah Brown, John Kilburn, E. F. Balch, and William P. McMullan, agents; and John Howard Fallon, trustee; and records (1848-1940) of the Danvers Bleachery and Dye Works, Peabody, Mass.
Inventory in the library.
Gift of Indian Head Mills, 1955.

13 Industrial workers of Brandywine Valley and Wilmington, Del., region, 1890's to 1920's: Hagley Museum oral history collection, 1955-70. 5 ft. (65 items)
In Eleutherian Mills Historical Library (Greenville, Del.)
Transcripts of tape-recorded interviews with persons employed in the Du Pont Company powder mills on Brandywine Creek, north of Wilmington, Del., and with persons living in the vicinity of these mills and nearby textile, paper, tanning, and car wheel manufacturers. Occupation of the interviewees include laborer, wagoner, machinist, powderman, pipe-fitter, bookkeeper, salesman, chemist, engineer, personnel man, and superintendent. Includes information on the operations of the powder mills, labor-management relations, wages, hours, and conditions of work, and patterns of life in the industrial communities of the area. Many of the transcripts include a photo of the interviewee.
Table of contents and card file of interviewees in the Research Division of the Hagley Museum.
The original tapes are retained in the repository's Pictorial Collections.
Permanent deposit.

14 Ingersoll, Arthur William, 1894-1969.
Papers, 1917-62. ca. 5 ft.
In Joint University Libraries (Nashville, Tenn.)
Professor of organic chemistry, Vanderbilt University, Nashville, Tenn. Correspondence, mss. of of papers published by Ingersoll in the Journal of the American Chemical Society on the electrolytic reduction of organic compounds, the resolution of amino acids, and other topics, research and lecture notes on various aspects of organic chemistry, and material relating to undergraduate and graduate chemistry programs at Vanderbilt University and other universities.
Unpublished register in the library.
Information on literary rights available in the library.
Gift of the Chemistry Dept., Vanderbilt University, and of Mrs. Ingersoll.

15 **Institute cf Women's Professional Relations.**
Records, 1928-41.
4 boxes.
In Radcliffe College Library, Women's Archives (B-5)
Minutes of the board, reports to the board, studies and interviews on women in banking, chemistry, dietetics, home economics, etc., together with studies and reports on Connecticut and Rhode Island State government in cooperation with W. P. A. project 2132 and the League of Women Voters.
Unpublished inventory in the library.
Acquired 1955.

16 International Education Board.
Records, 1923-1941. - 22 ft. and 1 v.
In Rockefeller Archives Center (Pocantico Hills, North Tarrytown, NY).
Correspondence, memoranda, reports, surveys, minutes, administrative, financial, and personnel records, fellowship and traveling professorship records, appropriations files, charts, and clippings. Includes material on agriculture, astronomy, biology, chemistry, dairying, farm demonstrations, forestry, geology, home economics, mathematics, physics, rural development, scientific publications, and veterinary medicine. Projects and organizations represented include: Booker Washington Agricultural and Industrial Institute of Liberia, College of West Africa, Monrovia, Liberia; Columbia University International Institute of Teachers College; International Institute of Agriculture; Mount Palomar telescope; Naples Zoological Station; National Research Council; New York State Colonization Society; Pan American Union; and universities around the world. Correspondents include Niels Bohr, James Breasted, Abraham Flexner, Simon Flexner, George E. Hale, and others.
Unpublished finding aid available.

17 International Union of Operating Engineers, Big Spring, Texas.
Records, 1946-1975. 10.5 feet
In University of Texas at Arlington Library, Department of Special Collections
Union representing workers at several companies in the fields of oil refining, petrochemicals, potash, and rubber production. - Correspondence, 1946-1975; arbitration cases and contract negotiations, 1956-1969; discharge cases; NLRB actions; organizing materials; clippings.
Permanent deposit.

Inyo Development Company.
Records, 1885-1916. - 12 ft.
In University of Nevada Library, Special Collections Department (Reno).
Records of an Inyo County, California company whose main business was the production and shipment of crude soda and soda products. - The Inyo Development Company was incorporated in Nevada in 1885 by H. M. Yerington, President. Its base of operation was at Keeler, California. Included in the collection are correspondence, financial statements, receipts, labor records, soda records, and legal documents from 1885 to 1916.
Gift, date unknown.
Unpublished finding aid available.

Ipatieff, Vladimir N., 1867-1952.
Papers, 1867-1950. - 3 cu. ft.
In Northwestern University Archives, (Evanston, IL).
Lecturer in Chemistry (1939-1945); Director, Ipatieff High Pressure and Catalytic Laboratory for Chemical Research (1939-1952). - Collection covers the entirety of his career. Material is arranged in the following categories: personal correspondence; diaries in English and Russian; professional articles by Ipatieff; hand- and typewritten manuscript materials; news clippings and tributes to Ipatieff; legal documents regarding the management of his and his wife's estates; and miscellaneous writings and memorabilia.
Finding aid available.

Ithaca Glass Works (Ithaca, NY). I10
Records, 1878-1891. - 1 ft.
In Cornell University Archives (Ithaca, NY).
Collection contains correspondence, committee reports, legal documents, bills, receipts, inventories of raw materials and equipment, sales records, price lists, fire insurance policies, printed circulars pertaining to the industry, and one photograph.

Ivanhoe Mining and Smelting Corporation, I11
Ivanhoe, Virginia.
Records, 1907-1929. - 1 cu. ft.
In Virginia Polytechnic Institute Library (Blacksburg).
Correspondence files, 1915-1929; iron ore analysis books, 1907-1915.

J1 Jackson, Charles Loring, 1847-1935.
Papers, 1884-1937.
In Harvard University Archives (Cambridge, MA).
Professor of Chemistry. – Small collection including miscellaneous printed material by and about Jackson; some undated personal manuscripts by him, mainly autobiographical; a research notebook begun in 1884; and a couple of writings by Jackson on topics in the field of literature.

J2 Jackson, Charles Thomas, 1805-1880.
Papers, 1829-1915. 65 items.
In Library of Congress, Manuscript Division.
Physician, chemist, and geologist, of Boston, who claimed the discovery of the basic principles of the telegraph, the discovery of guncotton, and the use of ether as an anesthetic. Letters (1829-59), and other papers. Letters written chiefly to Jackson's brother-in-law, Charles Brown of Brown & Ellis, merchants of Boston, relating to Jackson's student days in Europe, technical, topographical, and geological descriptions, the European cholera epidemic (1831), and Jackson's surgical anesthesia claims (1859) as opposed to those of William T. G. Morton. Other correspondents include his sister, Lydia (Jackson) Emerson.
Described in the Library's Information bulletin, v. 25, no. 22 (June 2, 1966).
Purchased from Franklin Book Shop, Philadelphia, 1920, and Goodspeed's Book Shop, Boston, 1966.

J3 Jackson, Charles Thomas, 1805-1880.
Papers, 1846-73. 92 items.
In Massachusetts Historical Society Library (Boston)
Chemist and geologist. Correspondence relating to Jackson's claim to have discovered the anaesthetic use of ether, together with diplomas and affidavits. Correspondents include Ralph Waldo Emerson, Edward Everett, Abbott Lawrence, and Dr. John C. Warren.
Information on literary rights available in the library.
Gift of heirs of Dr. Jackson, 1915.

J4 Jacobs, Joseph, pharmacist, of Atlanta, Ga.
Papers, 1880's-1940's. 1 ft.
In Atlanta Historical Society collections (Ga.)
Founder of Jacobs' Pharmacies, Atlanta, Ga. Formulae for cosmetics and medicinals, price lists, bottle labels, and sales information; and mss. of writings on Dr. Crawford W. Long (1815-1878), to whom Jacobs was apprenticed, and on Andrew Lipscomb, and articles by and about Jacobs.

J5 Jacobs, Walter Abraham, 1883-1967.
Papers, 1907-59. 2 ft.
In Rockefeller University Archives, Rockefeller Archive Center (Pocantico Hills, North Tarrytown, N.Y.)
Chemist at Rockefeller Institute for Medical Research, New York, N.Y. Chiefly 30 v. of laboratory record books (1914-36) signed by Jacobs, Michael Heidelberger, and others; together with correspondence with Donald Van Slyke, autobiographical sketch, biographical material by Jacob's wife, Laura F. Dreyfoos Jacobs, bibliography, collected reprints, and photos.
Unpublished finding aid in the repository.
Information on literary rights available in the repository.
Gift of the Rockefeller University administration.

J6 Jaffa, Meyer E., 1857-1931.
Papers, ca. 1896-1923. – 1 carton and oversized material.
In University of California, Berkeley, Bancroft Library.
Professor of Agricultural Chemistry and Nutrition, UC Berkeley; Director of State Food and Drug Lab. – Correspondence, subject folders relating to Jaffa's research, a scrapbook of clippings, and Central Valley and San Francisco newspapers containing mention of Jaffa's work and other agricultural news.
Partially processed collection; folder list available in repository.

J7 Jay, Philip, 1898-1974.
Papers, 1925-1942. – 4 ft.
In University of Michigan, Bentley Historical Library, Michigan Historical Collections (Ann Arbor).
Professor of Dentistry, University of Michigan. – Official and personal correspondence relating to Jay's teaching and research activities, particularly on fluoridation and administration of the School of Dentistry.

J8 Jeffress, Elizabeth Talbott (Gwathmey), b. 1900.
Papers, 1952-1972. – 26 items.
In Virginia Historical Society (Richmond).
Includes speeches concerning chemistry, delivered 1952-1961 at the Virginia Military Institute (Lexington) by Allan Talbott Gwathmey (1903-1963) Professor of Chemistry at the University of Virginia; letters, 1971-1972, written to Mrs. Jeffress of Richmond, concerning scholarships established at the Virginia Military Institute and the University of Virginia as a memorial to Allan Talbott Gwathmey; and miscellany concerning Allan Talbott Gwathmey (with photographs of Allan Talbott Gwathmey and Roberta Hollinsworth Gwathmey).
Gift of Mrs. Robert Miller Jeffress, Richmond, Va., December 29, 1972.

J9 Jenckes, Thomas Allen, 1818–1875.
Papers, 1836–78.
29 ft.
In Library of Congress, Manuscript Division.
Jurist and legislator. Correspondence, journals, ledgers, diaries, account books, patent papers, legal briefs and related papers, law library catalog, and photos. The papers deal with Jenckes' career from his days as a student at Brown University until his death, embracing his extensive law practice in the patent field, his services to Rhode Island as legislator, adjutant general, and secretary to the 1842 Constitutional Convention and to the Governor's council, and his four terms in the U. S. House of Representatives. Particular detail is found for the Congressional period, during which Jenckes was a pioneer in the civil service reform movement and a leader in the revision of the bankruptcy, patent, and copyright laws. A set of papers deals with the Crédit mobilier fraud prosecution, and others with patent cases, including the Corliss steam engine and Goodyear rubber controversies, and suits concerning ventilating and refrigerating patents. Individual correspondents include Benjamin Vaughan Abbott, Charles Adams, Henry B. Anthony, Hugh Burgess, Alexis Caswell, Horace H. Day, Ezra D. Fogg, George Gifford, R. Benton Hammond, Alexander Hay, Julius E. Hilgard, Charles R. Ingersoll, David Lyman, Dennis F. Murphy, Francis C. Nye, James H. Parsons, William Sprague, Henry E. Wallace, Augustus Woodbury, Charles C. Van Zandt, and others.
Indexed in part.
Unpublished finding aid in the library. Also described in the Library's Quarterly journal of current acquisitions, v. 7, no. 3 (May 1950) p. 27.
Gift of Thomas A. Jenckes (grandson), 1949.

J10 Jessop, Joshua, 1806–1869.
Dye recipe books, 1826–27; n.d. – 2 v.
In Maryland Historical Society Library (Baltimore).
Owner and operator of dye works in Baltimore County, MD. – Volumes contain "recipes" for dyes (including samples of dyed cloth and yarns). The volumes also contain Jessop's notes on the carding of wool and notes and a glossary on Edward Bancroft's Experimental Researches Concerning the Philosophy of Permanent Colors (1813).
Gift of Mrs. Nelson Carter, Palmetto, Florida, August 26, 1980.
Unpublished finding aid available.

J11 Jessup, Augustus Edward, 1797–1859.
Papers, 1822–1913. 121 items.
In Academy of Natural Sciences of Philadelphia collections (375)
Geologist, mineralogist and paper manufacturer, of Philadelphia. Correspondence, biographical material, data regarding the Jessup Fund (bequeathed to the Academy of Natural Sciences of Philadelphia to aid students in natural history and in support of publications) and applicants for its benefits, and articles published in the academy's journal. Some of the correspondence relates to Jessup's paper business.
Unpublished finding aid in the repository. Also described in Guide to the manuscript collections in the Academy of Natural Sciences of Philadelphia, compiled by V. T. Phillips and M. E. Phillips (1963) p. 190.
Gift of Mr. Jessup and his family.

J12 John L. Thompson, Sons & Company (Troy, N.Y.)
Records, 1818–1889. – 132 items.
In Albany Institute of History and Art (N.Y.)
Wholesale dealers in drugs, chemicals, and medicines, established in 1797. – Correspondence, bills, receipts, orders, and other material, including printed history of the firm, 1797–1972, and orders and receipts for merchandise purchased by the U.S. Arsenal at Watervliet, N.Y., 1827–1832.

J13 Johnson, Frederick, b. 1904.
Papers, 1948–1968. – 9 boxes.
In University of California-Los Angeles Libraries, Department of Special Collections.
Anthropologist, curator. – Materials, chiefly correspondence, of Johnson as Chair of the Committee on Radioactive Carbon-14, which was set up by the American Anthropological Association at the request of Willard F. Libby; correspondence and related materials about various conferences in the U.S. and elsewhere on radiocarbon dating which Johnson helped to organize; and organizational materials, correspondence and financial records relating to the Radiocarbon Dates Association Inc., of which Johnson was president.
Gift of Frederick Johnson, March, 1982.
Unpublished finding aid available.

J14 Johnson, Norman G.
Papers, 1946–67; n.d. – 80 items.
In Hagley Library (Greenville, DE) (acc. #1081).
Research chemist. – Johnson joined the laboratory of the Explosives Department of E. I. du Pont de Nemours & Co. at Gibbstown, N.J., in 1926 and retired in 1967. He held numerous patents, co-authored three editions of the blasters' handbook issued by the Du Pont Co., and made many contributions to the seismic explosives industry. Papers consist of notes and memoranda. A sample of subjects includes his intended revision of Arthur Pine Van Gelder and Hugo Schlatter, History of the Explosives Industry in America (New York, 1927); lamination; vibration and noise; seismic prospecting; Du Pont's contributions in World War II; competitive companies; black powder; company annual reports; and others.
Described more fully in the Supplement to A Guide to Manuscripts in the Eleutherian Mills Historical Library (1978).

J15 Johnston, James A.
Papers, 1960-1967. – 6 boxes.
In Rutgers University Library, University Archives Collections (New Brunswick, NJ).
Biochemist; Rutgers faculty member. – Papers from Johnston's tenure on the faculty, Department of Biochemistry, College of Agriculture, consisting mainly of correspondence, research notes, student papers, reprints, etc. Johnston was later Associate Provost, Rutgers University.
Unpublished finding aid available.

J16 Jones, Alfred Goldsborough, 1821-1868.
Papers, 1840-67. 15 v.
In New York Public Library.
Lawyer, of New York City. Journal covering Jones' education at Columbia and Harvard Colleges, his legal training and law practice, his service as an officer and director of the Sixth Avenue Railroad Company, his association with Theodorus Bailey Myers, social, political, and cultural life of New York City and New Brunswick, N. J., and his travels in the U. S., West Indies, South America, and the Pacific Ocean; and a chemistry notebook from Columbia College, later used as a scrapbook.
Described in Dictionary Catalog of the Manuscript Division, New York Public Library (1967) p. 458.
Purchase, 1945.

J17 Jones, Day, Cockley & Reavis, Cleveland, Ohio.
Records, 1917-59. 72 v.
In Western Reserve Historical Society collections (Cleveland)
Correspondence, memoranda, deeds, indentures, contracts, stock certificates, and other records, relating to various legal or financial transactions engaged in by different Cleveland businesses or industries for which the firm of Jones, Day, Cockley & Reavis, or one of its predecessors, served as legal counsel. Companies represented include Chesapeake and Ohio Railway Company, 1928-30, 1936; Cleveland Terminals Building, 1931; Goodyear Tire and Rubber Company, 1921, 1927; Lake Erie Chemical Company, 1933; Otis Steel Company, 1936, 1937, 1942; Pere Marquette Railway Company, 1930-32; Republic Steel Corporation, 1930, 1934-36, 1941; Van Sweringen Company, 1928, 1930; and White Motor Company, 1934.
Unpublished register in the repository.
Access restricted.
Gift of Jones, Day, Cockley & Reavis, 1967.

J18 **Jones, Grinnell,** 1884-1947.
Correspondence, 1912-47. 3 ft.
In Harvard University Archives.
Professor of chemistry at Harvard. Professional correspondence.

Open to investigators only upon prior application to the repository.
Information on literary rights available in the repository.

Jones, Meriwether, 1853-1936.   J19
Papers, 1817-1921. 3273 items and 14 v.
In Duke University Library (Durham, N.C.)
Mining engineer, of Richmond, Va. Correspondence, accounts, statistics, and other records, of companies owned or managed by Jones and of which Ferral C. Dininny, Jr. was owner or executive, including Alleghany Iron Company, operating pig iron furnaces and iron ore mines in Alleghany, Botetourt, and Rockbridge Counties, Va., coal mines in Chesterfield County, Va., and Old Dominion Tobacco Warehouse, Richmond, Va. Includes letters (1913-20) from members of the Class of 1874 at Virginia Military Institute, Lexington.
Card index in the repository.
Purchases, 1949-53.
This entry incorporates and replaces MS 68-1543.

Jones, Richard Uriah, 1877-1941.   J20
Papers, 1914-52. ca. 500 items.
In Macalester College, Weyerhaeuser Library (St. Paul, Minn.)
Presbyterian minister, professor of chemistry, and dean of Macalester College. Correspondence (personal and as dean and professor); mss. of Jones' book, "The scientific eye of faith" (1935) and of his M. A. thesis on chemistry from the University of Wisconsin (1916); sermon notes; materials in the Welsh language; and miscellaneous family papers.
Unpublished inventory in the library.
Gift of Mrs. Jone, 1964–

Joseph Bancroft & Sons, Company,   J21
Wilmington, Delaware.
Records, 1869-1969. – Ca. 90,000 items.
In Hagley Library (Greenville, DE) (Acc. # 1359, 1400, 1440).
Textile firm engaged in manufacturing, bleaching, dyeing, finishing, and licensing operations. – A full range of records including administrative, financial, production, and sales functions. Of particular interest to historians of chemistry are accessions 1400 and 1440 which contain material relating to dyeing, bleaching, and finishing research and processes.
Described more fully in the Supplement to A Guide to Manuscripts in the Eleutherian Mills Historical Library (1978) pp. 12-15 and the original Guide (1970) pp. 705-711.

Journal of Chemical Physics.
  Records, 1938-1957. - 0.3 ft.
  In American Institute of Physics, Niels Bohr Library (New York, NY).
  Correspondence files between AIP Director, Henry A. Barton, and successive editors of the Journal, Joseph E. Mayer and Clyde A. Hutchinson, pertaining to policy and administration; correspondence between Mayer and Margaret Griffin, the AIP publication manager, dealing with concerns such as scheduling and printing as well as other correspondence concerning the Journal.
  Deposited by J. W. Stout, editor of the Journal, 1967.
  Finding aid available.

Judge, John, 1830–1885. J23
  Business papers, 1852–81. 3 ft.
  In University of North Carolina Library, Southern Historical Collection (390)
  Army officer and businessman. Business correspondence and papers, chiefly 1860–73, relating to Judge's activities as owner of a paper mill in Wilmington, N. C., as operator of a yarn and sock factory at Columbia, S. C., and as a commission merchant. Includes account books, bankbooks, sales and daybooks, deeds, articles of agreement, specifications for factories and machinery, and receipts. Correspondents include Henry Lowe and P. H. Winston.
  Unpublished description in the library.
  Gift of Jane Judge, before 1940.

The Kodak Research Laboratories began operation in this building at Kodak Park in Rochester, New York, in early 1913. Courtesy Eastman Kodak Company.

K1 **Kaftan, Arthur,** 1912–
Papers, 1929–54. 1 box.
In State Historical Society of Wisconsin collections (Madison)
Lawyer, of Green Bay, Wis., chairman of the pollution committee of the Brown County Chapter of the Izaak Walton League of America and chairman of the Wisconsin division of the league. Correspondence, investigations of the committee on water pollution, reports of the State Water Pollution Committee and the State Board of Health, and other papers. The bulk of the collection is for 1949–54.
Gift of Mr. Kaftan.

K2 Kahlenberg, Louis, 1870-1941.
Correspondence, 1900-1939. - 12 boxes.
In State Historical Society of Wisconsin (Madison).
Professor (1895-1940) and Chairman of the Department of Chemistry at the University of Wisconsin. - Correspondence concerning problems of research, administration, publications, and placement of students.
Unpublished finding aid available.

K3 **Kalamazoo Paper Company.**
Records, 1866–1957. 1 item and 109 v.
In Western Michigan University, Regional History Collections (Kalamazoo)
Board of Directors' minute books (1866–92), stock certificate book (1868–92) and other record books of a paper manufacturing company of Kalamazoo, Mich. Includes a copy of "New horizons at ninety."
Gift of Kalamazoo Paper Company, 1963.

K4 Kamen, Martin David, b. 1913.
Papers, 1937-1954. - 2 cartons.
In University of California, Berkeley, Bancroft Library.
Professor of Chemistry at UC San Diego; formerly researcher at Radiation Laboratory of UC Berkeley. - Research notes and notebooks from research, much of which relates to the photosynthetic process. Also includes research notes by Samuel Ruben who, with Kamen, co-discovered carbon-14.
Partially processed collection; folder list available in repository.

K5 Kedzie, Frank Stewart, 1857-1935.
Papers, 1915-1946. - 2.3 cu. ft.
In Michigan State University, University Archives and Historical Collections (East Lansing).
Chemist; President of Michigan State University, 1915-1921. - Collection contains mainly correspondence. Also included are notes and papers by Kedzie on the history of Michigan Agricultural College and a booklet commemorating the establishment of the Michigan Agricultural College Union Memorial Building in 1923.
Gift of Madison Kuhn.
Unpublished finding aid available.

Keefer, Horace A K6
Papers, 1881–1916.
178 items.
In Pennsylvania Historical and Museum Commission collection.
Manager of Pine Grove Furnace. Correspondence and a history of the furnace.

Kefauver, Estes, 1903–1963. K7
Papers, 1935–63. 1204 ft.
In University of Tennessee Library (Knoxville)
U. S. Representative and Senator, of Madisonville, Tenn. Constituent, political, and personal correspondence, MSS. of Kefauver's books, Crime in America (1951) and A Twentieth Century Congress (1947), political files, speeches, films, tapes, legislative papers, clippings, and other material, relating to Kefauver's political career. Includes material on his work in the Congress with the Armed Services Committee, Senate Judiciary Committee, Senate Subcommittee on Antitrust and Monopoly's drug investigations (1962), and the Special Committee of the U. S. Senate to Investigate Organized Crime in Interstate Commerce.
Unpublished guide in the library.
Open to investigators under library restrictions.
Information on literary rights available in the library.
Gift of Mrs. Kefauver, 1963.

Keller, Anatole J. K8
Family papers, 1885-1919. - 2 linear ft.
In Louisiana State University, Department of Archives (Baton Rouge).
Sugar technologist. - Correspondence (1910-1916, n.d.) pertains to social life, Keller's work as a sugar technologist, and elementary school education in Louisiana. Other papers include student notes and assignments from LSU classes (1907-1911, n.d.) and copies of the Negro Advocate, with correspondence and other papers related to Black causes (1918-1919).

Keller, Arthur G., 1902-1968. K9
Papers, 1878-1967. - 19 linear ft.
In Louisiana State University, Department of Archives (Baton Rouge).
Professor of Chemical Engineering; specialist in sugar cane technology; consulting engineer. - Professional papers include correspondence; notebooks; teaching materials; and files, financial records, reports and other papers relative to a firm of consulting engineers run by Keller. Materials pertaining to sugar cane technology include reports on refineries and production in the U.S., Latin America and elsewhere, papers, and printed materials on sugar technology and production.

10 **Kelley, William J** 1886–1966.
Federal Hill Story papers, 1952–66. ca. 50 items.
In Maryland Historical Society Library (Baltimore) (1692)
Head, Real Estate Dept., Baltimore Chamber of Commerce. Ms., notes, and maps, for a book on industries located in the Federal Hill area of Baltimore. Includes chapters on the chemical, fertilizer, glass, and iron industries, and a list of Baltimore firsts in industry, 1757–1965.
Gift of the estate of William J. Kelley, through Richard H. Randall, Sr., 1968.

11 **Kelly, Wayne Clinton,** 1844–1923.
Papers, 1864–92. 13 v.
In University of Louisville, Medical School Library (Ky.)
Physician, of Louisville, Ky. Notes (1864–67) taken while a student at the Medical School of McGill University, Montreal, Can., and notes (1892) on Dr. Kelly's obstetric and gynecological patients. Subjects include chemistry, materia medica, theory and practice of medicine, medical jurisprudence, clinical medicine, surgery, and midwifery.

12 **Kennedy, James Arthur,** 1894–
Papers, 1953–57. ca. 3 ft.
In University of Louisville, Medical School Library (Ky.)
Professor and chairman of the Dept. of Microbiology at the University of Louisville Medical School. Correspondence and minutes relating to the organization of a program leading to a Ph. D. degree in the School of Medicine, University of Louisville. Includes a petition for the approval of a Ph. D. program in chemical engineering at Speed Scientific School, University of Louisville, and a petition (1953) for chapter status of the Sigma Xi Club of the University of Louisville.
Gift of Dr. Kennedy, 1966.

13 **Kerr, George Alexander.**
Papers, 1927–36. ca. 50 items.
In University of Virginia Library (Charlottesville) (9894)
Business executive. Correspondence relating to the Philippine Cutch Corporation and to the Philippine economy and government, financial papers, articles on vegetable tannin, and reprint of an address (1933) to the American Leather Chemists Association.
Gift, 1973.

14 **Kerr, W. W.**
Papers, 1913–1951. – Ca. 8,600 items.
In Louisiana State University, Department of Archives (Baton Rouge).
Consulting mechanical engineer specializing in sugar factory work, 1935–1949; efficiency engineer for the Cuba Cane Sugar Corporation of Havana, 1916. – The bulk of the collection is made up of correspondence, technical data and notes, blueprints and related printed items for the Cuba Cane Sugar Corporation and for sugar factories in southwest Louisiana. In addition, the collection contains family letters; personal correspondence relating to business matters; and printed material including articles from technical periodicals.

Kessler, Charles Nicholas. K15
Papers, 1865–1953. 288 ft.
In Montana Historical Society collections (Helena)
Businessman and brewer, of Helena, Mont. General correspondence, financial records, and miscellaneous papers. Includes business records of Kessler Brewery, Helena, Mont.

Kessler, Nickolas. K16
Family papers, 1856–1953. – 40 linear ft.
In Montana Historical Society (Helena).
Early Helena brewer and businessman; immigrated to U.S. from Luxembourg in 1854. Besides brewing, Kessler was involved in mining, real estate, farming, ranching, and brickmaking. The Kessler Brewing Company was the principal family interest. It formally incorporated in 1901 and remained in the family until it closed in 1958. – Collection contains 7 subgroups: Nickolas Kessler, Kessler Brewery, Kessler Brickyard, the Saloon, C. N. Kessler, Nick Baatz Company, and Miscellany. Correspondence, both personal and business; financial records; legal documents; subject files; minutes of Montana Brewers Association; photographs; printed materials; production records; reports; research notes; speeches; clippings; miscellany.
Gift of C. N. Kessler, Jr., Los Angeles, and of Helen Kessler Buterbaugh, Helena, November 13, 1957.
Unpublished finding aid available.

Kilgore, Benjamin Wesley. K17
Papers, 1892–1921. – Ca. 3,000 items.
In North Carolina State Office of Archives and History Collections (Raleigh).
Chemist. – Kilgore was Assistant Chemist of the North Carolina Agricultural Experiment Station from 1889 to 1898; State Chemist of Mississippi from 1898 to 1899; State Chemist of North Carolina from 1899 to 1919; Director of the North Carolina Agricultural Experiment Station and Extension Service from 1911 to 1921; and U.S. Department of Agriculture Director of Extension Service for North Carolina from 1919 to 1920. – Most of the items concern these professional activities and include research notes and reports; addresses; minutes of various committees and boards; and correspondence with research chemists in the United States, county agents in North Carolina, federal extension service officials, purchasers of cotton seed and other products, and sellers of farm equipment and supplies. There are also bulletins, conference proceedings, and other printed items relating to agriculture.

K18 Kimball, Philip J., 1894-1975
Papers, 1920-66. - 60 boxed items and 3 v.
In Hagley Library (Greenville, DE) (acc. #1489).
Du Pont Co. executive; Director of the Eleutherian Mills-Hagley Foundation, Inc. - Notebook on corporate organization and on foreign alliances of E. I. du Pont de Nemours & Co., 1926-29, 200 pp.; his reports on visits to Du Pont explosives plants, 1940; organization charts of the Explosives Department; correspondence and news items concerning Kimball's career with Du Pont; scrapbook of photographs and letters; and papers concerning the dedication of the Eleutherian Mills Historical Library, 1961.
Described more fully in the Supplement to A Guide to Manuscripts in the Eleutherian Mills Historical Library (1978).

K19 **Kimberly, John,** *d.* 1882.
Papers, 1821-80. ca. 1790 items.
In University of North Carolina Library, Southern Historical Collection (398)
Teacher in Hertford Co., N. C. and professor of chemistry at the University of North Carolina. Family correspondence, lecture notes, laboratory notebooks, and accounts. Consists mainly of letters written by Kimberly's wife, Bettie (Maney) Kimberly, to members of the Maney and Southall families in North Carolina, Tennessee, and Georgia; Kimberly's notes on Prof. Jean Louis Rodolphe Agassiz's lectures; his own teaching notes; and personal and farm accounts. Correspondents include James H. Otey, Charles Phillips, and James Woodrow.
Unpublished description in the library.
Gift of Rebecca Kimberly, before 1940; and Mary Kimberly, 1943 and 1948.

K20 **Kimberly-Clark Corporation.**
Records, 1880-1906, 1940-1952. - 2 reels microfilm (negative), plus unprocessed additions.
In State Historical Society of Wisconsin (Madison).
Minutes of the Board of Directors and stockholders' meetings (1880-1906) of the Kimberly and Clark Company, relating to incorporation and the formative years of the paper company and to the activities of the leaders in the Wisconsin waterpower and paper industries; and material used in preparing the company's 75th anniversary history in 1947 describing the development of the company's chief products and its cooperation with the war effort in the 1940's, including letters from retired employees, some copy for publicity and news releases, individual departmental histories and reports, a master's thesis by Melvin E. Bartz, "Origins and Development of the Paper Industry in the Fox River Valley (Wisconsin)," submitted to the History Department of the State University of Iowa, and an unpublished 80th anniversary history.

King, Franklin Hiram, 1848-1911.
Papers, 1899-1929. - 2 boxes.
In State Historical Society of Wisconsin (Madison).
Soil scientist, professor, and author. - Mainly correspondence, together with a volume of field notes on soils of the Goldsboro, North Carolina region. Letters relate chiefly to soil research, King's technical publications, and the extension of agricultural research through federal grants. Letters from King to his wife during his trip to the Orient in 1909 contain data collected for his book Farmers of Forty Centuries. Regular correspondents include Congressman Henry C. Adams; Eugene W. Hilgard of the University of California Agriculture Department; William D. Hoard; Milton Whitney, Chief of the United States Bureau of Soils; and Harvey W. Wiley of the United States Bureau of Chemistry.
Unpublished finding aid available.

King, Victor L.
Papers, ca. 1903-1918. - 1.5 lin. ft.
In Dartmouth College Library (Hanover, NH).
Chemist. - Letters and legal papers; notebooks containing notes on chemistry and material for Ph.D. dissertation; copies of the dissertation; notebooks with material pertaining to his work at Middlesex Chemical Company and other chemical companies. King attended Dartmouth College in 1907.
Unpublished finding aid available.

King, Ben H
Papers, 1864-1921. ca 150 items.
In State Historical Society of Colorado collections (Denver)
Records of military equipment issued to Civil War troops (1864-65), manual, invoices for equipment, map, memorabilia, and other items relating to World War I, and a booklet (1921) with notes, relating to Lincoln Co., Colo.

Kirkwood, Frank Coates, 1862-1945.
Papers, 1880-1945. ca. 3 ft.
In Maryland Historical Society Library (319, 515, 1501)
In part, transcripts (handwritten).
Farmer and naturalist, of Baltimore Co., Md. Correspondence, diaries, record book (1880-90) of drug prescriptions and formulas, field notes on Maryland birds, notes on birds' eggs, MS. of Kirkwood's "The List of birds of Maryland" (1895), typescript of his "A Partial list of the nesting birds in the proximity of Baltimore City, Maryland, 1892," logs of bicycling trips, farm and personal accounts, clippings, and other papers relating to birds, reptiles, fish, and conservation.
Gift of W. Bryant Tyrrell, Takoma Park, Md., 1946, 1960.

K25 Klein, J J
Papers, 1843–1902. 964 items and 5 v.
In University of South Carolina, South Caroliniana Library.
Druggist, of Walterboro, S. C. Correspondence, ledger, receipts, invoices, and bills. Includes business correspondence with Klatt & Company of Charleston, S. C., A. Loryea, and other business concerns and prescription and letter books (1883–1902) of Klein's son, John Marcus Klein (1859–1927) who took over his father's business.

K26 Kleinheksel, J. Harvey, 1900-1965.
Papers, 1929-1965. - 170 items.
In Hope College Archives (Holland, MI).
Professor of Chemistry at Hope College, 1928-1965. - Correspondence consisting mainly of recommendations for students; course examinations; record books; articles; clippings; photographs; and manuscripts: "Cooperation, Not Conflict," "The Role of Research at Hope College," and "Does Science Conflict With Christianity?" Also includes a student paper, "The Life of J. Harvey Kleinheksel," by Barbara Springer.

K27 Knight, Harry Granger, 1878-1942.
Diary, 1929-1942. - 2 v.
In National Agricultural Library (Beltsville, MD).
Chemist. - Knight became Chief of the USDA Bureau of Chemistry and Soils in 1927. He and his assistants were responsible for planning and building the four regional laboratories of the USDA during his term as Chief. His bureau eventually became Agricultural and Industrial Chemistry. The diary is a daily record of the activities of the Bureau of Chemistry and Soils during the main years of Knight's tenure as Chief.
Unpublished finding aid available.

K28 Knight, John.
Papers, 1902-18. ca. 50 items.
In State Historical Society of Colorado collections (Denver)
Mining entrepreneur. Correspondence, patents, deeds, bonds, and plats, relating to production of tungsten and molybdenum in Colorado.

K29 Kohler, Elmer Peter, 1865-1938.
Correspondence, 1912-38. 3 boxes.
In Harvard University Archives.
Professor of chemistry at Harvard University.
Professional correspondence.
Information on literary rights available in the repository.

Koops, Matthias. K30
Papers, 1801–04. 102 items.
In Dard Hunter Paper Museum collection (Appleton, Wis.)
Papermaker, of Westminster, Eng. Correspondence, contracts, inventories, agreements, wage and labor disputes, accounts, lists of stockholders, proceedings, price lists, indenture of shares, and other papers relating to Koops' papermaking venture and patents issued to Koops for making paper from wood, straw, and other fibers other than cotton and linen rags and for the de-inking of printed and written papers.
Described in unpublished Catalog of the Dard Hunter Paper Museum, by Dard Hunter.
Gift of Dard Hunter.

Kotch, Alex, b. 1926. K31
Letters, 1950-1976. - 11 boxes.
In University of Wisconsin Archives (Madison).
Professor of Chemistry. - Letters of recommendation, entrance evaluations to medical and graduate schools, draft board correspondence, industrial correspondence. Also minutes of the Chemistry Department meetings, executive, departmental, and finance committees.
Received in 1979.
Unpublished finding aid available.

Kraus, Edward Henry, 1875-1973. K32
Papers, 1910-1956. - 3 boxes.
In University of Michigan, Bentley Historical Library, Michigan Historical Collections (Ann Arbor).
Professor of Mineralogy; Dean of the summer session, 1915-1933; Dean of the College of Pharmacy, 1923-1933; and Dean of the College of Literature, Science and the Arts at the University of Michigan. - Contains correspondence, speeches and other papers relating to his professional career, including material, 1930-1938, concerning the Ann Arbor First Methodist Church, and papers relating to the University Research Club and to the activities of Phi Kappa Phi, 1918-1940.
Unpublished finding aid available.

Krauskopf, Francis Craig, 1877-1947. K33
Papers, 1917-1947. - 2 boxes.
In University of Wisconsin Archives (Madison).
Professor of Chemistry. - Correspondence, both to and from Professor Krauskopf. Letters of recommendation requested of him, as well as professional advice and correspondence from colleagues.
Received in October, 1967.
Unpublished finding aid available.

K34 Krebs, August Sonnin, 1877-1969.
Papers, 1899-1903; n.d. - 21 items.
In Hagley Library (Greenville, DE) (acc. #1276).
Chemical engineer. - Collection consists of personal correspondence received by Krebs, including letters from: The Pusey & Jones Co. of Wilmington, and William Sellers & Co. of Philadelphia, containing records of his employment with both companies; Jacob A. Riis (1849-1914), author, concerning pictures of Denmark taken by Krebs and Riis' proposal for their use by him for lecture and publication purposes; E. A. Wilson, classmate of Krebs at Cornell University, an employee of the Grasselli Chemical Co., Inc., in New Jersey. Krebs was the son of Henrik Johannes Krebs (1847-1929), a native of Denmark, who came to America and founded the Krebs Pigment & Chemical Co. at Newport, Del. in 1901. He succeeded his father as president of the company in 1918; the firm was acquired by E. I. du Pont de Nemours & Co. in 1929.

K35 Krebs, Henrik Johannes, 1847-1929.
Scrapbooks, 1899-1929. - 2 items.
In Hagley Library (Greenville, DE) (acc. #1297).
Krebs was the founder in 1886 of the Delaware Chemical Co., which was consolidated with other firms in 1888 to become the National Ammonia Co., important in the manufacture of ammonia for use in the production of ice; also founder in 1901 of the Krebs Pigment & Chemical Co., Newport, Del., purchased by E. I. du Pont de Nemours & Co. in 1929. Collection consists of two scrapbooks relating to the production of lithopone and other chemical pigments, with related articles from European and American periodicals.
Described more fully in the Supplement to A Guide to Manuscripts in the Eleutherian Mills Historical Library (1978).

K36 Krotz Springs Cycling Plant.
Blueprints, 1954. - 1 v.
In Louisiana State University, Department of Archives (Baton Rouge).
Cycling plant of Gulf Refining Company. - Blueprints consist of mechanical flow sheets and plans for the plant.

K37 Kunitz, Moses, 1887-1978.
Papers, 1913-1970. - 5 ft.
In Rockefeller University Archives, Rockefeller Archive Center (Pocantico Hills, North Tarrytown, NY).
Chemist. - Chiefly laboratory procedures, notes, and records, for the preparation of crystalline proteins, such as chymotrypsin, chymotrypsinogen, deoxyribonuclease, hexokinase, pyrophosphatase, ribonuclease, trypsin, and trypsinogen; together with reprints of articles (2 v.), negatives, and prints of various crystals.
Gift of Mr. Kunitz.
Information on literary rights available in the repository.
Unpublished finding aid available.

K38 Kunz, George Frederick, 1856-1932.
Papers, 1835-1938. ca. 600 items.
In Henry E. Huntington Library (San Marino, Calif.)
Gem expert for Tiffany & Company, mineralogist, geologist, numismatist, and adviser to government bodies. Includes correspondence with U. S. consuls abroad and relating to pearl culture, mining of precious stones, the diamond market, monetary systems, and politics. The bulk of the papers is dated 1900-28.
Purchase, 1964.

K39 Kunz, George Frederick, 1856-1932.
Correspondence, 1919-1922. - 137 l.
In New York Academy of Medicine Library.
Geologist and gemmologist. - Correspondence relating to Marie Curie's visit to the U.S. to raise money for her research in radium.

K40 Kunz, George Frederick, 1856-1932.
Papers, ca. 1880-1932, n.d. - 0.85 linear ft.
In Smithsonian Institution Archives (Washington, DC).
Mineralogist; authority on gems. - Throughout his career Kunz was associated with Tiffany & Company, the U.S. Geological Survey, U.S. Fish Commission; and the American Museum of Natural History. The papers consist primarily of correspondence with geologists, mineralogists, gemologists, jewelers, museums, government agencies, and acquaintances. Subjects include identification and acquisition of gems and mineral specimens; mineralogical research, especially the effects of radiation on the color of gems; and other topics. Also included is an address book kept by Kunz on a trip to Europe in 1881; statistics on American mining compiled by Kunz; manuscripts; and a few photographs.
Unpublished finding aid available.

L1 Labor education, 1946-72. 63 ft.
In Pennsylvania State University Library (University Park)
Records of institutes and conferences on labor education held at Pennsylvania State University, chiefly involving the Communications Workers of America, foreign labor teams, Gas, Coke, and Chemical Workers, International Ladies Garment Workers Union, International Union of Electrical, Radio, and Machine Workers, Pennsylvania AFL-CIO Community Services, Pennsylvania Federation of Labor, Retail Clerks, Teamsters, Union Advisory Committee, United Auto Workers, United Steelworkers of America, and the Upholsterers Union.
Information on literary rights available in the library.
Gift of the Labor Studies Dept., Pennsylvania State University.
Additions to the collection are anticipated.

L2 Ladd, Edwin Freemont, 1859-1925.
Family papers, 1890-1966. ca. 1 ft.
In North Dakota Institute for Regional Studies (North Dakota State University, Fargo)
Agricultural chemist and U.S. Senator, from Fargo, N.D. Correspondence, speeches, articles, biographical material, newspaper clippings, and subject files, relating to Ladd's life and career. Subjects include food adulteration and pure food laws. Includes Ladd family history and genealogy and letters (1890-95) to James B. Power and J.H. Bosard concerning a North Dakota dairymen's association.
Unpublished finding aid in the repository.
Presented by Milton Ladd, Washington, D.C., and others, 1953, 1956.

L3 Ladd, Edwin Freemont, 1859-1925.
Papers, 1905-1965. - 30 items.
In University of North Dakota, Chester Fritz Library, Orin G. Libby Manuscript Collection, (Grand Forks, ND).
Professor of Chemistry at North Dakota State University and Senator from North Dakota, 1921-1925. - Publications, speeches and news clippings on special topics in chemistry and politics as they relate to North Dakota, including laws governing pure food and drugs.

L4 LaForte, Benoist, b. 1761.
Archives of Benoist LaForte, 1784-97. ca. 8 ft. (ca. 3000 items)
In Cornell University Libraries, History of Science Collections (Ithaca, N.Y.)
French Government official. Letters (ca. 2200) from officials in Régie des poudres et salpêtres in Paris and the departments and districts of France, and saltpeter manufacturers, chiefly relating to gunpowder production. Correspondents include Jean Antoine Claude Chaptal de Chanteloup, Antoine François de Fourcroy, and Antoine Laurent Lavoisier.
Unpublished finding aids in the repository.
Information on literary rights available in the repository.
Gift of Arthur H. Dean, 1966.

L5 Laitinen, Herbert August, b. 1915.
Papers, 1949-1974. - 0.3 ft.
In University of Illinois Archives (Urbana).
Professor of Chemistry, University of Illinois. - Correspondence and related material (1949-1966) concerning the writing, publication, and promotion of Laitinen's book Analytical Chemistry; correspondence with Dr. Robert Rabin of National Science Foundation, together with notes and reports, concerning a National Science Foundation grant to the university for a study of environmental pollution by lead and other materials, directed by Laitinen (1971-1973); correspondence (1971-1974) concerning environmental pollutants and the state of environmental studies programs; and general research, professional societies, awards and honors, editorial duties, lectures and symposia, sabbaticals, job offers, and the scientific interests of colleagues (1959-1972).
Acquired 1974.
Unpublished finding aid available.

L6 Lamb, Arthur Becket, 1880-1952.
Papers, 1912-1952. - 22 ft.
In Harvard University Archives (Cambridge, MA).
Professor of Chemistry at Harvard. - Correspondence, diaries, articles, book reviews, lecture notes, and speeches.
Open to investigators only upon prior application to the repository.
Information on literary rights available in the repository.

L7 Lamb, George G., 1906-1977.
Papers, 1935-1977. - 4.4 cu. ft.
In Northwestern University Archives, (Evanston, IL).
Professor of Chemical Engineering, Northwestern University, 1946-1975. - Research interests in aviation fuels, lubricants, catalytic petroleum cracking. The papers include biographical materials; general and family correspondence; teaching files; research files; material on professional societies; papers presented at professional meetings; publications; administrative materials; and miscellaneous. The general correspondence consists primarily of letters dealing with Lamb's writings, research, attendance at meetings, and other educational and business matters. The teaching files include 1 folder of material concerning Northwestern's Center for the Interdisciplinary Study of Science and Technology. The papers presented at professional meetings include relevant correspondence, notes, drafts of the papers, and a final copy, if printed. Most of the papers prior to 1960 deal with technical or educational matters. Most of the papers after 1960 concern Lamb's studies on the

application of engineering concepts and techniques to the problems of society.
Donated to the University Archives by Mrs. George Lamb on August 25, 1977.
Finding aid available.

L8 La Motte, Arthur, 1871-1947.
Papers, 1896-1948. - 80 items.
In Hagley Library (Greenville, DE) (acc. #1345).
Chemist in the field of development and industrial application of commercial explosives; manager (1916-41) of the Technical Service Section, Explosives Department, Du Pont Co. - Papers include: notes on nitroglycerin, with correspondence, 1906; diary, 1896-1907; scrapbook containing articles written by La Motte, 1913-14; correspondence, 1903-48; correspondence and reports on safety precautions in European explosives plants, 1911, with a notebook and photographs on the subject; and extracts from experimental reports on explosive rivets, 1938-39.
Described more fully in the Supplement to A Guide to Manuscripts in the Eleutherian Mills Historical Library (1978).

L9 Lane, Samuel, 1718-1806.
Family papers, 1727-1861. 2 ft. (ca. 500 items)
In New Hampshire Historical Society collections (Concord)
Tanner, shoemaker, church deacon, and town official of Stratham, N. H., and member of the New Hampshire legislature. Biography, daybooks (1736-1810), ledger (1741-1802), appointment (1750) as assistant surveyor in New Hampshire, and other papers of Samuel Lane; ledger (1756-69) of Lane's uncle, Isaiah Lane, shoemaker of Hampton, N. H.; daybooks (1780-1810), ledger (1784-1830), tanyard journal (1780-1829), and other papers of Lane's son, Jabez Lane; ledger (1824-40) and daybooks (1825-61) of Jabez' son, Charles Lane, shoemaker and storekeeper of South Newmarket (now Newfields) N. H.; and will (1766) and journal (1727-55) of Lane's father, Deacon Joshua Lane, of Hampton, N. H. Other papers include 6 plans of the town of Bow and ca. 100 smaller surveys, and miscellaneous town and church papers.
Typewritten inventory in the repository.
Gift of Abby Lane, Stratham, N. H., 1914, and Henry C. Hines.

L10 Langley family.
Papers, 1856-1906. 73 items.
In University of Michigan Library, Dept. of Rare Books and Special Collections (Ann Arbor)
Forms part of the repository's D. E. Heineman bequest.
Chiefly correspondence and other papers of Samuel Pierpont Langley (1834-1906), astronomer, physicist, aeronautical engineer, and secretary of the Smithsonian Institution, Washington, D. C.; and papers of his brother, John Williams Langley (1841-1918), professor of chemistry and physics at University of Michigan, and of other family members, including Annie W. Ciocca, Mary W. (Langley) Herrick, Annie Langley, and Mrs. J. W. Levering.
Indexed in the repository's ms. catalog.
Bequest of David E. Heineman, 1935.

Langley, John Williams, 1841-1918. **L11**
Papers, 1859-1917. ca. 250 items and 2 v.
In University of Michigan, Michigan Historical Collections.
Professor of chemistry at the University of Michigan and Case Institute in Cleveland. Correspondence, mss. of Christmas stories, legal and patent papers, notes dealing largely with metallurgy, particularly the mercury pump and armor plate; the papers of Samuel Pierpont Langley (1834-1906), astronomer and physicist; and includes letters chiefly valuable for autographs from Alexander Agassiz, Edward E. Hale, Oliver W. Holmes, Rudyard Kipling, Robert T. Lincoln, Samuel S. McClure, Herbert Spencer, Gideon Welles, and Kaiser Wilhelm II.
Gift of Samuel P. Langley, 1937.

Langmuir, Irving, 1881-1957. **L12**
Papers, ca. 1871-1957. 43 ft. (ca. 32,000 items)
In Library of Congress, Manuscript Division (Washington, D. C.)
Chemist. Correspondence, diaries, experimental notebooks, articles, speeches, clippings, photos, awards, a card reference file, and printed matter. The notebooks (1894-1957) contain data which led to the development of the gas-filled incandescent lamp, the high vacuum power tube, atomic hydrogen welding, and screening smoke generators for the Armed Forces. Includes material on cloud seeding experiments and smoked bathythermograph records obtained at Lake George, N. Y. Correspondents include Niels Bohr, Vannevar Bush, Leopold Stokowski, and Willis R. Whitney.
Register published by the Library in 1962.
Bequest of Dr. Langmuir, 1958.

Laporte, Otto, 1902-1971. **L13**
Papers, 1926-70. 2 ft.
In University of Michigan, Bentley Historical Library, Michigan Historical Collections (Ann Arbor)
Professor of physics, University of Michigan, and specialist in the dynamics of fluids at high temperatures and atomic spectroscopy. Research notebooks, mss. of writings, and reprints of scientific articles.

Larsen, Gustive O. **L14**
Manuscripts. - 3 items.
In University of Utah Libraries, Special Collections Department (Salt Lake City).
Writer and Mormon educator. - Three papers written by Larsen, one of which is "The Early Beginnings of the Sugar Industry in Utah Territory (1852-1891)." Twelve pages in length, footnoted with a bibliography.
Gift.
Unpublished finding aid available.

L15 Lavoisier, Antoine Laurent, 1743-1794.
Papers, 1770-181-. ca. 10 ft.
In Cornell University Libraries, History of Science Collections (Ithaca, N.Y.)
French chemist and physicist. Letters, documents, and other papers, dating mostly from 1770-94, relating to Lavoisier's career and to Académie des sciences, Assemblée des notables (1787), Fermiers généraux des fermes royales unies, and Régie des poudres et salpêtres. Persons represented include Joseph Black, Claude Bourdelin, Henri Louis Duhamel du Monceau, and Marie Anne Pierrette (Paulze) Lavoisier.
Unpublished finding aid in the repository.
Access restricted.
Information on literary rights available in the repository.
Purchased in 1962 from H. P. Kraus, assembled by Denis I. Duveen, and was primarily from the Chazelles family, Lavoisier's heirs. Substantial additions of mss. in 1964.

L16 **Law, Thomas Cassels**, 1811-1888.
Papers, 1770-1899. 561 items.
In University of South Carolina, South Caroliniana Library.
In part, transcripts and photocopies.
Resident of Darlington District, S. C. Correspondence, business papers, plantation records, bills, receipts, legal papers, slave records, Bible records, photos, and other papers. Includes many letters and other papers relating to the establishment of the Presbyterian Church at Center Point and Hartsville, S. C.; chemistry notes from the lectures of Joseph LeConte at South Carolina College, and Civil War correspondence (1864-65) of Law's son, Hugh Lide Law; and genealogical material on the DuBose, Hart, Law, and Lide families. Correspondents include Law's brothers and other members of the family and the following Presbyterian ministers: William Brearley, Edward H. Buist, Thomas R. English, John F. Matheson, and William S. Plumer.
Gift of Mrs. Sarah Law Jones.

L17 Lawrence, Ernest Orlando, 1901-1958.
Correspondence and papers, ca. 1920-1968. - 48 cartons, 2 boxes.
In University of California, Berkeley, Bancroft Library.
Physicist, inventor of the cyclotron, Nobel laureate. - Included are the personal papers of Lawrence, administrative records for the University of California Radiation Laboratory, the Joseph C. Hamilton papers dealing with the Crocker Radiation Laboratory, and some papers of Lawrence's close associate, Donald Cooksey. They relate primarily to the Radiation Laboratory, University of California; relations with Office of Scientific Research and Development and the Atomic Energy Commission; development of cyclotron technology and establishment of cyclotrons at other institutions, etc.
Unpublished finding aid available.

Lawrence, Howard Cyrus, 1890-1961. L18
Papers, 1916-1966. - 25 ft. and 2 v.
In University of Michigan, Bentley Historical Library, Michigan Historical Collections (Ann Arbor).
Businessman and politician, of Grand Rapids, Michigan. - Correspondence, appointment books, business records, and other papers. Subjects include personal affairs; Grand Rapids Varnish Corporation; Michigan Trust Company of Grand Rapids; Ypsilanti-Reed Furniture Company, Ionia, Michigan; Albion College; the Methodist Church; the Ionia Free Fair; and the Republican Party in Michigan.
Unpublished finding aid in the repository.

LeBaron, Francis, 1781-1829. L19
Papers, 1809-26. 36 items.
In Pilgrim Society collections (Plymouth, Mass.)
Apothecary General of the U.S. Army, of Plymouth, Mass. Chiefly letters from LeBaron to his brother-in-law, Nathaniel Russell, of Plymouth, Mass., and to his parents, Isaac and Martha (Howland) LeBaron; and ms. of genealogy entitled Francis LeBaron and his Descendants.

Le Conte, Joseph, 1823-1901. L20
Papers, 1864-97. 95 items.
In University of North Carolina Library, Southern Historical Collection (2564 and 420)
Geologist, chemist, and professor at the University of California at Berkeley. Chiefly letters (1869-97) written by Le Conte to his daughter, Emma, wife of Farish C. Furman of Milledgeville, Ga., relating to scientific matters, his teaching, and family matters. Includes some letters (1870-94) written to Farish Furman about his agricultural experiments in growing cotton, and a diary (1864-65) of Emma Le Conte in South Carolina.
Unpublished description in the library. Emma Le Conte's diary was edited by Earl Schenck Miers and published as When the world ended: the diary of Emma Le Conte (1957)
Gift of Mrs. John R. L. Smith, 1943; and purchase, 1963.

LeConte-Furman family papers, 1810-96. 43 items. L21
In Georgia Historical Society collections (Savannah)
Chiefly correspondence between Emma (LeConte) Furman (b. 1847) and her husband, Farish Carter Furman (1846-1883) before and after their marriage in 1869; together with plantation and household accounts from their home in Scottsboro, Ga., pamphlets by and about Emma's father, Joseph LeConte (1823-1901), noted professor of physics and chemistry, and a letter from John Le Conte (1818-1891), geologist.
Unpublished finding aids in the repository.
Purchase, 1960.

L22 Lee, George F 1812–1894.
Business records of George F. Lee and Franklin Lee, 1820–92. 2400 items and 24 v.
In Historical Society of Pennsylvania collections.
Correspondence and other papers of Lee and Franklin Lee (1786–1861), both bricklayers, builders, and contractors of Philadelphia. Franklin Lee's business records cover 1835–46, while George F. Lee's cover 1820–73 and include receipt books, account books, notebooks, and cashbooks. The material reflects Lee's activities in organizing and building gasworks in all sections of the country (e. g. the Foundry pay book (1850) of the Chicago, Ill. Gas Works and the ledger (1842–47) of the St. Louis (Mo.) Gas Works. Lee's correspondence deals with the Chicago fire of 1871; the problems of building gasworks in Albany, Troy, and Utica, N. Y., as well as Chicago and St. Louis; and other business activities.
Gift of Mr. Harry E. Sprogell, 1964.

L23 Lehninger, Albert L., b. 1917.
Papers. – Ca. 4 linear ft.
In Johns Hopkins University, Alan Mason Chesney Medical Archives (Baltimore, MD).
Biochemist. – Collection consists of research notes, manuscripts, and departmental material. Lehninger was the author of a standard biochemistry text. His research includes studies of the understanding of the chemical route by which the biological oxidation of fatty acids occurs in animal tissues, cancer metabolism, ascorbic acid biosynthesis, thyroid hormones, and active transport mechanisms. In 1952 he became DeLamar Professor of Physiological Chemistry at the Johns Hopkins University School of Medicine.
Unprocessed collection; not open for research.

L24 LeMaistre, Frederic J., 1879–1944.
Laboratory notebooks, 1907–13. – 7 v.
In Hagley Library (Greenville, DE) (1709).
Chemical engineer for Du Pont Company. – LeMaister worked on explosives, artificial silk, pyralin and solvents. Several notebooks are indexed; frequently headings appear for nitrates and nitroglycerin. In addition to calculations, tables and notations, printed items are laid in, including companypolicyleaflets.

L25 Leonard Iron Works, *Taunton, Mass.*
Records, 1656–1876.
2 ft. (66 items)
In Old Colony Historical Society Library (Taunton, Mass.)
Fifty-six ledgers of James Leonard, his sons and grandsons, ironmasters, and several papers concerning the early iron industry at Taunton, Mass.
Unpublished guide in repository.
Open to investigators under the society's restrictions.
Gift, ca. 1880.

Lepine, J. Wilson. L26
Papers, 1890–1926. – 50 ft.
In Nicholls State University Library (Thibodaux, La.)
Owner and operator with Frank Barker of Laurel Valley and Melodia sugar cane plantations and Laurel Valley Sugar Refinery, Lafourche Parish, La. – Correspondence, letter books, ledgers, journals, payroll books, invoices, and other records, of Lepine and Barker's enterprise. Includes plantation diaries (1903–1915).
Gift of Mrs. J. Wilson Lepine, Jr., 1975.
Unpublished finding aid in the repository.

Leslie, Eugene Hendricks, 1892–1976. L27
Papers, 1916–1964. – 12 ft.
In University of Michigan, Bentley Historical Library, Michigan Historical Collections (Ann Arbor).
Chemical engineer and research scientist, pioneer in the refining of petroleum and synthesis of rubber. – Collection contains correspondence files, financial records, and published research materials.
Unpublished finding aid available.

Levene, Phoebus Aaron Theodor, 1869–1940. L28
Papers, 1905–1940. – 4.5 ft.
In Rockefeller University Archives, Rockefeller Archive Center (Pocantico Hills, North Tarrytown, NY).
Physician and chemist at the Rockefeller Institute for Medical Research, New York, N.Y. – Collection includes administrative and professional correspondence with Simon Flexner, Edric Smith, J. B. Conant, F. G. Hopkins, C. S. Hudson, John H. Northrop, A. Szent-Gyorgyi, and Otto Warburg. Also includes biographical material, collected reprints, and photos. Includes letters summarizing accomplishments of twenty-five researchers in Levene's laboratory (1929), commenting on developments in chemistry as a necessary tool in biology (1930), and relating to work in nucleic acids and with vitamins.
Gift of the Rockefeller University administration which acquired the collection as a gift of Dr. Levene's widow, Anne Erickson Levene, 1940.
Information on literary rights available in the repository.
Unpublished finding aid available.

Levison, Wallace Goold, 1846–1924. L29
Papers, 1866–1921. 17 boxes and 2 packages.
In New York Public Library.
Chemist. Personal correspondence (1874–1921), journal (1866–99) containing scientific propositions, experiments, researches, and notes, ms. of Levison's book on luminescence (1903), lectures, lecture

notes, miscellaneous notes and writings, patents, newspaper clippings, magazine articles, and other papers.

Gift of Mrs. Josephine Grimwood and Chancellor Levison, 1925-26.

L30 Lewis family.
Papers, 1755-1890.
1500 items and 16 v.
In Historical Society of Pennsylvania collections.
Mainly the papers of Mordecai Lewis, 1748-1799, a Philadelphia merchant, and of his sons Mordecai Jr. and Samuel N. Lewis, merchants and, after 1830, lead manufacturers, producing white and red lead and linseed oil. The earlier papers (1780-1830) pertain to trade with Europe, Madeira and the Far East. A few letters (1760-61) from H. Steele in London to his sister Elizabeth tell of his life abroad. Other papers include marine insurance ledgers (1755-59); ledgers and daybooks (1747-84) of Thomas Wharton, Philadelphia merchant; journal, daybook, and accounts current (1783-98) of James C. Fisher, Philadelphia merchant, and ledgers, waste book and schedule of property (1820-49) of trustees of Rebecca C. Lewis.
Presented by Leonard T. Beale, 1955.

L31 Lewis, Gilbert Newton, 1875-1946.
Papers. - 1 carton.
In University of California, Berkeley, Bancroft Library.
Professor of chemistry, UC Berkeley. - Correspondence and papers.
Finding aids available.

L32 Lewis, Howard Bishop, 1887-1954.
Papers, 1930-54. - 12 linear ft.
In University of Michigan, Bentley Library (Ann Arbor).
Biochemist and Chairman of the Department of Biological Chemistry at the University of Michigan. - Correspondence, reports, notes, and other papers dealing with departmental affairs, and Lewis' professional activities.

L33 Lewis, Richmond Addison, 1824-1900.
Student notebook, 1845-1847. - 75 [i.e., 70] p., 7 1/4 x 4 1/2 in. holograph.
In Virginia Historical Society (Richmond).
Bound volume. - Includes chemistry notes kept at Transylvania University, Lexington, Kentucky. Also includes accounts.
In the Holladay family papers, 1787-1968.

L34 Libby, Willard Frank, 1908-1980.
Papers. - Ca. 200 linear ft.
In University of California - Los Angeles Libraries, Department of Special Collections.
Chemist, 1960 Nobel laureate. - Correspondence, speech and travel files, and research materials reflecting Libby's work in the Manhattan Project, Carbon-14 research, service on the Atomic Energy Commission, and teaching at UCLA.
Finding aid in process; papers available for research.

Licking County papers, 1816-81.   ca. 10 ft.   L35
In Western Reserve Historical Society collections (Cleveland)
Records of the Licking County Pioneer, Historical, and Antiquarian Society, including correspondence and financial records of its secretary, Isaac Smucker, and reports and papers presented to the society; correspondence and business records of Dr. John N. Wilson (1802-1872), druggist, of Newark, Ohio, and David Wilson (d. 1870's), treasurer of Madison Township, Licking Co.; correspondence, minutes, and financial records of the Madison Township Farmer's Club; personal correspondence and business records of Bradley Buckingham, soldier in the War of 1812 and merchant, of Newark; History of Licking County (1881) by N. N. Hill; and miscellaneous personal papers of Amos Caffee, James Colville, John and William Cunningham, John and William Roe, and Adam and Mortimer Seymour; and other papers pertaining to Licking Co., Ohio.
Register in the repository.

Lillie, Foress B   1855-1926.   L36
Papers, 1872-1949.
9 ft.
In University of Oklahoma Library.
Pharmacist. Correspondence, diaries, legal papers, scrapbooks, newspaper clippings, material relating to the history of Guthrie, Okla., and photos. Includes correspondence relating to the passage of pharmacy laws in Oklahoma, a minute book of the Oklahoma Territorial Board of Pharmacy, the constitution (1907) of the Oklahoma Pharmaceutical Association, reports to the territorial governors from the Board of Pharmacy, Lillie's platform (1910) as candidate for mayor of Guthrie, Okla., a report of the city marshal of Guthrie, and city ordinances.

Linforth, F. A.   L37
Speeches, 1952-1953. - 0.1 linear ft.
In Montana Historical Society (Helena).
Assistant to the Vice-President of the Anaconda Copper Mining Company of Butte, in 1952 and 1953. Collection contains two speeches: "Resources and Operations of the Anaconda Copper Mining Company in Montana," and "An Introduction to the Richest Hill On Earth." The speeches provide operational statistics for the Anaconda Company, as well as some comment on planned future operations of the company.
Unpublished finding aid available.

Link, Louis, 1866-1934.   L38
Papers, 1883-1934. - 45 items, and 2 mss. v.
In Louisiana State University, Department of Archives (Baton Rouge).
Born Ludwig Link in Bavaria; immigrated to U.S. in 1882; Standard Oil employee for forty years in Texas and Louisiana, General Superintendent and member, Board of Directors; inventor and patent holder. - Collection consists of business letters, personal documents, printed materials, and photographs.

L39 Lipmann, Fritz Albert, 1899-
Papers, 1970. 2 items.
In Library of Congress, Manuscript Division (Washington, D.C.)
Biochemist. Two partial drafts of Lipmann's memoirs, Wanderings of a Biochemist (1971), relating to his life and work in biochemical research, including his Nobel Prize winning discovery of coenzyme A.
Gift of Dr. Lipmann, 1975.

L40 Little, Arthur Dehon, 1863-1935.
Papers, 1863-65, 1900-64. 1 ft. (572 items)
In Library of Congress, Manuscript Division.
Chemist, engineer, and inventor. Fan mail from business and professional associates, memorabilia, clippings, and other printed matter relating to Little, awards, testimonials, photos, prospectuses for his works, printed matter, progress reports, letters of condolence, necrology, and other papers.
Gift of Royal Little, and Arthur D. Little, inc., 1964.

L41 Lloyd, John Uri, 1849-1936.
Papers, 1879-1936. - ca. 36 boxes.
In Lloyd Library and Museum (Cincinnati, OH).
Manufacturing pharmacist; Professor of Chemistry at Eclectic Medical College; Professor of Pharmacy at Cincinnati College of Pharmacy. - Correspondence, clippings, miscellaneous experimental notebooks and notes taken during apprenticeship, 1866; undated manuscripts pertaining to pharmacy, drugs, chemicals, plants, etc.
Unpublished guide available.

L42 Lloyd, Malcolm, 1837-1911.
Papers, 1858-1911. 50 items.
In Historical Society of Pennsylvania collections (Philadelphia)
Industrialist, of Philadelphia, Pa. Leases to property, membership certificates, patent assignments from Richard and William Cox, agreements with petroleum brokers and shippers, other contracts, patents, and permits, and other business records of Lloyd's oil refinery, Gibson's Point, Pa., purchased 1888, by the Atlantic Richfield Company, of which Lloyd became an officer.
Information on literary rights available in the repository.
Gift of Mrs. Nathan Hayward, Wayne, Pa., 1968.

L43 Locke, William Lovering, 1853-1915.
Papers, ca. 1883-1913. 2 boxes.
In University of California, Berkeley, Bancroft Library.
Mainly letters (1897-99) to Locke from the London office of Pacific Borax and Redwood's Chemical Works, concerning processing and sale of borax; scrapbook (1883-97) of clippings on borax; and genealogical information on the Locke family.

L44 Loeb, Jacques, 1859-1924.
Papers, 1889-1924. ca. 11,000 items.
In Library of Congress, Manuscript Division (Washington, D.C.)
Physiologist and educator. General and professional correspondence (ca. 8000 items), family correspondence (ca. 1500 items), biographical data, speeches, awards, photos, and other material. Loeb's scientific writings include drafts of his books Forced Movements, Tropisms, and Animal Conduct (1918), Proteins and the Theory of Colloidal Behavior (1922), and Regeneration From a Physicochemical Viewpoint (1924); laboratory notebooks relating principally to his research on bryophytes, gelatin, and frogs, which led to his development of the tropism theory; and scientific articles (in English and German) on such topics as colloid chemistry, genetics, osmosis, and proteins. Correspondents include Svante Arrhenius, Bernard Berenson, James B. Conant, Paul De Kruif, Paul Ehrlich, Albert Einstein, Sigmund Freud, Julian Huxley, Ivan Pavlov, and Harlow Shapley.
Finding aid in the Library.
Information on literary rights available in the Library.
Gift of Dr. Loeb's children, Leonard B. and Robert F. Loeb and Mrs. Anne L. Osborne, 1960-63.

L45 Loeb, Jacques, 1859-1924.
Papers, 1906-24. ca. 2 ft.
In Rockefeller University Archives, Rockefeller Archive Center (Pocantico Hills, North Tarrytown, N.Y.)
In part, transcript (typewritten) of letter from Albert Einstein to Leonard Loeb.
Physiologist and educator at Rockefeller Institute for Medical Research, New York, N.Y. Administrative and other correspondence, bibliography, collected reprints (7 v.), and photos. Includes letters to Simon Flexner about coming to the institute, letter from Albert Einstein to Leonard Loeb asking for support to pay an assistant mathematician, and condolence letters from Einstein, Curt Herbst, Hans Meyer, Otto Myerhoff, and Ivan Pavlov.
Unpublished inventory in the repository.
Information on literary rights available in the repository.
Gift of the Rockefeller University administration.

L46 Long, Cyril Norman Hugh, 1901-1970.
Papers, 1920-70. - Ca. 10,000 items (12 linear ft.).
In American Philosophical Society Library (Philadelphia, PA).
Biochemist, physiologist. - In addition to extensive correspondence, there are numerous laboratory notes, lecture notes, drafts of published papers, reports. This material documents Long's career, with much on his work with diabetes, including the International Diabetes Federation. His 1950 trip to Japan, as advisor to Japanese medical schools on medical education, is

Long, William Lunsford, 1890-1964.
Papers, 1925-59. 6000 items.
In University of North Carolina Library, Southern Historical Collection (Chapel Hill) (3682)
Mining corporation executive, of Warrenton, N.C. Correspondence and other papers, chiefly 1951-58, relating to Haile Mines, Inc., of New York, and its subsidiaries, especially Tungsten Mining Corporation of North Carolina, and Manganese, Inc., of Nevada, involving properties in 10 Southeastern and Western States and Canada and Mexico; activities of the Tungsten Institute, of which Long was a director and president; and business, farming, personal, social, and political interests, in Warren, Vance, and Northampton Counties, N.C.
Unpublished description in the library.
Gift of Mrs. Long, 1964.

Longsworth, Lewis Gibson, 1904-1981.
Papers, 1930-1970. - 2 ft.
In Rockefeller University Archives, Rockefeller Archive Center (Pocantico Hills, North Tarrytown, NY).
Physical chemist at Rockefeller Institute for Medical Research, New York. - Professional correspondence, biographical information, bibliography, collected reprints, and photos, especially scientific photographs (primarily glass slides) on electrochemical research. Includes correspondence with A. V. Astin regarding National Academy of Sciences Committee on Tables and Constants and the Committee on Battery Additives; material on design and use of apparatus, experimental results, and calculations; and interpretations of diffusion and electrophoretic work.
Gift of Mr. Longsworth.
Information on literary rights available in the repository.
Unpublished finding aid available.

Longyear (E. J.) Company, Minneapolis.
Records, 1885-1949. - 192 boxes, including 243 v. and 21 oversize items.
In Minnesota Historical Society (St. Paul).
Drilling and mining engineering company. - Longyear formed Longyear and Hodge in 1903 with John E. Hodge. The company expanded in 1911 and was incorporated as the E.J. Longyear Company. Longyear retired in 1924. Company's business had become worldwide: drilling and shaft digging contracts; professional geological and mining engineering services; sale of drilling equipment. Minerals drilled for or analyzed include molybdenum, bauxite, tungsten, titanium, sulphur, clay, marble, granite, aluminum, silica, quartzite, lithium, chromium, trona, phosphates, and barium. Company also did soil sampling and foundation testing.
Unpublished finding aid available.

Lord, Nathaniel Wright, 1854-1911.   L50
Papers, ca. 1880-1933. - 6 feet.
In Ohio State University Libraries (Columbus, OH).
University professor. - Business and family correspondence, literary manuscripts and other papers. The business letters relate primarily to mining and metallurgical matters. Correspondents include James Hulme Canfield, Edward Orton, and William Henry Scott.

Lord, Richard Collins, b. 1910.   L51
Papers, 1946-1981. - 32 cu. ft.
In Massachusetts Institute of Technology, Institute Archives and Special Collections (Cambridge, MA).
Physical chemist; educator. - Includes correspondence with Optical Society of America, student and course materials and related correspondence, book drafts and notes including Practical Spectroscopy, records of grants and correspondence regarding research projects, and material about the M.I.T. Department of Chemistry and the M.I.T. Spectroscopy Laboratory.
Box list available.

Louisiana Sugar Planters Association.   L52
Records, 1877-1911. - 849 items and 1 v.
In Louisiana State University, Department of Archives (Baton Rouge).
Correspondence, papers and records of the Louisiana Sugar Planters Association. A bound volume contains the minutes of the Association, 1877-1891.
Subject summary: topics to be discussed at meetings; a new type of evaporation; a molasses test; a paper by N. A. Helmer, "Evaporation in Multiple Effects," and related papers, read before the meeting of the Association, June 1907; American Protective Tariff League; membership dues and resignations; data on cane growing; and the perfecting of a new cane cutter.

L53 Louisville Textiles, inc.
Records, 1890–1950. 12 ft.
In Merrimack Valley Textile Museum (North Andover, Mass.)
Journals, ledgers, cash, production, and day books, stock certificates, accounts, trial balances, cotton and synthetic yarn chemical and dye books, and fabric swatch books, of a cotton and synthetic textile manufacturing company in Louisville, Ky.
Unpublished listing in the repository.

L54 Lovering, Mary Campbell, 1868-1947.
Verses and essay, 1887-1889. - 1 folder.
In Massachusetts Institute of Technology, Institute Archives and Special Collections (Cambridge, MA).
Author; chemist. - Collection consists of a manuscript essay entitled "The Three Little Fairy Sisters, Chlorine, Bromide, Iodine," written in 1888 and a collection of poems entitled "Verses and Acrostics," dated 1887 to 1889.

L55 Lowry, Homer Hiram, 1898-1971.
Papers, 1930-37, 1941-53. - Ca. 200 items.
In American Philosophical Society Library (Philadelphia, PA).
Physical chemist at Bell Telephone Labs, 1925-30, and then Director, until 1953, of the Coal Research Lab at the Carnegie Institute in Pittsburgh. There is much correspondence concerning research at Bell Labs, and Carnegie Institute matters, relating to coal technology, machinery, and the steel industry. There are also several photographs, taken shortly after World War II showing the industrial destruction in Germany. Correspondents include: Thomas S. Baker, Robert M. Burns, Robert E. Doherty, George A. Hulett, Eliot Janeway, and John Johnson.
Accessioned, 1981
Unpublished finding aid available.

L56 Lucas, John R., ca. 1866-1917.
Papers, 1903-1916.
In Montana Historical Society (Helena).
Assistant Superintendent of Hope Mining Company's mine at Philipsburg, Montana from 1892 to 1903, and Superintendent of Granite-BiMetallic Consolidated Mining Company from 1903 to 1916. Also managed Henderson, Combination, Sapphire, and Basin Mines. - Primarily correspondence with Paul A. Fusz and Charles D. McLure of St. Louis concerning operation of mines. Also considerable personal correspondence concerning Democratic Party politics and veterans' affairs. Fills a gap from 1903 to 1912 in the Granite-BiMetallic Consolidated Mining Company correspondence (MC 17, Montana State Historical Society), but covers many other companies as well. Arranged into two subgroups: John R. Lucas and American Gem Mining Syndicate.
Unpublished finding aid available.

L57 Lucas-White collection, 1806-1911. ca. 1 ft.
In Maryland Historical Society Library (Baltimore) (2277)
Correspondence of the related Lucas and White families, of Baltimore, Md., including correspondence (1807-11) of Fielding Lucas, Jr. (1781-1854), stationer, with his wife, Eliza M. Carrell Lucas, and friend, Eliza Slough, relating to Lucas' settling in Baltimore, his children, and family and other affairs; correspondence (1855-57) between Kate Butler Lucas and her mother while she was away at school at Eden Hall, Torresdale, Pa.; correspondence (1872-73) between Thomas Hurley White and his father, Ambrose Abram White, concerning the firm White and Elder, agents of Chesapeake Sugar Refinery, Baltimore, buying and selling of sugar, operation of boilers at the refinery, coffee trade, and fertilizer business; letters (1870-1902) from Thomas Hurley White to his wife, Kate Butler (Lucas) White, written while on business trips in the U.S., including references to the Centennial Exhibition, 1876; and other White family correspondence including references to the Civil War, Southern campaigns of Grant and Sherman, Lee's surrender, and Lincoln's death. Other correspondents include Catharine Lucas, Edward C. Lucas, Ethel White, and John Charles White.
Gift of Ernestine K. Payan, Sacramento, Calif. 1977.

L58 Luck, James Murray, b. 1899.
Papers, 1930-1979. - 5.5 linear ft.
In Stanford University Archives (Stanford, CA).
Professor of Biochemistry at Stanford. - Primarily correspondence and subject files relating to professional activities, the Chemistry Department, and his position as science attache at American embassies in London, Stockholm, and Bern.

L59 Luckey, Thomas D., b. 1919.
Papers, 1946-ca. 1980. - 3 ft. + 13 ft. of unprocessed material.
In University of Missouri-Columbia Library, Western Historical Manuscript Collection and State Historical Society of Missouri Manuscripts.
Biochemist. - Correspondence with other scientists and with chemical and pharmaceutical and animal feed companies, relating to Luckey's research on animal care and zoo animal nutrition, nutrition and metabolism of germ-free vertebrates, folic acid and related compounds in chick nutrition, and additives and the use of antibiotics in animal feed. Unprocessed material includes research data, reference

files, class lecture materials, and some correspondence.
    Gift of Mr. Luckey, 1969. Unprocessed material transferred to repository on January 22, 1985.
    Entire collection available for research.
    Unpublished finding aid available.

Lukens Iron and Steel Company, Coatesville, Pennsylvania. L60
    Records, ca. 1798-1917. - 1,420 v. and 170,000 boxed documents.
    In Hagley Library (Greenville, DE).
    Company archives including correspondence, operating reports, and financial records. Described more fully in the Supplement to a Guide to the Manuscripts in the Eleutherian Mills Historical Library by John B. Riggs (1975).

Lutz family. L61
    Papers, 1785-1874.
    ca. 100 items.
    In Historical Society of Pennsylvania collections.
    Miscellaneous papers relating chiefly to Lutz family interests in chemicals and dyeing; also documents on Bedford County politics.
    Presented by William Filler Lutz, 1939.

Lyman, Rufus Ashley, 1875-1957. L62
    Papers, 1895-1958. - 47 boxes.
    In State Historical Society of Wisconsin (Madison).
    College dean, founder of the American Journal of Pharmaceutical Education, and active member of pharmaceutical organizations. - Collection includes correspondence; speeches; articles; organizations' minutes, reports, convention materials, and financial records; and other documents concerning the field of pharmacy.
    Gift of the American Institute of the History of Pharmacy, Madison, Wisconsin, July 27, 1970.
    Unpublished finding aid available.

Lyon, James E L63
    Business records, 1865-67. 3 v.
    In State Historical Society of Colorado collections (Denver)
    Mining entrepreneur and pioneer promoter of the smelting industry, of Central City, Colo. Journal of business transactions, cashbook, and record book.

M1 McAdams, William Henry, 1892-1975.
Student notes, 1916. - 1 folder.
In Massachusetts Institute of Technology, Institute Archives and Special Collections (Cambridge, MA).
Chemical engineer; educator. - Problem sets with solutions for Course X, 245, Heat Engineering which McAdams submitted to Professor Charles William Berry.
Gift of Marshall T. Sanders, 1968.

M2 McBain, Bertram Telfar.
Papers, 1913-36. ca. 2 ft.
In University of Oregon Library (Eugene)
Paper mill manager and consultant. Ten proposals for pulp mill sites, mostly located in the Pacific Northwest (one in Alaska), ms. of a history of the pulp and paper industry in the Pacific Northwest, and photo albums relating to the Crown Pulp Mill, Oregon City, Or., (1914) and the construction of the Oregon City-West Linn bridge (1922).

M3 McBryde, John McLaren, 1841-1923.
Papers, 1883-1945. 12 v.
In University of South Carolina, South Caroliniana Library.
Confederate officer and educator. Chiefly correspondence and business papers connected with McBryde's position as president of the University of South Carolina, including reports to him as director of agricultural experiment stations in the State.

M4 McChesney, Joseph Henry, 1828-1895.
Papers, 1859-75. 73 items.
In University of Illinois at Urbana-Champaign, Illinois Historical Survey collections.
Paleontologist, professor of chemistry and geology, and U.S. Envoy to Great Britain. Correspondence, bills, and other papers, chiefly relating to McChesney's activities as U.S. consul at Newcastle-on-Tyne, Eng., together with papers pertaining to his work as Illinois State geologist. Includes material on the creation of a scientific exchange program between the U.S. Dept. of Agriculture and groups in England and Germany, and comments on the Civil War and politics in the U.S. and in Great Britain.

M5 McCoy, Herbert Newby, 1850-1945.
Papers, ca. 1930-1944. - 2 boxes. Manuscripts and printed material.
In University of California - Los Angeles Library, Department of Special Collections.
Professor of Chemistry. - Contains laboratory notes and articles by Professor McCoy concerning his research on the rare earths.
Gift of the Estate of Ethel Terry McCoy, December 5, 1966.
Catalog card entry.

M6 McCrae, James Archibald.
Manuscript, 1926. - 1 item.
In University of North Dakota Library, Orin G. Libby Manuscript Collection (Grand Forks, ND).
Manuscript of The Relation of Chemistry to Agriculture, submitted to the American Chemical Society.

M7 McCulloch, Warren Sturgis, 1898-1969.
Papers, ca. 1935-1968. - Ca. 30,0000 items (30 linear ft.).
In American Philosophical Society Library (Philadelphia, PA).
Neurologist, psychologist. - Collection is composed of correspondence and papers which center on McCulloch's study of the functional organization of the central nervous system and the related field of cybernetics. There is much on computers as well, including "biological computer" studies. Other topics include the brain or neural studies, biological psychiatry, chemical warfare, space biology, and U.S. Army studies. His participation in the American Society of Cybernetics is extensively documented. There are numerous papers and notes on conferences attended, etc.

M8 McCutcheon, Samuel.
Papers, 1832-1874. - 82 items and 8 v.
In Louisiana State University, Department of Archives (Baton Rouge).
Sugar planter of Ormond Plantation, St. Charles Parish, Louisiana, and manager for Young, Toledo and Company of Regalia Estate, a sugar plantation and sawmill near Belize, British Honduras. - Collection contains diaries; plantation registers; record books; supplementary plans for Regalia Estate; and miscellaneous printed material pertaining principally to sugar manufacturing. Diaries contain some detailed information on operation and management of the sugar plantations and the sugar and sawmills, especially the processing of sugar.

M9 McFarland, David Ford, 1878-1954.
Papers, 1810-1963. 3 ft.
In Pennsylvania State University Library (University Park)
In part, transcripts.
Chemist, metallurgist, and educator. Correspondence, biography, papers, theses, and an article, by various authors, notes and sketches, 2 maps, photos and some negatives of furnaces with an accompanying description, pamphlets and government reports, magazine and newspaper clippings, and other papers, relating principally to early mining and iron production in Pennsylvania, mainly

the counties of Blair, Centre, Clinton, Cumberland, Fayette, Huntingdon, Lebanon, Lycoming, and York. Most of the papers date from the time McFarland became professor and head of the Dept. of Metallurgy, Pennsylvania State University, in 1920, and after his retirement in 1945.

Information on literary rights available in the library.

Gift of Mr. McFarland.

10 McGill, John Thomas, 1851-1946.
Papers, 1876-1947. 7 ft.
In Joint University Libraries (Nashville, Tenn.)
Professor of pharmacy and organic chemistry and dean of the School of Pharmacy, Vanderbilt University. Correspondence, articles by McGill, notes, reports, newspaper and magazine clippings, programs, announcements, photos, and other papers relating to Vanderbilt University, its history, Alumni Association, Barnard Club (an astronomy club), Pharmacy Club, Chancellor Garland Memorial, and administration of the School of Pharmacy. Includes material concerning air pollution of the Ducktown Sulphur, Copper and Iron Company, and the Pharmaceutical Association.

Unpublished register in the library.
Gift of Dr. McGill.

11 MacInnes, Duncan Arthur, 1885-1965.
Papers, 1926-65. 8 ft.
In Rockefeller University Archives, Rockefeller Archive Center (Pocantico Hills, North Tarrytown, N.Y.)
Chemist. Personal and professional correspondence, journals, diaries of travels and camping trips, biographical and bibliographical material, illustrations for scientific publications, reprints, travel maps, and photos. Includes material on MacInnes' interests in the American Alpine Club, American Soviet Science Society, Appalachian Mountain Club, and national parks, limited attendance conferences he instigated under the auspices of the National Academy of Sciences, and his membership on the American Philosophical Society's committee on research grants, and records of arrangements and personnel for research work during World War II. Correspondents include S. F. Acree, of National Bureau of Standards, Herbert Spencer Harned, professor of chemistry at Yale University, and Edgar Reynolds Smith.

Unpublished finding aid in the repository and in Rockefeller University Archives, New York, N.Y.

Information on literary rights available in the repository.

Bequest of Mr. MacInnes, 1965.

McKay, Frederick Sumner, b. 1874. M12
Correspondence, 1908-1954. - 2 boxes plus unprocessed additions.
In State Historical Society of Wisconsin (Madison).
Dentist of Colorado Springs, Colorado. - Letters received by McKay, a major figure in the campaign for fluoridation of municipal water supplies as a means of reducing dental cavities. Correspondence of the 1930's includes an exchange with H. V. Churchill, chemist of the Aluminum Company of America, who in 1931 discovered the presence of fluorine in municipal water supplies. The letters of the years 1940-1954 are mainly requests for advice and information on fluoridation techniques.

McKim, William Duncan, 1855-1935. M13
Papers, 1708-1937. - 2 boxes; ca. 400 items.
In Maryland Historical Society Library (Baltimore).
Physician, of Washington D.C. - Letters, medical drawings, medical tracts, commonplace book, indentures, plats, genealogical notes, notebooks, daguerreotypes, and ambrotypes.

A section of the papers (1825-1846) deals with the formation of the Maryland Chemical Works in 1827. This section includes legal documents, plan of the works, and letters from Richard Caton to David T. McKim (fl. 1832-1866) about property for the works; also stock certificates and insurance policies.

Gift of Mrs. William D. McKim, 1941.

Mackintosh, James Buckton, 1856-1891. M14
Papers, 1885-1891. - 2 boxes.
In Columbia University, Rare Book and Manuscripts Library (New York City).
Research chemist (chemistry of minerals); instructor; Columbia alumnus. - The collection consists of letters written to Mackintosh chiefly on scientific subjects from his colleagues. Correspondents include Thomas Sterry Hunt; William Earl Hidden; Thomas Egleston, founder of the Columbia University School of Mines; R. S. Penniman; and Pierre Eugene Marcelin Berthelot. Included also are miscellaneous documents relating to Mackintosh and about 15 printed articles and brochures.

Gift of James H. Mackintosh, 1958.
Unpublished finding aid available.

M15 McLachlan, Dan, Jr., 1905-1982.
Papers, 1957-1958. – 10 in.
In American Institute of Physics, Niels Bohr Library (New York, NY).
Crystallographer; President of the American Crystallographic Association. – Letters selected by McLachlan for preservation: one for each of ca. 1,000 correspondents and records of the American Crystallographic Association, dating from the time of his vice-presidency and presidency of the Association in 1957-58.
Gift of Dan McLachlan, 1975.
Finding aid available.

M16 Maclean family.
Papers, 1771-1902. ca. 1000 items and 40 v.
In Rutgers University Library (21, 254, 346, and 805)
Correspondence, bills, receipts, and other papers of John Maclean, Jr. (1800-1886), chiefly relating to his position as president of the College of New Jersey (now Princeton University); an incomplete history of the College; a biography and chemistry notes of John Maclean, Sr. (1771-1814); correspondence, medical notes, reports, and other papers of Dr. George M. Maclean (1806-1886); and lecture notes, indentures, legal documents, and other papers of members of the Maclean family of Mercer Co., N. J.

M17 McMillan, Edwin Mattison, b. 1907.
Papers, ca. 1927-1972. – 26 cartons, oral history, and diary.
In University of California, Berkeley, Bancroft Library.
Professor of nuclear physics; winner (with Glenn Seaborg) of 1951 Nobel Prize in Chemistry for discovery of plutonium and research on the transuranium elements. – Collection contains synchrotron notebooks, files, research data, diagrams and plans. Also contains oral history for the American Institute of Physics, including discussion of the development of radar and the work of the UCB Radiation Laboratory and the Manhattan Engineer District; and a diary (1927) kept by McMillan during a prize tour in Europe.

M18 MacNider, William DeBerniere, 1881-1951.
Papers, 1901-52. ca. 24 ft.
In University of North Carolina at Chapel Hill, Library, Southern Historical Collection (837)
Professor of pharmacology, dean of the Medical School, and research professor at the University of North Carolina at Chapel Hill. Personal, professional, and family correspondence, relating to scientific research, medical organizations and publications, local affairs in Chapel Hill, university activities, and personal and professional friends.

M19 McReynolds family.
Papers, 1794-1965. 105 items.
In Western Kentucky University, Kentucky Library (Bowling Green)
In part, photocopies made in 1945 from originals in the Library of Congress, Manuscript Division, Washington, D. C. (1 item).
Correspondence, receipts, recipes for food, medicine, and dyes, genealogical and biographical material, and autograph album, chiefly of Susan (Edwards) Reeder Phillips, her husband Thomas Phillips (1774-1843), and their daughter, Mary Jane (Phillips) Reeves McReynolds (1814-1860), of Elkton, Todd Co., Ky. Includes correspondence from their son, Benjamin E. Phillips, from Mississippi and Texas; receipts of his brother G. E. Phillips, of Texas; questions answered by Senator John Edwards, of Kentucky, father of Susan Phillips, before the Democratic Society of Washington, D. C., about the free navigation of the Mississippi (1794); correspondence (1945) between Mary (Taylor) Leiper Moore, librarian of the Kentucky Library, Western Kentucky University, and Justice James Clark McReynolds; genealogical material on the Edwards, McReynolds, Phillips, and Reeves families. Topics include travel in Texas (1837-43), especially Nacogdoches and Galveston, political relationships between the U. S. and Mexico, Mexican invasions in Texas (1842-43), and social life in Kentucky and Texas.
Unpublished guide, descriptive inventory, and card catalog in the library.
Gifts of Justice James Clark McReynolds, Washington, D. C., and Robert Phillips McReynolds, Los Angeles, Calif., 1945.
The library also has copies (typewritten and photocopy) of the genealogical and biographical material.

M20 Mah, Richard S. H., b. 1934.
Papers, 1980-1981. – 0.15 linear feet (1/2 box).
In Northwestern University Archives, (Evanston, IL).
Professor of Chemical Engineering. – In addition to 1 folder of biographical materials, the collection includes 2 folders of correspondence, mailing lists, papers, and related materials pertaining to a conference on computer-aided design held July 6-11, 1980 at New England College, Henniker, New Hampshire, and sponsored by the Engineering Foundation and the American Institute of Chemical Engineers. Much of the conference material concerns program planning and publication of papers.
Transferred to the Archives on July 18, 1983.
Finding aid available.

M21 Mallmann, Paul.
Papers, 1917-1924. - 5 boxes.
In New York Public Library, Rare Books and Manuscripts Division (New York).
Chemist, metallurgist, and engineer. - Includes correspondence with British authorities concerning the development of the "Death Ray" for use in World War I; correspondence dealing with Mallmann's law suit against the British government after the war; writings about steel production, metallurgy, and engineering; correspondence and notes concerning other engineering works abroad; papers discussing Mallmann's theory entitled "Mallmann Continuous Steel Process"; and printed ephemera on real estate.
Unpublished finding aid available.

M22 Margaretta Furnace, Margaretta, Pa.
Records, 1841-64. ca. 450 items.
In Historical Society of York County collections (Pa.)
Correspondence (1843-64) and account books (1844-47) during the period when the furnace was operated by James Curran after its failure under the Slaymaker family.
Unpublished inventory in the repository.
Gift.

M23 Marine Paint and Varnish Company, inc., *New Orleans*.
Records, 1917-59. 48,527 items.
In Tulane University Library.
Correspondence (1924-45), three books of minutes of the meetings (1920-59) of the board of directors and the stockholders, tax papers (1917-44), papers (1936-44) dealing with renegotiation of war contracts, inventories (1946-52), trial balances (1935-46), balance sheets (1935-39 and 1945-46), weekly inventories (Dec. 1940-Dec. 1947) for liquids, rosins, and fish oils, stock certificates (1919-59), weekly payroll time books (July 1919-Dec.1927), factory timecards (1952-53), office timecards (Jan. 1952-Jan. 1954), dept. payrolls (Dec. 28, 1951-Jan. 3, 1957), office payroll checks (1952-54), factory payroll checks (1953-57), voucher records (1952-53), daily delivery reports (1941-45), daily truck operation reports (1941-45), cashbooks (1952-56), salesbooks (1952-58), and cancelled checks (Dec. 1951-Dec. 1957)
Information on literary rights available in the library.
Gift of H. Stanley Butterworth and Paul B. Lemann, 1959 and 1964, the liquidators of the company.

M24 Maron, Samuel Herbert, 1908-1975.
Papers, 1946-1974. - Ca. 9 linear ft.
In Case Western Reserve University Archives (Cleveland, OH).
Professor of Chemistry and Polymer Science. - Correspondence, grant proposals, reports, publications, research and lecture notes, course outlines, and lab manuals; relating to research and consulting for the synthetic rubber industry, his textbook Principles of Physical Chemistry, and the physical chemistry of polymers.
Gift of Mrs. Maron, 1976.

M25 Marshall, Christopher, 1709-1797.
Diaries, 1773-93.
9 v. and 1 package.
In Historical Society of Pennsylvania collections.
Philadelphia pharmacist and Revolutionary patriot. Diaries, kept in Philadelphia and Lancaster, with details of events during the Revolution; also information on pharmacy. Includes miscellaneous notes, a letter book (1773-78), and Waste book: accounts with the Continental Congress (1776)
Published in part in Extracts from the diary of Christopher Marshall ... 1774-1781, edited by William Duane (published 1877) Four letters to Peter Miller of Ephrata (1773-77) are published in Pennsylvania magazine of history and biography, v. 28 (1904) p. 71-77.
Gift of Charles Marshall.

M26 Martin, Sumner Leroy, b. 1887.
Papers, 1917-36. ca. 75 items.
In Ohio Historical Society collections (Columbus) (MSS 695)
Methodist minister. Correspondence, diary, writings, notes, lists, orders, and other papers, relating chiefly to Martin's service as chaplain for 7th Ohio Regt. during World War I, establishment of a social center at Camp Sheridan, Ala., for the regiment, and his advocacy of the abolition of religious deferments from the draft. Includes notes on gas defense school.
Unpublished inventory in the repository.
Gift of Mary Anne Martin Wyall, 1977.

M27 Maryland Chemical Works Account Books.
1827-1833, 1 v.
In Maryland Historical Society Library (547.2) (Baltimore).
Payroll record and journal of the Maryland Chemical Works, of Baltimore, of which David T. McKim was president.

M28 Mason, William Pitt, 1853-1937.
Papers, 1877-1924. - 1.5 linear ft.
In Rensselaer Polytechnic Institute Archives and Dept. of Special Collections (Troy, NY).
Professor of Chemistry and head of RPI Chemical Engineering Department, 1875-1925. - The papers consist of letterpress books of scientific and business correspondence, investigation results and records of water analyses, bibliographical notations, notes by RPI students on experiments and water investigations, and published books and articles by Mason.
Unpublished finding aid available.

M29 Massachusetts Institute of Technology. Department of Chemical Engineering.
Records, 1917-1971. - 87 cu. ft.
In Massachusetts Institute of Technology, Institute Archives and Special Collections (Cambridge, MA).
Records include bound student reports from the Practice School stations, 1917-1961, bound course notes, 1923-1971, and photograph albums from the Mt. Katahdin trips, 1930-1935 and from the Practice School, 1958-1972.
Box list available.

M30 Massachusetts Institute of Technology. Department of Chemistry.
Organic chemistry seminar reports, 1946-1980. - 6.25 linear ft.
In Massachusetts Institute of Technology, Institute Archives and Special Collections (Cambridge, MA).
Papers presented to weekly seminars on organic chemistry.

M31 Massachusetts Institute of Technology. Department of Chemistry.
Records, 1971-1979. - 0.35 cu. ft.
In Massachusetts Institute of Technology, Institute Archives and Special Collections.
The collection consists of grant records of John Deutch, head of the Chemistry Department, and includes projects on the theory and exploration of new applications for laser light scattering from fluid systems; and statistical mechanical investigations of various aspects of structure and relaxation in fluids.
Restricted; for further information consult the Institute Archivist.

M32 Massachusetts Institute of Technology. Department of Nuclear Engineering.
Course notes, 1953-1976. - 33 cu. ft.
In Massachusetts Institute of Technology, Institute Archives and Special Collections (Cambridge, MA).
Consists of bound course notes, including reading lists and problem sets, from the Department of Nuclear Engineering. Covers the early years, when the department had first evolved from the Chemical Engineering Department.
Box list available.

M33 Massachusetts Institute of Technology. Department of Physics.
The Laboratory Circular, 1887-1878. - 1 folder.
In Massachusetts Institute of Technology, Institute Archives and Special Collections (Cambridge, MA).
Circulars (#1-4) describing laboratory apparatus and processes in physics and chemistry which "you have found convenient and think not in _general_ use in other laboratories. Published jointly with the Chemistry Department.

M34 Massachusetts Institute of Technology. Office of the President, 1897-1930.
Records, 1897-1931. - 14 linear ft.
In Massachusetts Institute of Technology Libraries, Institute Archives and Special Collections (Cambridge, MA).
Subject files of administrative records of the President's Office from 1897 to 1930. Includes correspondence from all the Presidents in this period: James Mason Crafts (1897-1900), Henry Smith Pritchett (1900-1907), Arthur Amos Noyes (Acting President, 1907-1909), Richard Cockburn Maclaurin (1909-1920), Administrative Committee (1900-1923), Elihu Thomson (Acting President, (1920-1923), and Samuel Wesley Stratton (1923-1930). Crafts, Noyes, and Thomson were all chemists. Most of the correspondence is with Presidents Maclaurin and Stratton.
Box list available.

M35 Massachusetts Institute of Technology. Office of the President: Richards, Ellen Henrietta (Swallow).
Records, 1907-24. - 28 items.
In Massachusetts Institute of Technology, Institute Archives and Special Collections (Cambridge).
Richards (1842-1911) was a chemist, professor, and first President of the American Home Economics Association. Letters she wrote while she was a faculty member at MIT are addressed to MIT administrators and concern ventilation and pure water, women students at MIT, and the Association. Also includes correspondence about her death, a resolution passed by the faculty concerning her service to the Institute, and a transcript of an interview Richards conducted with Margaret E. Dayton Stinson, who was hired as a chemistry assistant at MIT in 1865, becoming the first woman employee at the institution. Stinson held the chemistry assistant's position until 1911. In the interview the two women talk about their experiences as women and as associates of the MIT Chemistry Department.

M36 [Massachusetts Institute of Technology] Ten Club.
Ledgers, 1930-1969. - .3 cubic feet.
In Massachusetts Institute of Technology Libraries, Institute Archives and Special Collections (Cambridge).
Ledgers record the names of speakers and information about their speeches at meetings of the Ten Club, a secret and select club of the top ten students in the Department of Chemical Engineering.

M37 Massachusetts Institute of Technology. Women's Laboratory.
Records, 1875-1922. - 0.35 cu. ft.
In Massachusetts Institute of Technology, Institute Archives and Special Collections (Cambridge, MA).
Collection consists of correspondence, reports, and unidentified writings concerning the initiation, funding, and operation of M.I.T. Women's Laboratory in cooperation with the Women's Educational

Association of Boston. Principal correspondent is Ellen Swallow Richards, Head of the Women's Laboratory. Other correspondents include Francis Amasa Walker and Susan Minns.
Folder list available.

M38 Mathews, Joseph Howard, 1881-1970.
Papers, 1932-1948. - 2 boxes.
In University of Wisconsin Archives (Madison).
Professor of Chemistry. - Committee reports, questionnaires, memoranda, correspondence while Chairman of the University Committee on Research in National Defense during World War II, and 4 folders of general correspondence, including correspondence with Farrington Daniels, 1943-1947.
Received in October, 1967.
Unpublished finding aid available.

M39 Maver, Mary Eugenie, b. 1891.
Autobiography, 1965. - 78 p.
In National Library of Medicine (Bethesda, MD).
Biochemist at the National Cancer Institute. - Typescript. Includes some related papers.

M40 Maxim, Hudson, 1853-1927.
Papers, 1889-1925. 147 items.
In Eleutherian Mills Historical Library (Greenville, Del.) (509, 655)
Inventor and mechanical engineer. Correspondence, agreements, patent material, and other papers, relating to Maxim's gunpowder inventions and his work as a consultant in explosives to E. I. du Pont de Nemours and Company. Correspondents include Pierre Samuel du Pont and William George Armstrong, Baron Armstrong of Craigside, Northumberland, Eng.
Described in A Guide to the Manuscripts in the Eleutherian Mills Historical Library, by John B. Riggs (1970) p. 872-875.

M41 Maxim, Hudson, 1853-1927.
Papers, 1883-1927. 9 cartons.
In New York Public Library.
Inventor, mechanical engineer, and explosives expert. General correspondence, writings, speeches, and other papers, relating especially to Maxim's inventions, armaments, explosives, nutrition, the soybean as a food product, peace, prohibition, and legal and business matters, including the development of his property at Lake Hopatcong, N. J. Includes correspondence with Maxim's father-in-law, William Durban, of London, Eng. Other correspondents include Luther Burbank, Gutzon Borglum, Francis I. Dupont, Hiram Percy Maxim, Frank A. Tichenor, and George Sylvester Viereck.
Gift of Michael F. Dee, 1953.

Mayer, Ralph, 1895- M42
Papers, 1929-64. 3 reels of microfilm.
In Archives of American Art.
Microfilm made in 1965 from originals owned by Ralph Mayer.
Artist. Correspondence, notebooks, business records, scrapbooks, photos, and publications, reflecting Mayer's activities as a painter and a chemist interested in the qualities of paint and color, who conducted technical classes for artists, as well as his work as a conservator and restorer. Correspondents include George Biddle, Peter Hurd, Abraham Rattner, John Sloan, and David Smith.

Maynard, Leonard Amby, 1887-1972. M43
Papers, 1940-1963. - 6 ft.
In Cornell University Libraries, Dept. of Manuscripts and University Archives (Ithaca, NY).
University professor. - Administrative papers pertaining to Maynard's positions as director of the U.S. Plant, Soil, and Nutrition Laboratory, New York State College of Agriculture; director of the School of Nutrition, Cornell University; and head of the Department of Biochemistry and Nutrition and of the Laboratory of Animal Nutrition, New York State College of Agriculture. Collection includes correspondence, notes, and interviews.
Gift of Mr. Maynard, 1964; interview deposited by the Cornell University Oral History Program, 1963.
Unpublished finding aids available.

Mead Johnson and Company, Evansville, Ind. M44
Mead Johnson and Company collection, 1895-1971. ca. 55 ft., ca. 280 items, and 9 boxes.
In Indiana State University, Evansville Campus Library, Special Collections and University Archives.
Historical documents relating to the growth and development of a pharmaceutical company engaged in manufacture and research; papers concerning administration and organization, industrial and public relations, marketing and merchandising, product histories of Mead products, and research and development; Johnson family photos; product samples and memorabilia; and audiovisual material.
Unpublished guide in the repository.
Gift of the company, 1973 and after.
Additions to the collection are expected.

Means family. M45
Business records, 1840-1954.
ca. 3 ft. (3792 items, 40 v., and 1 reel of microfilm)
In University of Kentucky Library.
In part, microfilm (negative)
Correspondence, financial records, journals, ledgers, and scrapbooks of John Means (1821-1910), Harriet Hildreth Perkins Means (1826-1895), his wife, Ellison Cooke Means (1864-1956), his son, and William Biggs Seaton (1855-1927), his son-in-law, pioneer industrialists of Ashland, Ky., dealing with the iron industry in eastern Kentucky and southern Ohio, rail and water transportation, and the development of Ashland, Ky. as an industrial city. Includes genealogical material relating to the family on microfilm. The material is related to the library's Seaton family papers.
Indexed in manuscript catalog.
Acquired from Mrs. Harriett Means Witt, 1956.

M46 Media-Related Perceptions of Contemporary Problems.
Oral history, 1975. - 20 tapes and ca. 500 pp. of transcripts.
In Indiana University, Oral History Research Project (Bloomington).
Includes interviews with five retired professional women: Theodora Allen and Sada Murayama, who were social workers; Elizabeth Cleland, a chemist and homemaker; Agnes Newton, a teacher and homemaker; and Eunice Roberts, an administrator who was involved in women's education and women's rights at Indiana University. The women discuss current events and their relationship to them. They also discuss their careers and their activities since their retirement.
Closed.
Unpublished guide available.

M47 Medical history collection, ca. 1584-1979. ca. 30 ft.
In Smithsonian Institution, National Museum of American History, Division of Medical Sciences (Washington, D.C.)
In part, photocopies.
Account books, prescription forms, anatomical charts, public health posters, newsclippings, advertisements for patent medicines, diplomas and certificates, and other medical materials, and photos, prints, and other graphic material concerning medical equipment, hospitals, pharmacies, pharmaceutical manufacturing, and materia medica.
Gifts from various sources.

M48 Medical lecture notes, 1746-1865. ca. 50 v.
In National Library of Medicine (Bethesda, Md.)
Students' notes on lectures on materia medica, surgery, anatomy, midwifery, and the theory and practice of medicine, given in Edinburgh, London, and Philadelphia, by Francis Boott, Sir Robert Christison, William Cullen, John Syng Dorsey, John Eberle, George Fordyce, John Hunter, Adam Kuhn, Alexander Monro, Donald Monro, Joseph Parrish, Philip Syng Physick, Francis Henry Ramsbotham, Benjamin Rush, William Shippen, and George Bacon Wood. Persons taking notes include Sir Thomas Lauder Brunton, Jonathan Elmer, William Elmer, William Hewson, Hugh Lenox Hodge, Samuel Poultney, John Redman, Elihu Hubbard Smith, and John Todd. Includes clinical cases and reports taken at the Royal Infirmary of Edinburgh from Dr. William Cullen; and notes on chemistry lectures given in 1834 by Franklin Bache.
A composite collection consisting of groups of papers cataloged separately in the library.
Additions to the collection are anticipated.

M49 Mees, Charles Edward Kenneth, 1882-1960.
Experimental notebook, 1912-1920. - 1 reel microfilm, neg.
In American Institute of Physics, Niels Bohr Library (New York, NY).
Physical chemist. - Notes on experimental work relating to photography.
Originals in International Museum of Photography, George Eastman House, Rochester, New York.
Finding aid available.

M50 Meggers, William Frederick, 1888-1966.
Papers, 1917-66. 12 ft.
In American Institute of Physics, Center for History and Philosophy of Physics collections (New York City)
Physicist and spectroscopist. Correspondence, reports, MSS. of writings, financial records, photos, and other papers relating to Meggers' career from laboratory assistant to chief of the Spectroscopy Section of the National Bureau of Standards, his work on the Welch periodic table of elements, the analysis of complex atomic spectra, the mercury 198 lamp, rare earth spectra analysis, rare gas analysis, infrared spectroscopy, standards of wave lengths, scientific and learned societies, national and international conferences on spectroscopy, Meggers' collection of portraits of Nobel laureates, his preparations for publications, and other activities.
Unpublished description in the repository.
Information on literary rights available in the repository.
Gift of Mrs. Edith R. Meggers, 1967.

M51 Mell, Patrick Hues, 1850-1918.
Scrapbooks, 1869-97. 8 v.
In University of Georgia Library.
Scientist and educator. Letters, articles from Engineering and mining journal, class notes, lists, invitations, programs, clippings, obituaries, and other material pertaining to Mell's college career, family, church affiliation, interests, and activities while he was a student at the University of Georgia, State chemist of Georgia, and professor of geology and botany at Alabama Polytechnic Institute, Auburn, Ala.
Unpublished list in the library.
Purchased from Patrick Mell, 1957.

M52 Mendel, Lafayette Benedict, 1872-1935.
Papers, 1879-1941. - 6.5 linear ft.
In Yale University Library, Department of Manuscripts and Archives (New Haven, CT).
Physiological chemist; Professor at Yale, 1897-1935; served on the Inter-Allied Scientific Food Commission during World War I; author. - Family and professional correspondence, writings, diaries (1884-1935), scientific notebooks (1884-1909), research files, photographs, and memorabilia largely relating to Mendel's research on nutrition and growth. His professional correspondents include Yale colleagues, as well as scientists and nutritionists from around the world, e.g., James R. Angell, F. C. Bing, W. B. Cannon, Russell H. Chittenden, Harvey Cusing, R. G. Hubbell, Jacques Loeb, Graham Lusk, Anson Phelps Stokes, Carl Voegtlin, William H. Welch, and R. S. Woodward.
Gift of Richard L. and Robert F. Herrmann, 1980.
Unpublished finding aid available.

M53 Michigan College Chemistry Teachers Association.
Records, 1925-69. 3 v. and 1 box.
In Detroit Public Library, Burton Historical Collection.
Correspondence, minutes (1952-68), register (1939-64), and other papers.

M54 Michigan State University. Department of Chemistry.
Records, 1889-1963. - 20 cu. ft.
In Michigan State University, University Archives and Historical Collections (East Lansing).
Records of MSU Chemistry Department. - Correspondence, 1889-1957. Includes letterpress books and index. Subjects are sugar beets, Michigan Agricultural College Alumni Association, R. G. Kedzie, and F. S. Kedzie. Also includes departmental administrative correspondence. Other series are faculty minutes; departmental information book; student grade books; inventories; alumni files; publications; building plans; Alpha Chi Sigma theses; lab manuals, textbooks, and bulletins.
Received March 23, 1982.
Unprocessed collection; access to some boxes restricted.
Unpublished inventory available.

M55 Michigan State University. Miscellaneous Papers.
Research papers, ca. 1983. - 0.1 cu. ft.
In Michigan State University, University Archives and Historical Collections (East Lansing).
Research papers on and draft of Robert Clark Kedzie's biography. Collected and written by Sister Mary Grace Waring.
Gift of P. Hudson, August 26, 1983.
Unprocessed collection.

M56 Michigan State University. Miscellaneous Records.
Student notebook, ca. 1888. - 0.1 cu. ft.
In Michigan State University, University Archives and Historical Collections (East Lansing).
Notebook containing class notes of Louis Bregger, an 1888 graduate of Michigan Agricultural College. Professor Kedzie's and Cook's classes are included.
Gift of Doris Finn, August 16, 1983.
Unprocessed collection.

M57 Michigan Sugar Company, *Bay City, Mich.*
Records, 1898-1914. 100 items and 14 v.
In University of Michigan, Michigan Historical Collections.
Bylaws, prospectus, records of shipments, records of sugar beet purchases, payroll accounts, general accounts, miscellaneous papers; and a few records of the Iowa Sugar Company, of Waverly, Iowa, successor of the Michigan Sugar Company.
Gift of Dan Gutleben, 1957.

M58 Michigan Sugar Company, Bay City.
Records, 1892-1903. 2 v.
In Michigan State University, Archives and Historical Collections (East Lansing)
Minutes of stockholder and board meetings and newspaper clippings pertaining to developments in the sugar beet industry in the U.S.
Gift of William Mantley.

M59 Middle and west Tennessee papers, 1803-1916. 1 box.
In Knoxville-Knox County Public Library, McClung Historical Collection (Tenn.) (MSC 14)
Correspondence, medical records, memorabilia, and other papers, of the families of James Coleman, of Memphis, Tenn., George D. Holmes, of Covington, Tenn., and William King, of Sumner County, Tenn. Includes material on drugs and medicine and on the Civil War, and genealogical information.
Described in A Guide to the Manuscript Collections of the Calvin M. McClung Historical Collection, by Linda Langdon Posey (1974) p. 34.
Gift of Mary U. Rothrock, 1929.

M60 Midwest Oil Corporation.
Records, 1895-1954. - Ca. 85 ft.
In University of Wyoming Archives-American Heritage Center (Laramie).
Correspondence, minutes, financial statements, reports, dispositions, depletion accounts, incorporation papers, powers of attorney, proofs of discovery, income tax papers, stocks, and other papers of the following companies: Belgo-Americaine, Central Salt Creek Company; Central Wyoming Oil and Development Company, Franco-Petroleum Company, Midwest Oil Company, Midwest Refining Company, Mountain Producers Corporation, Natrona Pipe Line and Refining Company, New Bradford Oil Company, Parkman Oil Company, Petroleum Maatschappij, Salt Creek Consolidated Oil Company, Salt Creek Producers Association, Saltmount Oil Company, Wyoming Associated Oil Corporation, Wyoming Associated Royalty Company, and Wyoming Oil Fields Company.

M61 Mikkelsen, Niels, 1876-1955.
Papers, 1897-1955. ca. 600 items.
In Nebraska State Historical Society collections.
Pharmacist, of Kenesaw, Neb. Correspondence, diaries (1906-09), speeches, biographical and historical material, and printed matter, relating to Mikkelsen's business with pharmacy suppliers, membership on the Pharmacy Board, and his duties as a justice of the peace. Correspondents include H. H. Antles, Charles A. Binderip, Edward R. Burke, Hugh A. Butler, Carl T. Curtis, Dwight Griswold, Donaldson W. Kingsley, Rufus A. Lyman, Samuel R.

McKelvie, George W. Norris, Val Peterson, Fred Seaton, A. C. Shallenberger, Robert G. Simmons, and Kenneth S. Wherry.
Gift of Max Mikkelsen, Kenesaw, Neb., 1955.

M62 Miller, Alamby M b. 1848.
Diary and student notebooks, 1866–71.
18 v.
In University of Virginia Library (38–580)
Notes on lectures (1866–68) at the University of Virginia on chemistry, French, geology, Hebrew, history, Latin, mathematics, and moral and natural philosophy, under Professors William H. McGuffey, Socrates Maupin, and others. The diary, in German, tells of Miller's experiences at the University of Heidelberg in 1868 and 1869.
Described in the Annual report of the archivist, University of Virginia Library, 1933–34, p. 5.

M63 Miller, Hugo W 1889–1963.
Records of the Hugo W. Miller and Miller and Cox Assay Offices, Nogales, Ariz., 1902–58. 5 ft.
In University of Arizona Library.
Assay and ore records, and miscellaneous financial and administrative papers. Includes mill, smelter, and converter reports (1902) for the Pride of the West Mining and Milling Company.
List of contents in vol. 1.

M64 Mills family.
Papers, 1895–1970. 484 items.
In University of South Carolina, South Caroliniana Library (Columbia)
Chiefly family correspondence, together with research notes and papers, of James Edward Mills (b. 1876), professor of chemistry, University of South Carolina, and other family members. Includes material on gas warfare and on Wilson Plumer Mills' experiences in China until his internment by the Japanese and repatriation in 1943.
More fully described in University South Caroliniana Society Program (1980).

M65 Mills, Frederick Ira, 1807–1830.
Lecture notebook, 1826–27. – 1 v. (87 pages).
In American Philosophical Society Library (Philadelphia, PA).
Notes on Benjamin Silliman's lectures on chemistry and pharmacy, Yale College. Volume begins with lecture number 23 (Dec. 13, 1826) and ends with number 59 (Feb. 21, 1827). Mills was in the Yale class of 1827.
Accessioned, 1982.

M66 Mineral Industries.
Records, ca. 1820. – 46 cu. ft.
In National Museum of American History Archives Center (Washington, DC).
Original and photocopied annual reports, government reports, blueprints, photographs, maps, trade literature, and publications concerning mining operations, equipment, and companies, chiefly in the United States. Also included are microfilm copies of some mining-related newspapers.
Unpublished finding aid available.

Mirsky, Alfred Ezra, 1900–1974.
Papers, 1913–1984. – 38 ft.
In Rockefeller University Archives, Rockefeller Archive Center (Pocantico Hills, North Tarrytown, NY).
Biochemist and physiologist. – Professional and personal correspondence; biographical material; manuscripts of published and unpublished papers, lectures, and conferences; laboratory notes and notebooks; collected reprints (6 v.) and bibliography; laboratory memorabilia; engraving plates from publications from Proceedings of the National Academy of Sciences; slides (6 boxes) with index; electron photomicrographs and negatives; scientific photo prints and negatives; and other photos. Includes information on the Christmas Lectures for High School Students which Mirsky established at Rockefeller University; also includes student materials from the Ethical Culture School in New York City (1913–1918), Harvard University (1918–1922), and Cambridge University (1923–1925).
Gift of Sonya Wohl Mirsky, 1980–1984.
Information on literary rights available in the repository.
Unpublished finding aid available.

Missoula Brewing Company.
Records, 1915–65. ca. 5 ft.
In University of Montana Library (Missoula)
General correspondence (1934–66), financial records (1915–19, 1934–64), legal documents (1933–64), and miscellaneous papers.
Unpublished guide in the repository.
Gift of Mrs. Anita Taylor.

Mitchell, Harold Hanson, 1886–1966.
Papers, 1906–66. 3 ft.
In University of Illinois Archives (Urbana)
Chemist and professor of animal science at the University of Illinois. Correspondence, alphabetical subject file, photos, 80 unpublished MSS., and 158 publication reprints, relating to the administration of research for the saltpeter investigation in a laboratory of physiological chemistry, apparatus used in the saltpeter experiment and in metabolism, digestion, perspiration, and respiration studies, protein and mineral nutrition and metabolism, comparative nutrition, and climatic stress research.
Unpublished finding aid in the repository.
Information on literary rights available in the repository.
Acquired, 1967.

Moeller, Therald.
Papers, 1937–69. 2 ft.
In University of Illinois Archives (Urbana)
Professor of chemistry at the University of Illinois. Correspondence; Moeller's textbooks on inorganic chemistry (1952), quantitative analysis (1958), and general chemistry (1965), his laboratory manual (1965), and his book on the chemistry of lanthanides

(1963); reprints of articles by Moeller on rare earths, especially gallium, indium, thallium, and thorium, radio tracer techniques, chelate chemistry, sulfamide properties, inorganic polymers, physical methods in inorganic chemistry, and spectrophotometry; lecture notes for chemistry and thorium analysis; abstracts of theses done under Prof. Moeller; instructional films for which Moeller was an educational collaborator; symposia proceedings on rare earth ions; and Montana State University chemistry institute syllabi (1963-64) and Air Force contract research reports (1948-69) concerning the rare earths, compounds of bromine, chlorine and fluorine polymers and sulphur-nitrogen compounds. Includes correspondence relating to Moeller's lectures and visits with the College of St. Teresa, Winona, Minn., Institute for Continuing Education in Engineering and Applied Science, University of Missouri at Rolla, University of São Paulo, and Western Michigan University.

Unpublished finding aid in the repository.
Information on literary rights available in the repository.
Acquired, 1969.

Monsanto Chemical Company.
Questionnaires and reports on employee activities, 1951, 1956.
3 ft.
In Harvard Business School, Baker Library.
Replies to questionnaires and reports submitted by about 500 employees on their day's activities, as of May 9, 1951, together with additional reports, five years later, by about forty of these. Material was submitted for a proposed book about Monsanto, which was not published.
Indexed in the library's catalog.
Access restricted until May 9, 2001.
Gift of the company, 1960.

Montana Ore Purchasing Company.
Records, 1900-1910. - 1 linear ft.
In Montana Historical Society (Helena).
Incorporated in Butte, Montana in March, 1893. In 1894, began offering low-priced custom smelting work to small mining companies and thus placed itself in direct competition with Anaconda Copper Mining Company. Bulk of property sold in 1906 to Butte Coalition Mining Company, a division of Amalgamated. - Collection contains Montana Ore Purchasing Company records, including incoming correspondence (1900-1910), 2 letterpress books of outgoing correspondence (May 1, 1905-April 30, 1906), court papers, 2 account journals, legal documents, and miscellaneous materials. In addition, there is a subgroup for Butte Plumbing Company, including incoming, outgoing, and interoffice correspondence.

A portion of this collection was donated by William L. Wallace of Butte, Montana, in June, 1972. The remainder of the collection was separated from Alice Gold and Silver Mining Company Records (MC 57, Montana State Historical Society Archives).
Unpublished finding aid available.

Montgomery Court House ledgers, 1792-1893. M73
3 ft.
In University of Virginia Library (38-67)
Letter books, ledgers, daybooks, expense books, and journals of stores, hotels, blacksmith shops, banks, farms and other businesses mainly in Blacksburg, Christiansburg, and New River White Springs, Va. Includes records of D. Baldwin, Baldwin and Gibboney, Bank of the Valley, Christiansburg, B. Lipscombe and Co., Montgomery Hotel, W. H. Peck and Co., B. W. Stewart, West Virginia Arsenic Bromine Lithia Co., and the minutes and membership list of the S. S. S. Society (a young women's sewing circle)
Indexed in part.
Described in the Annual report of the archivist, University of Virginia Library, 1937-38, p. 30.

Moore, Edward Warren, 1906- M74
Papers, 1940-50. 1 ft.
In Harvard University Archives (Cambridge, Mass.)
Teacher of sanitary chemistry at Harvard University. Correspondence and other papers.
Access restricted.
Information on literary rights available in the repository.

Morfit, Campbell, 1820-1897. M75
Papers, 1855-57. -ca. 100 items.
In University of Pennsylvania Libraries, E.F. Smith Collection (Philadelphia).
Industrial chemist and author of several books on chemistry and pharmaceuticals. - Correspondence with James Curtis Booth, S. Muspratt, and E.W. Horsford among others. Some of the letters concern the publication of Morfit's Chemical and Theraputic Manipulations.

Morfit, Campbell, 1820-1897. M76
Manuscripts, 1862-1865.
In Library of Congress, Manuscript Division (Washington, DC).
Chemist. - Treatises on the manufacture of soaps, 1862-1865, phosphate of lime, and potential fertilizers.

Morfit family. M77
Papers,1815-1941. - 54 items, 28 v.
In Maryland Historical Society Library (994, 1374) (Baltimore).
Correspondence, deeds, contracts, chemical treatises, notes, essays, and other business and professional papers of Campbell Morfit (1820-1897), Professor of Applied Chemistry at the University of Maryland; his mother, Catherine (Campbell) Morfit; his father, Henry M. Morfit, lawyer and member of the Maryland legislature; his grandfather, John Campbell; and his daughter, Marie C. C. Morfit. Miss Morfit's correspondence relates to the construction of a bakery as envisioned by her father.
Some of the volumes are indexed.
Gift of Louis H. Dielman, 1934, and Mrs. H. T. Bredehorn, 1934.

M78 Morgan, Agnes Fay, 1884-1968.
Papers, ca. 1905-62. - 11 boxes.
In University of California, Berkeley Library, Manuscripts Division.
Chemist, nutritionist, and professor at the University of California, Berkeley. - Correspondence, drafts, reprints, manuscripts of articles and speeches, concerning home economics, nutrition, and biochemistry.
Published guide available.

M79 Morland and Warwick, Villaldama, Mexico.
Records, 1899-1907. 20 v.
In State Historical Society of Colorado collections (Denver)
Letter press books (1904-07), journals of accounts, mine expenses, and ore shipments, timebooks, and receipt book, of operators of mines and smelters. Includes liquidation journals (1904-06) of American Smelting and Refining Company, Monterrey, Mexico.

M80 Morley, Edward Williams, 1838-1923.
Papers, 1851-1922. - Ca. 3 ft.
In Case Western Reserve University Archives (Cleveland, OH).
In part, photocopies of originals in the Library of Congress, Washington, D.C.
Chemist, physicist, and Professor of Natural History and Chemistry at Western Reserve University. - Personal correspondence with American and European scientists; family correspondence; 22 notebooks (1890-1900) relating to experiments and calculations of the amount of oxygen in the earth's atmosphere and the relative atomic weights of hydrogen and oxygen, and photos. Also includes reprints of lectures and articles by Morley.

M81 Morley, Edward Williams, 1838-1923.
Papers, 1833-1923. ca. 1200 items.
In Library of Congress, Manuscript Division (Washington, D.C.)
Chemist and physicist. Personal, professional, and family correspondence, and miscellaneous certificates and printed material, chiefly 1863-99, consisting almost entirely of correspondence from members of Morley's family, personal friends, and fellow scientists. Includes personal letters from Myron A. Munson, some written while Munson was in the Union Army in 1864, and extensive correspondence with European and American scientists. Correspondents include Henry E. Armstrong, H. Brereton Baker, Richard Börnstein, Carl Wilhelm Böttger, Charles F. Brush, Frank W. Clarke, Edward S. Dana, James Dwight Dana, Harold Dixon, Hugo Wilhelm Erdmann, Phillippe Guye, Edward Hart, Walter Hempel, Francis Herrick, William M. Hicks, Sir William Higgins, Frank F. Jewett, Lord Kelvin, Samuel P. Langley, Sir Joseph Larmor, Thomas C. Mendenhall, Albert A. Michelson, Dayton C. Miller, Charles E. Munroe, William A. Noyes, Wilhelm Ostwald, Henry S. Pritchett, Frederic W. Putnam, Sir William Ramsay, Lord Rayleigh, Ira Remsen, William A. Rogers, Frederick Soddy, and William F.G. Swann.
Unpublished finding aid in the repository.
Gift of Howard R. Williams, 1956.
This entry replaces MS 59-213.

Morley family.
Edward Williams Morley collection, 1829-1926. 10 boxes.
In California Institute of Technology Library (Pasadena).
Letters to Edward Williams Morley (1838-1923), chemist and physicist, from his father, Sardis Brewster Morley (1804-1889), Congregational clergyman in Massachusetts and Connecticut, his mother, Anna Clarissa (Treat) Morley, brothers John Henry Morley, Congregational clergyman in the Midwest, and Frank Gibson Morley (d. 1875), and sister, Lizzie Morley; together with correspondence between the other members of the family; letters to Sardis Brewster Morley from classmates at Williams College, many relating to theological questions; other family papers; and reprints of Edward Williams Morley's scientific writings, including those relating to the Michelson-Morley experiment. Includes Civil War letters of Sardis Brewster Morley (a chaplain), John Henry Morley, and Frank Gibson Morley; letters by Lizzie Morley written from Mount Holyoke Female Seminary and Dio Lewis's school for young ladies at Lexington, Mass.; letters from John Henry Morley in Magnolia, Iowa, Sioux City, Iowa, and while superintendent of Congregational Home Missionary Society in Minneapolis, Minn.; family chronology (1864-1885) kept by Anna Clarissa (Treat) Morley; and Treat family papers.
Gift of Mr. and Mrs. William K. Morley, Altadena, Calif., 1977.
Finding aid in the repository.

Morris family.
Papers, 1684-1935. 16 ft.
In Eleutherian Mills Historical Library (Greenville, Del.) (721, 944, 951, 1004)
Chiefly the William Henry Russell collection of Morris family papers consisting mostly of correspondence, ledgers, accounts, deeds, receipts, bills, and genealogical data; together with other business and personal papers of a family of brewers, merchants, sugar refiners, real estate investors, and participants in public affairs, founded in Philadelphia by Anthony Morris (1654-1721). Most of the papers relate to Samuel Morris (1734-1812), his son, Isaac Wistar Morris (1770-1831), and their descendants, with some concerned with the Paschalls and Wistars, with whom they intermarried. Includes papers relating to Samuel Morris' service as Commissary General of the Middle District during the Revolution and to Bolton Farm, Bucks Co., Pa., and other property in Philadelphia and western Pennsylvania. Other persons include Anthony Morris (1681-1763), Anthony Morris (1705-1780), Anthony Morris (1766-1860), Anthony P. Morris (1798-1873), Dr. Benjamin Morris (1725-1755), Benjamin Wistar Morris (1762-1825), Caspar

Wistar Morris (1764-1828), Catharine Wistar Morris (1772-1859), Isaac Paschall Morris (1803-1869), Israel Morris (1741-1806), James Pemberton Morris (1795-1834), John Thompson Morris (1847-1915), Joseph Paschall Morris (1803-1869), Luke Morris (1707-1793), Luke Wistar Morris (1768-1830), Lydia Thompson Morris (d. 1932), Mary (Wells) Morris, Paschall Morris (1813-1875), Phineas Pemberton Morris (1817-1888), Rebecca (Thompson) Morris, Samuel Morris (1711-1782), Sarah Paschall Morris (1815-1905), and William Hudson Morris (1753-1807).
In part, described in A Guide to the Manuscripts in the Eleutherian Mills Historical Library, by John B. Riggs (1970) p. 878-880.
Gift, 1965; purchase, 1967-68.

Morton, William Thomas Green, 1819-1868.
Family papers, ca. 1849-1911. 2 ft.
In Smithsonian Institution, National Museum of American History, Division of Medical Sciences (Washington, D.C.)
Dentist and anesthetist. Correspondence, copies of diary entries, publications, government documents, genealogical notes, and other papers, of and about Morton, mostly collected by his son, William James Morton (1845-1920), in defense of Morton's character and priority of his discovery of the use of ether as an anesthetic.
Gift.

Moss, Frank E., b. 1911.
Papers, 1958-1956. - 380 ft.
In University of Utah Libraries, Special Collections Department (Salt Lake City).
U.S. Senator, judge, and attorney. - This collection is concerned with Moss's years in the U.S. Senate, 1958-1976. Included is a large section with several sub-sections concerning mining and minerals in the West.
Gift of Frank E. and Phyllis Moss, 1976.
Unpublished finding aid available.

Mrozowski, Stanislaw W   1902-
Papers, ca. 1950-74. 27 ft.
In State University of New York at Buffalo Archives.
Physicist and professor of physics at the University of Buffalo and State University of New York at Buffalo. Papers pertaining to Mrozowski's carbon research and, to a lesser extent, his work in spectroscopy, and papers as editor of Carbon International Journal and as director of Carbon Research Center at State University of New York at Buffalo.
Unpublished guide in the repository.
Information on literary rights available in the repository.
Gift of Mr. Mrozowski, 1974.

Mulliken, Robert S., b. 1896.   M87
Papers, 1958-1982. - 19.5 linear ft.
In University of Chicago Library.
Chemical physicist; Professor, Department of Physics (1928-56); Ernest DeWitt Burton Distinguished Service Professor, Departments of Chemistry and Physics (1956-date). Director, Informational Division, Metallurgical Project (1941-45). Nobel Prize, 1966, for molecular orbital theory. - Biographical and bibliographical materials (1958-81); expired grants and contracts files (1958-74); manuscripts, illustrations, and figures for works, including Molecular Complexes; undated lecture notes; personnel files (1962-71); payroll registers (1964-74); and general office records (1965-74).
Deposited by Mr. Mulliken.
Unprocessed and restricted.

Mummery, Arthur E.   M88
Papers, ca. 1884-1906. - 150 items and 18 v.
In University of Michigan, Bentley Historical Library, Michigan Historical Collections (Ann Arbor).
Ann Arbor pharmacist. - Correspondence and legal and business papers, including personal accounts and accounts of organizations.
Gift of Mr. Mummery, 1948.

Murphree, Egar Vaughan, 1898-1962.   M89
Papers, 1932-63. 1000 items.
In Library of Congress, Manuscript Division (Washington, D.C.)
Engineer, scientist, and businessman. Correspondence, speeches, memoranda, reports, newspaper clippings, photos, and biographical material, chiefly 1957-62, relating primarily to Murphree's activities as president of Esso Research and Engineering Company (research and development branch of Standard Oil of New Jersey) and as an official of the 5th World Petroleum Congress, New York, 1959. Includes correspondence concerning admission of Soviet scientists into the U.S. Correspondents include William B. Lacy, John A. McCone, Richard Nixon, Glenn Seaborg, Lewis L. Strauss, and Edward Teller.
Unpublished finding aid in the repository.
Gift of Esso Research and Engineering Company, 1966.

N1 Nason, Henry B., 1858-1894.
Scrapbooks and publications, 1858-1896. - 1 linear ft.
In Rensselaer Polytechnic Institute Archives and Dept. of Special Collections (Troy, NY).
Professor of Chemistry, Mineralogy, and Geology. - Two scrapbooks, one of Nason's and one of his wife's. Nason's scrapbook contains newspaper clippings, printed invitations, programs and some personal correspondence. Very little about his research or professional writing. Pertains mostly to life at RPI and in Troy. Mrs. Nason's scrapbook contains newspaper clippings, letters of condolence upon her husband's death, programs for his memorial services, and memorabilia. Collection also contains 4 of Nason's publications: his doctoral dissertation on the formation of ether; a Handbook of Mineralogy; a translated and revised edition of Elderhorst's Blowpipe Analysis; and the Biographical Sketches of the Graduates and Officers of RPI 1824-1886.
Gift of William E. Tinney of Glenmont, New York.
Unpublished finding aid available.

N2 **National Dye Works.**
Records, 1917-27. 10 items and 13 v.
In Duke University Library (Durham, N. C.)
Ledgers, journals, cashbooks, trial balances, and inventories of a firm which dyed, bleached, and finished seamless hosiery.
Card index in the library.
Gift, 1965.

N3 National Paint, Varnish & Lacquer Association, Inc., Washington, D.C.
Records and papers, 1860-1957. - 33 items.
In Hagley Library (Greenville, DE) (acc. #881).
Collection includes shipping journal, 1898-1915, containing record of paint and varnish materials, with names and addresses of companies, 1 vol.; material relating to the Paint & Oil Club of New England; correspondence relating to historical data on paint and varnish clubs in St. Louis, Pittsburgh, Chicago, and New York, with related memoranda and news items; photographs; card subject index to materials in the National Paint, Varnish and Lacquer Association Library, Washington, D.C.; price book of paint and varnish supplies kept by E. T. Trigg, undated, with printed price lists.
Described more fully in the Supplement to A Guide to Manuscripts in the Eleutherian Mills Historical Library (1978).

National Sugar Manufacturing Company. N4
Records 1900-1967. - 70,000 items.
In Colorado Historical Society (Denver).
Sugar City, Colorado enterprise. — Collection contains correspondence and business papers: beet receipts, sugar production records, invoices of shipments, sales data, time cards, and payrolls.
Unpublished finding aid available.

Nef, John Ulric, 1862-1915. N5
Papers, 1884-1915. 1 ft.
In University of Chicago Library.
Chemist and professor at the University of Chicago. Personal and professional correspondence, memoranda, reports, articles, reports and indexes of experiments, a series (1904-13) of "sugar experiments," and biographical material, including accounts of Nef's life and work written by his son, John Ulric Nef, Jr. (b. 1899).
Unpublished guide in the library.
Gift of John Ulric Nef, Jr., 1952.

Nef, John Ulric, 1899- N6
Manuscripts of writings, 1935-57. 5 ft.
In University of Chicago Library.
Educator and historian. Drafts of articles and books by Nef, including The United States and civilization (1942), War and the early industrial revolution, War and human progress (1950), The genesis of industrialism and of modern science, Industrial Europe at the revolution, Industry and government in France and England (1940), and Mining and metallurgy in mediaeval civilization.
Unpublished guide in the library.
Open to investigators having the permission of Mr. Nef.
Information on literary rights available in the library.
Gift of Mr. Nef, 1953 and 1959.

Nelson, Oscar M. F. N7
Papers, ca. 1925-1931. - 1 carton.
In University of California, Davis, Shields Library, Special Collections.
Manuscripts and mimeographed papers concerning sugar and methods of refining sugar.
Unprocessed collection.

Nelson, Wilbur Armistead, 1889- N8
Papers, 1913-37. 1 ft. (ca. 1000 items)
In University of Virginia Library (6835)
State geologist of Tennessee (1918-25) and of Virginia (1925-28) and professor of geology at the University of Virginia. Correspondence, reports, geological surveys, and ore assays for Alabama, Tennessee, and Virginia, drafts of clippings of articles, geological notebooks, U. S. Dept. of Labor industrial surveys of Virginia and West Virginia, blueprint maps and aerial photos of Tennessee, clippings, and information on the American Institute of Mining Engineers, the Engineering Association of Nashville, and the Nonacid Fertilizer and Chemical Company, Lakeland, Fla. Other subjects include the Scopes trial for which Nelson testified in defense of evolution, the development of Muscle Shoals, Ala., earthquakes in Tennessee, the Appalachian National Park, and the Temple bill on completion of topo-

graphic maps of the U. S. Correspondents include John Sharshāl Grasty, Albert Louis Kreiss, David A. Shepherd, and the patent attorney firm of Shepherd and Campbell.
Gift, 1962.

Neuberg, Carl, 1877-1956.
Papers, ca. 1929-56. - Ca. 10,000 items (12 linear ft.).
In American Philosophical Society Library (Philadelphia, PA).
Biochemist; Collection consists of correspondence, lab notebooks, documents, photographs, reprints, maps. Most of the correspondence is post-1940, when Neuberg arrived in New York and made contact with American chemical manufacturers and industries involved in fermentation. A notable exception is the correspondence with Kurt Jacobsohn in Portugal, 1929-56. There are many contacts with Japanese scientists, and significant involvement documented about the American Society of European Chemists, later, the Carl Newberg Society for International Scientific Relations. Scholars interested in the mechanics or politics of government grants will find the American Cancer Society, U.S. Public Health Service files of interest.
Presented by Dr. Irene Forrest, 1980.
Unpublished finding aid available.

Neurath, Hans, b. 1909.
Papers, 1947-1975. - 60 lin. ft.
In University of Washington Libraries, Archives and Manuscripts Division (Seattle).
Professor of Biochemistry. - Correspondence, lecture texts, notes, manuscripts and subject files reflecting his career at Duke University (1947-1950) and as Chairman of the Biochemistry Department at the University of Washington (1950-1975) and his membership in various professional organizations.
Access restricted. Contact repository for further information.
Unpublished inventory available.

Nevada Mining Companies.
Records, 1862-1901. - Ca. 12,000 pieces.
In The Huntington Library, (San Marino, CA).
Records of Savage, and Hale and Norcross Silver Mining Companies. - Letters and documents, including 27 bound volumes, documenting legal and business affairs of the mines. Receipts and disbursements, payrolls, bank books, time books, stockholders' statements and accounts, reports on ore extracted, and assays, vouchers, etc. Also includes letters to mining superintendents, a few pertaining to the mining town of Bodie, California, and some material concerning operations of the U.S. Mint in Carson City, Nevada.
Purchased from William P. Wreden, August, 1961.
Unpublished finding aid available.

Nevada Salt and Borax Company. N12
Records, 1882-1900. ca. 310 items.
In University of Nevada Library.
Chiefly monthly reports from the superintendent of the company to its secretary, G. F. Ford; together with correspondence. Names represented include H. M. Yerington, treasurer of the firm.
Formerly part of the records of the Virginia & Truckee Railroad.
Gift of Gordon A. Sampson, 1952.

New Almaden Mine, California. N13
Records, 1845-1973. - 85 l.f.
In Stanford University Libraries, Special Collections Department (California).
An important, long operated mercury mine in Santa Clara County. - Correspondence, reports, research notes, legal and financial documents, ledgers, photographs, maps and other materials relating to the operation of the mine and the Quicksilver Mining Company.
Unpublished finding aid available.

New York Brewery, Spokane, Washington. N14
Records, 1887-1904. - 6 ft.
In Washington State University Library (Pullman).
Letterbooks, ledgers, accounts, reports and other papers.
Unpublished finding aid available.

New York State Agricultural Society. N15
Records, 1792-1840.
58 items.
In Albany Institute of History and Art.
Letters and treatises of an agricultural nature, relating to animals, chemicals, produce, scientific observations, and machinery.

New York State College of Agriculture agricultural N16
  leaders project: oral history interviews,1962-69.
77 items.
In Cornell University Libraries, Dept. of Manuscripts and University Archives (Ithaca, N.Y.) (O.H. 1, 2, 4-6, 8-21, 23-30, 32-36, 38-41, 43-46, 49-52, 54, 62, 64-65, 79, 81, 101-102, 105, 107, 110, 114, 117-118, 121-124, 126-129, 133-137, 142, 144-146, 173, 200, 497)
Transcripts of tape-recorded interviews with farmers, officials of the New York State Grange, New York State and American Farm Bureau Federations, Dairymen's League, Grange League Federation, other agricultural businesses, county agents, extension specialists, faculty and staff members of College of Agriculture and Agricultural Experiment Station at Geneva, N.Y., and others involved in

agriculture, including Cornell-Los Baños Project at University of the Philippines; Rockefeller Foundation sponsored crop improvement programs (1940-62) in Mexico, Colombia, and the Far East; agricultural missionary work in China and the Agricultural Economics Dept. at University of Nanking; Bureau of Agricultural Economics, U.S. Dept. of Agriculture; Federal Land Bank of Springfield, Mass.; Long Island Agricultural and Technical Institute in Farmingdale, N.Y.; formation (1923) of New York Seed Growers' Cooperative and New York Metropolitan Milk Marketing Area. Subjects discussed include history of 20th century agriculture, student life at the college, agricultural production and marketing, farm practice and management, experiences in organizing farmers, fruit marketing cooperatives, impressions of early county agents, farmers' reaction to tuberculosis and brucellosis control, the seed business in New York State, and teaching, research, and extension work at the college and Geneva Station in agricultural economics, agricultural engineering, agronomy, animal husbandry, bacteriology, biochemistry, botany, conservation, entomology, farm practice, pomology, and vegetable crops. Persons interviewed or mentioned are listed in the index.

Described, with a complete list of persons interviewed, in Report of the Curator and Archivist, 1962-1966, Collection of Regional History and the University Archives (1974) p. 155-158.

Partially restricted.

N17 **New York. Agricultural Experiment Station,** *Geneva.*
Records, 1868-1925. ca. 13 v. and 1 storage case.
In Cornell University Library, Collection of Regional History and University Archives (22/1/31)

Records relating to the beginnings of the experiment station, the setting up of its Board of Control, the interest in its activities by individual and organized farmers, the discussions over conflicting operating and experimental plans, and the adoption of general policies, the aims of the various directors, W. A. Armstrong, E. Lewis Sturtevant, and others, current operating problems, and the general problems of scientific agriculture. Includes minutes of the meetings of the Board of Control, fiscal and operating reports of the director, experiment books, and 8 vols. of clippings relating largely to agricultural problems in New York State, and a long series of sketches, photos, and other biographical materials concerning staff members.

N18 **Nicklès, François Joseph Jérôme,** 1821-1869.
Correspondence, 1847-69. ca. 1000 items.
In Temple University Libraries (Philadelphia)
Professor of chemistry at the Université de Nancy. Correspondence from French and German chemists concerning social and professional matters, chemical experiments and work with metals, and engineering problems. Correspondents include Jean-Baptiste Dumas, A. Laurent, Herman Laurent, William Odling, Alexander W. Williamson, and other scientists and scholars.
Index in preparation.
Information on literary rights available in the library.
Purchased from Alain Brieux, of France, 1964.

Nieuwland, Julius Arthur, 1878-1936. N19
Papers, 1904-1936. - 6.1 linear ft.
In University of Notre Dame Archives, (Notre Dame, IN).
Catholic clergyman, chemist, botanist, editor of the American Midland Naturalist, a monographic series, and professor at the University of Notre Dame. - Letters and notes.

Nobel Laureates on scientific research: oral history collection, 1963-64. 37 items. N20
In Columbia University Libraries (New York City)
Transcripts of tape-recorded interviews with Nobel Laureates in science, discussing their relations with co-workers, their associations with Nobel Prize winners and other eminent scientists, descriptions of each prize winner's discovery, the events leading up to it, and the parts played by others in this process. Persons interviewed are listed in the index.
Described, with a complete list of persons interviewed, in The Oral History Collection of Columbia University (1964) p. 150, and Supplement (1966) p. 27-28.
Open to investigators under restrictions accepted by the library.

Norman, George Miller. N21
Records, 1898-1967. - 9 items.
In Hagley Library (Greenville, De) (acc. #1073 and 1150).
Chemist, of Hercules Powder Co., Wilmington, DE. - Notebook of chemical formulae, undated; notebook on water analysis, 1899; notebook of chemistry notes, 1902; notebook for a college course in German, 1898; journal of a trip to Europe, 1926, giving details of visits to chemical and industrial firms in England, France, Germany, and Italy; letter of Henry N. Lyons to Norman, 1921, concerning employment; commemorative medal for Nobel's invention of dynamite; news item. In addition, a notebook containing: correspondence, 1912-13; production statistics; O. A. Pickett, "Hercules Research, a critical review," 1933, on developments by Hercules Powder Co.; L. N. Bent, annual report of the Naval Stores Department, Hercules Powder Co., 1932-33; laboratory records and correspondence concerning the Repauno plant of E. I. du Pont de Nemours & Co., 1904-22.
Described more fully in the Supplement to A Guide to Manuscripts in the Eleutherian Mills Historical Library (1978).

North Carolina Academy of Science. N22
Records, 1902-1978, 1981. - Ca. 1,200 items and 2 v.
In East Carolina University Library, East Carolina Manuscripts Collection (Greenville, NC).
Records include minutes, correspondence, programs, N.C.A.S. newsletters, reports, and other papers.

Described in "East Carolina Manuscripts Collection, Bulletin No. 6."

Gift of Dr. John Yarbrough, Raleigh, NC, 1975; Dr. Randolph Ferguson, Beaufort, NC, 1977; and Dr. John W. Nowell, Winston-Salem, NC, 1982.

Northwestern University, *Evanston, Ill. Archibald Church Medical Library*, *Chicago*.

Medical lectures and notes, ca. 1760-1906. ca. 60 items and 34 v.

In Northwestern University, Archibald Church Medical Library (Chicago)

Lectures on various medical topics and notes from lectures given at the Chicago Medical College, Johns Hopkins Medical School, Northwestern University Medical School, the University of Edinburgh, and elsewhere. Subjects include the ear, the eye, obstetrics, pathology, organic chemistry, surgery, anatomy, medical jurisprudence and the practice of physic. Names represented include William Cullen, John Wesley Dal, Sir W. B. Dalby, Joseph Bolivar De Lee, George Fordyce, John Gregory, P. W. Hays, Ferdinand Hebra, A. I. Kendall, Herman Louis Kretschmer, Sir William Lawrence, Aaron Lufkin, Alexander Monroe, James Wilson, and G. E. Wire.

A composite collection consisting of single items and groups of papers cataloged separately in the library.

Northwestern University. Technological Institute. Dept. of Chemical Engineering.

Departmental files, 1951 - 1976; 1969 -1982 (addition). - 4 boxes.

In Northwestern University Archives, (Evanston, IL).

School of engineering established at Northwestern in 1940. - Bulk of the records of the main collection cover the years 1963-1973. Consists mostly of memos and descriptive accounts of programs and projects pertaining to various subject categories: environmental engineering; honorary societies; professional societies; placement information; visiting academicians; faculty; students; and courses. The addition, which concentrates on the years 1978-1981, includes general departmental correspondence, planning and faculty files, graduate student files, and subject files.

Main collection transferred to the archives by William F. Stevens, department chairman, in May, 1977. Addition transferred by Libbie Rubinoff, department secretary, on July 18, 1983.

Finding aid available.

Norton, John Pitkin, 1822-1852.
Papers, 1837-52. 3 ft.
In Yale University Library (New Haven, Conn.)

Professor of agricultural chemistry at Yale University. Correspondence, including letters concerning the publication of Norton's works; 10 vols. of diaries (1838-47, 1851-52); scientific papers, including Norton's additions to Henry Stephens' Book of the Farm, and numerous articles and letter essays contributed to the agricultural press, especially to the Albany "Cultivator"; addresses on scientific agriculture; ca. 33 lectures given by Norton in courses at Yale, 1847-51; notes, including a catalog of minerals, subject guide to his readings, and notes on various soils; agricultural and geological map (1844) of New York State; scrapbook of newspaper clippings; and other papers.

Unpublished register in the library.

Information on literary rights available in the library.

Gift of Mary DeWitt Pettit, 1969.

Noyes, Bradford, b. 1860. N26
Recollections, 1948. - 46 p.
In West Virginia University Library (Morgantown).

Charleston businessman. - Uncorrected typescript copy concerning Charleston, Kanawha County, 1866-1948, and recollections related to Noyes by early settlers or their descendants, 1790-1865. Subjects include Indian attacks, turnpikes and taverns, the first telegraph system, natural gas illumination, Civil War manufacture of saltpeter, schools and economy in post-Civil War Charleston, salt and chemical industries, carrier pigeons, steamboats on the Kanawha River, coming of the railroad to Charleston.

Loaned for photoduplication by Mrs. John J. D. Preston, 1966.

Noyes, William Albert, 1857-1941. N27
Papers, 1870-1942. - 9 ft.
In University of Illinois Archives (Urbana).

Professor of Chemistry at the University of Illinois. - Correspondence, mss. of writings, essays, photos, notebooks, and other papers. Includes correspondence with his colleagues on chemical problems; laboratory notebooks, 1895-1934; correspondence and mss. on early investigations into the structure of camphoric acid and the camphoric series, and of his later studies on valence, electron theory, ionization and their relationship to organic chemistry; notebooks and correspondence from Noyes' student years at Iowa College and Johns Hopkins University; files concerning Noyes' editorship of several journals relating to chemistry; correspondence and mss. (1920-1940) on war, the economy, fascism, interventionism, religion, and similar subjects; and personal correspondence with family and friends. Correspondents include Roger Adams, Marston Bogart, Arthur Capper, Arthur Compton, Charles Jackson, Arthur B. Lamb, Gilbert Lewis, Charles Marie, Robert S. Mulliken, Linus Pauling, Ira Remsen, Theodore Richards, Albert Shaw, Charles M. Sheldon, Alexander Smith, Julius Stieglitz, and Wilhelm Willstaetter.

Unpublished finding aid available.

N28 **Noyes, Winthrop Gilman,** 1869–1931.
    Letter books, 1906–10.
    2 v.
    In Minnesota Historical Society collections.
    Businessman. Copies of letters written by Noyes as executor of the estate of his father, Daniel R. Noyes, and as chairman of the Building Committee of the YMCA, St. Paul. A few of the letters relate to the St. Paul Society for the Relief of the Poor and to the Minnesota pure food and drug law, in which Noyes was interested as a member of the wholesale drug firm of Noyes Brothers and Cutler, St. Paul.
    Presented by Levi T. Jones.

N29 <u>Nycum, John and John Q.</u>
    <u>Papers, 1825-1900.</u> - 8998 items and 59 v.
    In Duke University Library (Durham, NC).
    Correspondence and mercantile papers (chiefly 1840-1890) of John Nycum and John Q. Nycum, of Ray's Hill, Pennsylvania, and of other members of the Nycum family. The collection relates to general merchandising, tanning, the sale of mineral water, and other matters, and includes letters from members of the family in Pennsylvania, Virginia, and other places; papers of Bedford Co., Pennsylvania, teachers (1850's); and personal letters written during the Civil War.
    Card index in the library.
    Acquired, 1951.

Oenslager, George, 1873-1956.
    Papers, 1850-1972; 1929-1933. - 5 boxes.
    In University of Akron, American History Research Center (Akron, OH).
    Rubber chemist; recipient of Charles Goodyear and Perkin Medals. - Collection is in 2 parts. Largest part (4 boxes) is General Materials Collection (1850-1972), and contains business and personal correspondence; personal financial records; biographical material; family photographs; travel memorabilia; and Oenslager's 4 U.S. patents. Smaller part (1 box) is Personal Correspondence Collection (1929-1933), and contains correspondence pertaining to an article written by James Lawrence, "Early History of Organic Accelerators," as well as correspondence pertaining to Oenslager's receipt of the Perkin Medal in 1933.
    Unpublished finding aid available.

Ohio State University. Department of Chemistry.
    Records, 1925-1942. - 4 cu. ft.
    In Ohio State University Archives (Columbus).
    General correspondence, 1929-36; Chemistry Department National Youth Administration and FERA materials, ca. 1935; bibliography of publications of the chemistry department, 1872-1941; yearbook, 1937-1941; budget materials, 1940; Committee on Deanship of Arts and Sciences, 1936; William Lloyd Evans, 1931; minutes, 1925-1928; scrapbook, ca. 1931-1941; scrapbook, 1933-1937.

Oil, Chemical and Atomic Workers International Union.
    Records, 1948-75. ca. 8 ft.
    In University of Texas at Arlington, Library, Dept. of Special Collections.
    Correspondence, agreements, arbitrations, reports, and studies.
    Unpublished finding aid in the repository.
    Information on literary rights available in the repository.
    Gift.

Oil, Chemical, and Atomic Workers International Union. Local 4-228, Port Neches, Tex.
    Records, 1934-69. 4 reels of microfilm (negative) (ca. 840 items)
    In University of Texas at Arlington, Texas Labor Archives.
    Microfilm made in 1970 from originals owned by Local 4-228.
    Minutes.
    Information on literary rights available in the repository.
    Permanent deposit, 1970.

Olin, Hubert Leonard, 1880-1964
    Papers, 1880-1964. 9 v.
    In Iowa State Historical Dept., Division of Historical Museum and Archives (Des Moines)
    Professor of chemical engineering at the University of Iowa, Iowa City, and chairman of Iowa State Mining Board. Correspondence; biographical material, typescripts, outline, and clippings, relating to Olin's writings; mining reports; correspondence, financial reports, and board meeting reports, of Iowa State Mining Board; county coal histories; mine inspectors' reports (1880-93); and maps, articles, stories. tape recording, photos, and other papers, relating chiefly to coal mining in Iowa.

Oliver Iron and Steel Company, *Pittsburgh.*
    Records, 1866-1922. 4 ft.
    In Archives of Industrial Society (Pittsburgh, Pa.)
    Minute book of the board of directors; charters; agreements; patents; partnership agreements (1863-91) of Lewis, Oliver and Phillips; mortgages and patents (1891-97) of the Oliver Coke and Furnace Company; miscellaneous records (1888) of the Hainesworth Steel Company; trade catalogs; and scrapbooks on the life of Henry W. Oliver (1840-1904).
    Inventory in the repository.
    Gift of Henry Oliver Rea, 1964.

Oliver-Gowen Collection.
    Papers, 1833-1928. - Ca. 1,100 pieces.
    In The Huntington Library, (San Marino, CA).
    Frederick Oliver (d. 1883) was Secretary and Treasurer of the Megantic Mining Company and was involved, along with Hammond Gowen, James Douglas, and others, in developing the mines and mineral resources of Quebec. At his death his widow, Dame Charlotte Maria (Gowen) Oliver moved to Toronto, where her business affairs were handled by her son-in-law, Edward Holloway. - Collection contains chiefly documents and letters, including 6 volumes of memoranda, diaries, and letterbooks, and a 1833 journal of a trip across New York. Also includes a few mining sketch maps and some ephemera. Subjects are the business affairs of the Olivers, Douglas, Gowen, and Holloway; the sale, purchase, or lease of lands involved in mining; mining and assay reports; and private financial and legal affairs of the Gowen and Oliver families.
    Acquired from R. R. Oliver, 1957.
    Unpublished finding aid available.

Omansky, Morris, 1889-1974.
    Papers, 1918-1972. - 6.5 linear ft.
    In Massachusetts Institute of Technology, Institute Archives and Special Collections (Cambridge, MA).
    Chemical engineer; consultant. - Collection includes material documenting research and teaching activities, but consists mainly of material documenting activities as a consultant on patent and insurance litigation concerning rubber.

09 Onsager, Lars, 1903-1976.
Papers, 1926-1977. Microform, 16 reels.
In Yale University Library, Department of Manuscripts and Archives (New Haven, CT).
Chemist and university professor; taught at Johns Hopkins, 1928; at Brown University, 1928-1933; at Yale, 1933-1972; won the Nobel Prize in Chemistry, 1968; author. - Correspondence, research materials for articles and lectures, teaching material related to Yale University and the University of Miami, and citations for awards received by Onsager.
Deposited by Margarete Adleter Onsager, 1977. (Originals held by the Onsager family.)
Donor must be notified whenever the microfilm is used.
Index available.

010 Oral history collection, 1926, 1951-68. 32 items.
In National Library of Medicine (Bethesda, Md.)
Transcripts or tape recordings of interviews with physicians and other persons connected with medical matters; addresses; and a tape recording of the William Stewart Halsted Centenary program, Baltimore, Md., 1952. Three of the oral histories were conducted by the American Gastroenterological Association. Topics covered include medical education, military medicine, public health, social medicine, costs of medical care, blood banks, drugs, quackery, cancer treatment, medical research, biochemistry, gastroenterology, immunology, psychiatry, and surgery. Persons interviewed include J. Arnold Bargen, Stanhope Bayne-Jones, Daniel L. Borden, Arthur Carlisle Christie, Burrill B. Crohn, Ward Darley, Raymond Osborne Dart, Michael Marks Davis, John Miller Train Finney, Warfield Monroe Firor, Robert Philipp Fischelis, Morris Fishbein, Gilbert S. Goldhammer, Albert Baird Hastings, Michael Heidelberger, Lister Hill, Andrew Conway Ivy, Boisfeuillet Jones, George P. Larrick, Zigmond Meyer Lebensohn, Warren Egbert Magee, Rudolph Matas, William Shainline Middleton, Wilder Graves Penfield, Herbert Percy Ramsey, Harvey Brinton Stone, Albert Szent-Györgyi, Vivien Thomas, Rexford Guy Tugwell, Donald Dexter Van Slyke, and Lawrence Richardson Wharton. Other persons and organizations mentioned include Alfred Blalock, Simon Flexner, William Stewart Halsted, Sir

Transferred to the Institute Archives by Frieda Omansky Cohen in July 1975 and May 1983.
Unpublished finding aid available.

William Osler, Franklin D. Roosevelt, American College of Surgeons, American Medical Association, Committee on the Costs of Medical Care, Johns Hopkins University School of Medicine, Rockefeller Institute for Medical Research, New York, U. S. Food and Drug Administration, and U. S. National Institutes of Health.
Cataloged and indexed individually in the library, where most of the original tapes are retained.
In part, restricted.

011 **Oregon. Agricultural Experiment Station,** *Corvallis.*
Records, 1890-1935. ca. 5 ft.
In Oregon State University Archives (Corvallis)
Daybook (1890-96, 1902-13) and plot records (1888-1910) relating to daily activities in experimenting with varieties of crops; and general correspondence, records of analyses, soil records, reports, experiment notebooks, and miscellaneous papers of the station's Chemical Laboratory.

012 Oregon Environmental Council.
Records, 1952-76. 38 ft.
In Oregon Historical Society Library (Portland)
Correspondence, environmental impact statements, hearing transcripts, newsletters, clippings, and other records, relating to nuclear power, mining, oil, timber, recreation, wildlife, environmental education, rivers, land, government agencies, and air, water, pesticide, solid waste, and noise pollution.
Unpublished finding aids in the repository.
Gift of the council, 1976.

013 Oregon State Pharmaceutical Association.
Records, 1934-37. 52 items.
In Oregon Historical Society Library (Portland)
Correspondence, convention and membership information, and other records, comprising papers of Roy A. Perry, president of the association.
Gift of Gene A. Westberg, 1978.

014 **O'Shaughnessy, Ignatius Aloysius,** 1885-1973.
Papers, 1917-1973. - 17 ft.
In Minnesota Historical Society collections (St. Paul)
Oil company executive and philanthropist, of St. Paul, Minn. - Correspondence and other papers relating to O'Shaughnessy's interest in educational, religious, cultural, and civic organizations, including College of St. Thomas in St. Paul, Notre Dame University, and other Catholic institutions; and business records relating to the petroleum industry, including Globe Oil and Refining Company, and O'Shaughnessy's work as director of several banks and other companies.
Gift of I.A. O'Shaughnessy Foundation, 1978.
Unpublished inventory in the repository.

15 Orton, Edward, 1863-1932.
  Papers, 1880-1932. 7 ft.
  In Ohio Historical Society collections (Columbus) (360)
  Geologist, dean of the College of Engineering, Ohio State University, Columbus, and ceramic engineer and manufacturer, Standard Pyrometric Cone Company. Business, personal, and academic correspondence, diaries, account books, and other papers. Includes material relating to Orton's military career during World War I, genealogy, geology, family affairs, and European travels, and letters of condolence to Mina A. Orton on the death of her husband.
  Gift of Mr. Orton's niece, Mrs. Stanley Robarge, Columbus, Ohio, 1971.

Commercial-scale esterification equipment used by Pfizer for ascorbic acid production in the late 1940s. Courtesy Pfizer Inc.

P1 **Pacific Guano Company.**
Pacific Guano Company records and Crowell family papers, 1860-1912. ca. 4 ft.
In Harvard University, Graduate School of Business Administration, Baker Library (Boston, Mass.)
Records of a company with offices in Boston and plants at Woods Hole, Mass., and Charleston, S.C., which manufactured fertilizer from guano obtained from Pacific and Caribbean islands; together with records of various fishing business in the Woods Hole, Mass., area, with which the Crowell family was also associated. Includes letter books and other correspondence, account books, trade catalogs, photos, and other papers. Persons represented include Prince S. Crowell (1813-1881) and Azariah F. Crowell (1846-1918). Fishing interests represented include Mitchell Fish Company, Boston, Mass., Isaiah Spindel & Company, Trap Fishermen's Association, and the Woods Hole Weir.
Unpublished inventory in the repository.
Gift of Olive C. Beverly, 1974.

Pacific Northwest Pollution Control Association.
Records, 1946-76. ca. 7 ft. (ca. 8500 items)
In Washington State University Libraries (Pullman)
Correspondence; administrative, financial, membership, committee, and section records; and photos and other newsletter editorial material, of an association affiliated with Water Pollution Control Federation.
Unpublished finding aid in the repository.
Gift of the association, 1977-78.

P3 **Palmer, Arthur William,** 1861-1904.
Papers, 1893-1903. - 0.3 ft.
In University of Illinois Archives (Urbana).
Professor and head of the Chemistry Department at the University of Illinois. - Letter books concerning needs of the Chemistry Department for buildings, chemicals, and laboratory equipment, analyses of chemical compounds for manufacturers and public officials, applications for academic positions, instruction in chemistry, pharmacy training, water survey, a pure food conference (1898), and other matters; and copies of Palmer's Chemical Survey of the Water Supplies of Illinois (1897) and Chemical Survey of the Waters of Illinois for 1897-1902 (1903).
Information on literary rights available in the repository.
Acquired, 1963.

P4 **Panknin, Charles F**
Business records, 1844-1909.
1050 items.
In University of North Carolina Library, Southern Historical Collection (3084)
Pharmacist, of Charleston, S.C. Account books and prescription books (1844-72), business correspondence (1881-88), and other papers, of Panknin's drug store.
Unpublished description in the library.
Purchase, 1955.

P5 **Parchen, Henry Martin,** 1839-1925.
Business records, 1906-27. 9 ft.
In Montana Historical Society collections (Helena)
Businessman, of Helena, Mont. Records of businesses in which Parchen was involved: Helena Cab Company (1912-19), Helena Motor Company (1919-21), Lakeview Land and Boating Company (1906-21), Montana Phonograph Company (1916-27) and Parchen Drug Company (1915-21). Includes correspondence and financial records.

P6 **Parke, Davis & Company.**
Records, 1862-1969. - 37 boxes.
In Detroit Public Library, Burton Historical Collection (Mich.).
Pharmaceutical company; headquarters in Detroit, Mich. - Correspondence, research reports, history of the company and its products, biographies of personnel, publications, and photos.
Finding aid in the repository.

P7 **Parrish, Maxfield,** 1870-
Papers, ca. 1888-1950.
ca. 60 items.
In Haverford College Library, Quaker Collection.
Artist. Letters, drawings, printed articles about Parrish, and reproductions of his work. Includes 13 letters (1891-1954) concerning Haverford College affairs; sketches for Haverford College alumni reunions and a classbook; and sketches in a German reader and chemistry and physics notebooks used by Parrish at Haverford College (1888-91)
Indexed in the library's card catalog.
Gift, ca. 1941-56.

P8 **Parsons, Henry Betts,** d. 1885.
Papers, 1876-1882. - 2 ft.
In University of Michigan, Bentley Historical Library, Michigan Historical Collections (Ann Arbor).
Pharmacist. - Report of faculty activities of Parsons at the University of Michigan; manuscript of a book on chlorophyll; pharmaceutical notes; and 14 v. containing personal accounts, notes, and data on scientific experiments.

P9 **Paschall family.**
Papers, 1705-70.
6 v.
In Historical Society of Pennsylvania collections.
Commonplace book of John Paschall consisting mainly of notes from the alchemic and philosophical writings of Thomas Vaughan; iron account book (1735-56) of Stephen Paschall; malt and barley book (1705-28) of Thomas Paschall.
Presented by Ellison J. Morris, 1944.

P10 **Pasteur, Louis,** 1822-1895.
Papers, ca. 1844-90. - Ca. 100 items.
In National Library of Medicine (Bethesda, MD).
Chemist. - Collection of Pasteur manuscripts (notes, texts, and letter drafts) covering Pasteur's work in crystallography, fermentation, vinegar, diseases of wine, spontaneous generation, diseases of silkworms, studies on beer, the

germ theory of disease, vaccination, and rabies. Also includes material pertaining to opponents of Pasteur, the Pasteur Institute, Pasteur the teacher, Pasteur the patriot, and Pasteur's speeches and honors.

Patterson family.
Papers, 1775-1854.
ca. 700 items.
In American Philosophical Society Library.
Correspondence; reports to the American Philosophical Society; notebooks on biology, chemistry, agriculture, mineralogy and physic; lectures; articles presented to the society for publication; and other material. Major figures in the collection are three members of the society: Robert Patterson, 1743-1824; Robert Patterson, 1819-1909; and Robert Maskell Patterson, 1787-1854. Correspondents include Andrew Ellicott, Nicolas Fuss, Ferdinand R. Hassler, Thomas Jefferson, William Lambert, Benjamin H. Latrobe, David Rittenhouse, John Vaughan and Jonathan Williams. Part of the collection is in the society's Archives.
Cataloged individually in manuscripts catalog of the library.
Purchase, and gift of James O. Patterson.

Pauling, Linus Carl, b. 1901.
Papers, 1923-1929. – 3 reels microfilm.
In American Institute of Physics, Niels Bohr Library (New York, NY).
Chemist; Nobel Prize winner. – Research notes, letters, lecture notes. Originals at the Gates and Crellin Laboratories of Chemistry, California Institute of Technology.
Finding aid available.

Peale family.
Peale-Sellers papers, ca. 1707–
ca. 4000 items and 18 ft.
In American Philosophical Society Library (Philadelphia)
Papers of Charles W. Peale include 18 volumes of letter books (1767-1826), 25 volumes of diaries (1765-1826), an autobiography, a two volume daybook (1805-24), 8 volumes of lectures on natural history and other subjects, farm accounts (1816-22), and 13 volumes of notebooks on portraits. Papers of Rembrandt Peale include letters, palettes and sketchbooks. Papers of Rubens Peale include correspondence (1826-49), letter books (1802-24), an autobiography and a sketchbook. Papers of Titian R. Peale include correspondence with Franklin Peale, a scrapbook and sketches. Papers of Coleman Sellers include personal and business letter books (1828-84), expense books and receipt books. Papers of John Sellers, Jr., include 7 volumes of diaries (1808-46), daybooks (1785-1817), and receipt books (1821-52). Papers of Nathan Sellers include letters and account books on papermaking, diaries (1773-1829) and notebooks (1784-1821) Also included are minutes (1821-35), stockbooks (1827-41), accounts and miscellaneous material on Peale's Museum; and 7 volumes of Sellers family letters valuable for social history of Philadelphia and Ohio in the 19th century.
Cataloged in manuscripts catalog of the library.
Purchases and gifts from members of the Peale and Sellers families, 1945-

Pearce, Charles Chester, 1888-1974.
Papers, 1907-73. 6 ft.
In State Historical Society of Wisconsin collections (Madison)
Lawyer. Correspondence and other personal and professional papers of a Wisconsin-born lawyer who practiced in Washington, D.C., and New York City and served as special assistant to the U.S. Attorney General, 1938-46. Includes private legal practice files; material relating to Pearce's work with Antitrust Division of U.S. Justice Dept., a fertilizer industry investigation (1939-40), and small loan investigation (1943-44); and files relating to his private work through Office of Alien Property and War Claims on behalf of German expatriates making claims for property confiscated by the Nazis.
Unpublished finding aid in the repository.
Gift of Mr. Pearce's estate through University of Wisconsin Foundation, 1976.

Pearson, Eliphalet, 1752-1826. P15
Papers, 1769-1857. 250 items.
In Phillips Academy, Dept. of Archives (Andover, Mass.)
Congregational clergyman and educator. Correspondence, diaries, teaching notes, sermons, biographical sketches and appraisal by Josiah Quincy, notebooks, personal account books, and other papers, dealing with Pearson's Hancock Professorship of Hebrew and Oriental Languages at Harvard College, religion, teaching, the American Society for Educating Pious Youth for Gospel Ministry, Dummer Academy, Massachusetts Society for Promoting Christian Knowledge, New Jersey College, Phillips Academy, and the Society for Promoting Christian Knowledge. Includes correspondence with his son, Henry B. Pearson, while a student at Phillips Academy (1804-05) and at Yale College (1810-15), and papers (1775) on making saltpeter for the gunpowder factory of Lt. Gov. Samuel Phillips.

Peckham, Stephen Farnum, 1839-1918. P16
Papers, 1865-1885. – 0.25 linear ft.
In Stanford University Libraries, Department of Special Collections (Stanford, CA).
Chemist. – Collection reflects Peckham's interests in the oil refining process, and particularly the occurrence of petroleum in Southern California. Includes journal (1865-1866) kept by Peckham during the period he live in San Buenaventura, California, and worked for the California Petroleum Company; and correspondence (1865-1866) between Peckham and geologists then working in California on the state survey—William Henry Brewer, William More Gabb, Benjamin Silliman, George M. Wheeler, and Josiah Dwight Whitney.
Gift of Mary W. Peckham, ca. 1920 and 1932.
Unpublished finding aid available.

Peduto, Michael L P17
Cayuga County sugar beet industry records, 1958-66. 415 items.
In Cornell University Libraries, Dept. of Manuscripts and University Archives (Ithaca, N. Y.) (2562, O. IL 139, 148, 155, 160, 165, 170)
In part, photocopies.
Correspondence, notes, lists of growers and potential growers, memoranda, reports, testimony, minutes, pamphlets, clippings, and other papers, chiefly relating to the efforts of the Cayuga County Redevelopment Area Organization of the Auburn Labor Market, Chamber of Commerce of Auburn and Cayuga County, Finger Lakes Sugar, Inc., Industrial Development Foundation of Auburn and

Cayuga County, faculty members of the New York State College of Agriculture at Cornell, and others, in promoting the sugar beet industry during the sugar shortage caused by the Cuban crisis; and 6 oral history interviews (1965-66) with Peduto and businessmen Paul W. Lattimore and Henry Stack, and New York State College of Agriculture professors C. Delmar Kearl, Joseph F. Metz, Jr., and Thomas W. Scott, relating to acreage allotment, formation and administration of a growers' Association, selection of the Pepsi-Cola Company as a refiner, acquiring background knowledge about raising sugar beets and production of beet sugar, and relations of producers to the processor and to the growers' association. Includes papers of Peduto, executive manager of the Chamber of Commerce and secretary of the Industrial Development Foundation, and of Paul Aron, economic consultant of the Redevelopment Area Organization, James F. Hoobler, and Paul W. Lattimore, chairman of the Redevelopment Area Organization.

Name card index to oral history interviews in the library.

Oral history interviews restricted, in part.

Information on literary rights available in the library.

Gift of Mr. Peduto and deposited by the Cornell University Oral History Program, 1965-66.

Additions to the collection are expected.

P18 **Pemberton family.**
Papers, 1800-1900.
10,000 items.
In Historical Society of Pennsylvania collections.
Papers of John Pemberton, Philadelphia merchant, 1783-1847, his wife Rebecca Clifford Pemberton, and their decendants. Papers of John Pemberton include several letters from Andrew Jackson. Papers of Israel Pemberton, 1813-1885, who surveyed routes for several early railroads in Pennsylvania, Delaware, Tennessee, and Ohio and also worked in the coastal surveys of Florida and Louisiana, reveal the difficulties under which the early engineers worked. His later letters (1870-85) show his interest in art and particularly in the Spanish artist Mariano Fortuny y Carbo. Letters (1830-81) of John C. Pemberton, Israel's brother, to his family give details of an army officer's life. Correspondence of another brother, Henry, a chemist and inventor, relates to salt manufacturing. Other members of the family are represented by less important papers.
Presented by Henry R. Pemberton, 1952.

P19 **Pence, Edward H**
Papers of Edward H. and John W. Pence, 1846-72. ca. 3 ft.
In Ohio Historical Society collections (Columbus) (Vol. 65-78)
Ledgers and daybooks relating to the operation of a flour mill and distillery in Mt. Holly, Ohio, owned by the Pences.

P20 **Penington family.**
Papers, 1764-1882.
ca. 5000 items.
In Historical Society of Pennsylvania collections.
Account books (1769-1826) of the sugarhouse of Edward Penington and son Edward; a volume of observations on the making of sugar; correspondence and business papers (1840-62) of John Penington, rare book dealer, and a diary (1827-41) of Henry I. Baird.
Purchase, 1946.

Penn Chemical Company. P21
Records, 1893-1950. - 3 v.
In Cornell University Libraries, Department of Manuscripts and University Archives (Ithaca, NY).
Ledger, 1911-1914; Board of Directors minute book, 1893-1950; and financial record book, 1920-1927.

Pennsylvania State University. P22
Agricultural Experiment Station.
Records, 1862-1930. - 12 ft.
In Pennsylvania State University Libraries (University Park).
Correspondence of Henry Prentiss Armsby and William Frear; financial records and experimental records, especially fertilizer experiments.

Pennsylvania State University. Petroleum P23
Refining Laboratory.
Records, ca. 1929-1972. - 92.5 cu. ft.
In Pennsylvania State University Libraries (University Park).
Pennsylvania State University's Petroleum Refining Laboratory was established in 1929 to investigate the components of Pennsylvania crude oil and to study the chemical and engineering aspects of its refining and use. The lab grew rapidly, providing research for the Penn Grade Crude Oil Association, Esso Research and Engineering, and many other companies. During World War II, it conducted war-related research for the Army and Navy. The Lab became part of the Chemical Engineering Department in 1959 and was phased out ca. 1966. The bulk of the collection consists of research reports for various clients, correspondence, and materials relating to the university Cryogenics Lab, naval diesel engineering courses, patents, and the administration of the Lab and the School of Chemistry and Physics.
Unprocessed; access restricted until processing completed.

Pennsylvania State University. School of P24
Agriculture.
Records, 1886-1907. - 4.5 ft.
In Pennsylvania State University Libraries (University Park).
Collection contains correspondence of Henry Prentiss Armsby and William Frear.

P25 Pennsylvania. Historical Society.
Forges and furnaces account books, ca. 1720–1832.
ca. 250 v.
In Historical Society of Pennsylvania collections.
Ledgers, daybooks, journals, receipt books, account books, waste books, store books, time books, and other business records of forges, furnaces, and ironmasters of Pennsylvania. Many accounts relate to the production of ammunition, cannon, and other war materials used by the American forces during the Revolution. Furnaces and forges represented include Coventry Iron Works, Chester Co.; Colebrookdale Furnace, Berks Co.; Warwick Furnace, Chester Co.; Pine Forge, Berks Co.; Pool Forge, Berks Co.; Popadickon Furnace, Berks Co.; Sarum Forge, Chester Co.; Mount Pleasant Furnace, Berks Co.; Potts Grove (later Pottstown) Montgomery Co.; Dale Furnace, Berks Co.; Hopewell Furnace, Berks Co.; Tulpehocken Forge (later Charming Forge); Elizabeth Furnace, Lancaster Co.; Mary Ann Furnace, York Co.; Berkshire Furnace, near Wernersville; New Pine Forge, Berks Co.; and Valley Forge, Montgomery and Chester Cos. Also, Manheim Glass Works, and George Ross and Co. Individual ironmasters represented include Samuel Nutt, Alexander Woodrop, Caspar Wistar, Thomas Rutter, Richard Lewis, John Mickle, Samuel Morris, John Potts, Samuel Potts, Robert Hobart, William Bird, William Stiegel, Jacob Huber, Thomas May, Robert May, and members of the Taylor, Potts, and other families.
Many records are gifts of the Potts family.

P26 Perlmann, Gertrude Erika, 1912–1974.
Papers, 1935–74. 33 ft.
In Rockefeller University Archives, Rockefeller Archive Center (Pocantico Hills, North Tarrytown, N.Y.)
Chemist at Rockefeller University and specialist in protein chemistry. Correspondence, mss. of lectures and published papers, collected reprints, biographical material, Rockefeller University subject files, laboratory notes, course notes, photos, and other papers. Includes programs and photos (1953–74) from the Gordon Conferences. Correspondents include Ephraim Katzir-Katchalski, Aharon Katzir-Katchalski, and K. Linderstrøm-Lang.
Unpublished finding aid in the repository and in Rockefeller University Archives in New York City.
Gift of Miss Perlmann's brother, Dr. Peter Perlmann.

P27 Perot Malting Company, *Philadelphia.*
Papers, 1818–1956. 2000 items and 80 v.
In Historical Society of Pennsylvania collections.
Records of the company, which was founded in 1687 and sold in 1963, together with some Perot family papers. Business papers include account books, cashbooks, cash ledger, private ledger, trial balances, rents and interests, personal expense book, blotter, salesbooks, daybooks, receipt books, minute books, expense book, barley book, malt book, tests of malt, receipts for malt, brewing book, accounts receivable and payable, order book, and contracts. Family papers include Pennsylvania deeds; papers pertaining to the Association of Centenary Firms; the Santo Domingo Silver Mining Company of Batopilas, Chihuahua, Mexico; Thomas Morris Perot (1828–1902); Thomas Morris Perot, Jr. (1872–1945); the Society of Friends; the Welsh Society of Philadelphia; the State in Schuylkill; and the estates of Elizabeth Marshall (1862–1883) and Mary Ann Marshall (1881–1913). Includes correspondence (1914–18, 1939–43) of Sara T. Hallowell (d. 1924) and her niece, Harriet Hallowell (d. 1943), Americans resident in France, containing information about the World Wars.
Gift of the company, 1963.

P28 Person, Alice (Morgan) 1840–1913.
Papers, 1872–1972. 10 items and 5 v.
In University of North Carolina at Chapel Hill, Library, Southern Historical Collection (3987)
In part, copies.
Ms. autobiography of Person, of Franklin County, N.C., manufacturer of Mrs. Joe Person's Remedy, a proprietary medicine; account book (1910–11); scrapbook of clippings; and diary (2 v., 1872–73, 1913–16) of her sister, Lucy (Morgan) Beard, teacher, of Hickory, N.C.

P29 Peru Steel and Iron Company, Clintonville, N.Y.
Records, 1800–89. 16 items and 11 boxes.
In State University of New York, College at Plattsburgh, North Country History Center.
Correspondence, memoranda, inventories, cost reports, records of orders received and agreements, orders for steel and coal, company store accounts, affidavits, contracts for charcoal and company owned land, company held mortgages, court orders, account book of Flack & Ransom store, legal papers of justice of the peace Alexis Ransom, papers pertaining to Clintonville, N.Y., and Francis Saltus and family, map of Livingston and Maul patents, drawings of charcoal kilns and a blast furnace, and other papers, of a company in Clinton County, N.Y.
Unpublished finding aid in the repository.

P30 Peter, Robert, 1805–1894.
Papers, 1828–1905.
ca. 2 ft. (339 items and 28 v.)
In University of Kentucky Library.
Physician, chemist, and university professor. Correspondence, laboratory notebooks, and field notes made by Peter for the first and second geological surveys of Kentucky, and for surveys of Indiana and Arkansas, together with a catalog of plants found in the vicinity of Pittsburgh (1828–33) and Lexington, Ky. (1833–35) The bulk of the material is for the first Kentucky survey in 1854. The unbound material is largely chemical analyses.
Indexed in manuscript catalog.
Acquired, 1947.

P31 Peters, John Punnett, 1887–1955.
Papers, 1909–1973. – 10 linear ft.
In Yale University Library, Department of Manuscripts and Archives (New Haven, CT).
Physician; A.B., Yale, 1908; M.D., College of Physicians and Surgeons (Columbia), 1913; Instructor in Clinical Medicine, College of Physicians and Surgeons, 1916–1917; Fellow, Russell Sage Institute of Pathology, 1917–1920; Associate Professor of Medicine, Vanderbilt University, 1920–1921; Yale, 1921–1927; John Slade Ely Professor of Medicine, Yale, and Associate Physician, New Haven Hospital and New Haven Dispensary, 1927–1955. – The papers contain correspondence, reports, writings and personal papers of John P. Peters, especially reflecting his work in clinical chemistry and with insulin.
Gift of Alice Peters Irwin, 1983.
Unpublished finding aid available.

P32 **Pettee family.**
Papers, ca. 1805–1957. ca. 2 ft.
In Salisbury Association collections (Conn.)
Papers relating to Joseph Pettee (1781–1838), ironmaster at Mt. Riga, Conn., and to the Mt. Riga Furnace (earliest 1813); papers relating to William E. Pettee (1845–1924), Litchfield Co., Conn., surveyor and civil engineer, including account books (1867–93), diaries (1863–1913), maps, and surveys; ms. of Julia Pettee's book "The Reverend Jonathan Lee and His 18th Century Parish of Salisbury" (1957), with notes for the same; mss. of Miss Pettee's articles on Mt. Riga Furnace and Village, local stories, and accounts of outstanding individuals and families of Salisbury; genealogical records, accounts, deeds, and wills.
Gift of Miss Julia Pettee, 1966.

P33 **Pettee, William Henry, 1838–1904.**
Papers, 1865–1883. – 17 items.
In University of Michigan, Bentley Library (Ann Arbor).
Professor of Mineralogy at the University of Michigan. – Correspondence, reports, and miscellaneous documents relating to his scholarly interests and activities at the University. Correspondents include Alvah Bradish, Frank W. Clarke, Henry S. Frieze, Israel Hall, and George S. Morris.

P34 **Pharr family.**
Papers, 1848–1934. – 85,762 items, 220 mss. v., and 141 printed v.
In Louisiana State University, Department of Archives (Baton Rouge).
Papers and account books of Captain John Newton Pharr (1828–1903), sugar planter and manufacturer of St. Mary Parish; and, after 1903, the records of J.N. Pharr and Sons, Limited. – Papers and volumes cover the operation of all the Pharr holdings, including the several plantations, Glenwild Sugar Refinery, the Fairview Dairy, and the Pharr Steamboat Line.

P35 **Philadelphia Quartz Company.**
Records, 1818–1981. – Ca. 40 linear feet; 15 photographs and imprints.
In Hagley Museum and Library (Wilmington, DE).
Nineteenth century soap and candle-making firm, founded in Philadelphia by Joseph Elkinton in 1831. In the twentieth century the firm became an important producer of detergents and industrial silicates. Name officially changed to PQ Corporation in 1973. – Collection consists of 7 series. Early Records of the Elkinton Company, 1831–88, include financial documents, common place books and research notebooks, letter books and inventory statements. Administrative Records cover the founding and dissolution of a copartnership, 1863–68, and meetings of the Executive Committee, 1913–16. Financial Records for Philadelphia Quartz and Subsidiary Companies include journals, cashbooks, and sales ledgers for the parent company, 1888–1949. Research and Development Records and Sales Department Records cover the period 1853–1930. Personnel Records, 1840–1953, include labor records, job descriptions, and conditions of employment. Records of Acquired Companies include legal, administrative, financial and construction records. The Elkinton family Personal Papers include official family genealogy, correspondence (1830–1940), and biography of Joseph Elkinton. In addition, there are imprints and photographs.
Unpublished finding aid available.

P36 **Phoenix Glass Company,** *Boston.*
Records, 1867–71. ca. 500 items.
In Massachusetts Historical Society Library (Boston)
Correspondence, receipts, price lists, sales and promotion information, and other records concerning the manufacture and sale of lamp chimneys and other glassware. Includes agreements and partnership contracts between Francis Cabot Lowell, Francis Lowell Hills, and William Cains for the formation of the company, which succeeded the firm of Hills, Turner and Harmon in 1869, the subsequent split, and the dissolution of the company in 1870.
Unpublished guide in the library.
Information on literary rights available in the library.
Gift of Harriet Ropes Cabot, Boston, 1967.

P37 Phoenix Steel Corporation.
Records, 1827–1963. 335 ft. and ca. 25,000 drawings.
In Eleutherian Mills Historical Library (Greenville, Del.) (683, 909, 916, 1179)
Letter books, daybooks, journals, ledgers, cashbooks, records of expenditures, minute books of directors, orders, requisitions, deeds, property and mine maps, and photos, relating to a Pennsylvania corporation and its predecessors, Reeves & Whitaker, Reeves, Buck & Company, Reeves, Abbott & Company, Phoenix Iron Company, and to the Spring Mill Furnace, Conshohocken, Pa. Two-thirds of the collection consists of the records of an allied firm, the Phoenix Bridge Company, Phoenixville, Pa., and its predecessor, Clark, Reeves & Company, including ca. 25,000 drawings (1905–62), correspondence, minute books, ledgers, contract orders, equipment records, and records of bridge erection. Includes correspondence concerning orders of iron and deliveries for railroads, bills, scale drawings and calculations, and certificates for Civil War prize money.
Unpublished inventory in the library. Also described in part in A Guide to the Manuscripts in the Eleutherian Mills Historical Library, by John B. Riggs (1970) p. 913.
Open to investigators under library restrictions.
Gift, 1966; purchase, 1964; and deposits, 1966–67, 1969 by the Phoenix Steel Corporation.

Physical Sciences History Collection.
Papers, 1533-1919. - 5 cu. ft.
In National Museum of American History Archives Center (Washington, DC).
Contains original and photocopied documents, publications, prints, drawings, photographs, and maps, documenting the history of the physical sciences, especially astronomy, chemistry, and physics, and the history of surveying. Included are graphic materials illustrating the history of scientific instruments and apparatus; photographs and portraits of prominent scientists; and a large number of printed maps. Individual items of note include two 17th or 18th century drawings of a salt works and water purification plant.
Unpublished finding aids available.

Physical Sciences Information Files.
Papers, n.d. - 10 cu. ft.
In National Museum of American History Archives Center (Washington, DC).
Contains copies of patents, trade literature, and illustrations of chemical apparatus, surveying instruments, astronomical instruments, and Da Vinci models. Also included are photocopies of old prints dealing with chemistry.

Piccard family.
Papers, 17th cent.-1968. 45 ft. (ca. 37,000 items)
In Library of Congress, Manuscript Division (Washington, D. C.)
Family, general, and personal correspondence, diaries, notebooks, biographical material; writings and speeches, financial papers, scrapbooks, blueprints, patent specifications and descriptions, newspaper clippings, printed matter, and other papers, chiefly 1926-66, documenting the varied careers of Jean Felix Piccard (1884-1963), Jeannette Ridlon Piccard (b. 1895), and their families, in the fields of chemistry, aeronautics, ballooning, and education. Includes Swiss legal and genealogical documents of the 17th and 18th centuries and material relating to Miss Akiko Tokuza, a student befriended by the Piccards, and to Jeannette Piccard's activities as a member of the Episcopal church and her interest in women's rights. Family correspondents include Auguste and Jacques Piccard, twin brother and nephew, respectively, of Jean Piccard, who shared common interests in chemistry, ballooning, and undersea bathyscaphe exploration. Other correspondents include Albert Einstein, Robert Millikan, the National Research Council, A. N. Stevens, and William Swann.
Finding aid in the Library.
Open to investigators under restrictions accepted by the Library.
Gift of Jeannette Piccard, 1969.

Pickel, J. J., 1850-1921. P41
Papers, 1896-1920. - 62 items.
In North Carolina State Office of Archives and History Collections (Raleigh).
Professor of Chemistry. - Collection contains seventeen roll books (1896-1913), which include student information; course outline; list of experiments; and comments on laboratory, oral examinations, and cheating. Also brochure of Shaw University (1898), where Pickel taught, and catalogue (1902-1903) listing faculty and graduates of the Leonard Schools of Medicine and Pharmacy at Shaw University; notes on additives in feed, testing of feeds (1903-1918), and salaries (1919) at the N.C. Department of Agriculture, where Pickel was also employed; an argument to the Board of Agriculture on the need for better feed inspection (1918); and letters about the manufacture of a fat extractor developed by Pickel (1919-1920).

Piedmont Guano Company (Charleston, S.C.) P42
Records, 1893-1895. - 1 v.
In South Carolina Historical Society collections (Charleston) (23-355)
Letter book of superintendent Rene Ravenel. Correspondents include Read Fertilizer Company, Charleston, S.C., and firms and individuals in Maryland, New York, and Mount Pleasant, Moncks Corner, and elsewhere in South Carolina.

Pierson (Isaac G.) and Brothers. P43
Records, 1795-1865.
28 ft.
In Harvard Business School, Baker Library.
General accounts, correspondence, and day-to-day accounts with employees of an iron firm, which operated a steel furnace at Hoboken, N. J., and an iron foundry at Ramapo, N. Y. Includes material on the manufacture of nails and screws and their sale through a New York outlet. Other enterprises recorded include cotton goods manufacture (the 1820's), a store for employees, a flour mill, a farm, and a woodlot.
Inventory in the library. Described in part in List of business manuscripts in Baker Library, compiled by Robert W. Lovett (2d ed., 1951)
Deposited 1932, by J. Fred Pierson.

Pinckney family. P44
Papers, 1735-1922. 315 items and 7 v.
In University of South Carolina, South Caroliniana Library.
Correspondence, accounts, bills, receipts, and other business, legal, and personal papers of a South Carolina family: Charles Pinckney (1699-1758), Charles Pinckney (1757-1824), Charles Cotesworth Pinckney (1746-1825), Charles Cotesworth Pinckney (1812-1899), Charles Cotesworth Pinckney (1839-1909), Harriott Pinckney (1776-1866), and Thomas Pinckney (1750-1828). Includes material relating to plantation affairs and to the Ashley Phosphate Company of Charleston. Thomas Pinckney's letter book (1779-98) contains correspondence with Edward Rutledge relating to the military defense of South Carolina.

P45 Pine Grove Furnace collection, 1789-1914. 40 ft.
In Pennsylvania Historical and Museum Commission collections (Harrisburg)
Business records of the Pine Grove Furnace and related companies, including Laurel Forge, South Mountain Mining Company, Fuller Brick and Slate Company, Cumberland Furnace, Hunter's Run and Slate Belt Railroad, and Gettysburg and Harrisburg Railroad. Persons represented include Jonathan M. Butler, Michael Ege, Peter Ege, Horace A. Keefer, J. D. North, M. Watts, and Edward V. William.
Unpublished listing in the repository.

P46 Pinkham (Lydia E.) Medicine Company.
Records, 1776, 1859-1968. 218 ft.
In Radcliffe College, Schlesinger Library on the History of Women in America (Cambridge, Mass.)
Chiefly financial, advertising, and general records of the company at Lynn, Mass., and its branches in Mexico and Canada. Includes correspondence, research studies, manufacturing, labeling, and packaging records, journals, inventories, payrolls and tax records, contracts, litigation records, court suits, and other legal papers, advertising material, family papers, and photos. Persons represented include Lydia (Estes) Pinkham, her sons, William, Daniel, and Charles, and her daughter, Aroline (Pinkham) Gove and grandson, Arthur Wellington Pinkham.
Access restricted.
Information on literary rights available in the library.
Gift of Charles H. Pinkham, 1968 and 1971.

P47 Pioneer Reduction Works, Nevada County, Calif.
Records, 1898-1926. ca. 3500 items.
In California State Library (Sacramento)
Correspondence and receipts of a gold and silver ore processing company.
Unpublished finding aid in the repository.
Gift of Sven Skaar, Nevada City, Calif., 1950.

P48 Plattner, Karl Friedrich, 1800-1858.
Papers, 1854-1855. - 0.5 linear ft.
In Yale University Library, Department of Manuscripts and Archives (New Haven, CT).
Metallurgist. - Three volumes of manuscripts by Plattner on metallurgy and smelting. One manuscript, "Vorlesurgen uber allgemeine Huttenkunde," was published in 1860.
Unpublished finding aid available.

P49 Playfair, Lyon, Baron, 1818-1898.
Lectures, 1861-62. - [14], 368, [3] p.; 19 cm.
In National Library of Medicine (Bethesda, MD).
Lectures on organic chemistry, Edinburgh, 1861-62. Signature of T. L. Brunton on flyleaf.

P50 Polanyi, Michael, 1891-1976.
Papers, 1913-1975. - 22.5 linear ft.
In University of Chicago Library.
Physical chemist, writer on social thought and economics.- Manuscripts, notes and correspondence with leading chemists, most notably Fritz Haber, Georg Bredig, Kasimir Fajans, Herbert Freundlich, Arthur Allmand, Sir Christopher Ingold, Sir Cyril Hinshelwood, Sir Erick Ridial, Juro Horiuchi, Alexander Frumkin, and Nicolai Semenoff.
Unpublished guide.

P51 Pontiac, Michigan. Weed's Drug Store.
Records, 1891-1908. - 2 v.
In University of Michigan, Bentley Historical Library (Ann Arbor).
Business records of Weed's Drug Store, Pontiac; prescription file book, miscellaneous memoranda, and scrapbook of recipes for patent medicines, dyes, etc.

P52 Port Stanley Kelp Plant, Washington.
Manuscript. - 0.1 cu. ft.
In Lopez Island Historical Society (WA).
An unpublished history of the Port Stanley kelp plant on Lopez Island, which operated in 1917-18 and produced potash for gunpowder in World War I; this historical sketch was compiled from the memory of L.C. Norderer, son of the plant's builder.

P53 Porter, G. Harvey, 1893-1978.
Papers, 1919-1977. - 22 boxes; ca. 500 items.
In Maryland Historical Society Library (Baltimore).
Executive officer of the Industrial Corporation of Baltimore, a firm specializing in financing, reorganization, and management counseling of small businesses and industry. - The collection consists of miscellaneous articles, speeches, and reports written or collected by Porter. The reports were prepared for over 200 companies, including rubber companies; metal companies; breweries; oil and stearic acid manufacturers; pottery and concrete companies; paper manufacturers; iron and steel companies; glass

manufacturers; paint manufacturers; food processing industries; plastics industries; canning companies; and enameling companies.
Gift of Benenia S. Porter, March 1, 1979.
Unpublished finding aid available.

54 Potosi Brewing Company, Potosi, Wis.
Records, 1848-1973. 38 ft.
In State Historical Society of Wisconsin, Area Research Center at Platteville.
Correspondence, minutes, ledgers, advertising material, legal records, tax returns, financial records, general reference files, photos, and other records, of the brewer of Potosi, Holiday, Alpine, and other beers.
Unpublished finding aid in the repository.
Gift of Edward Ragatz, Potosi, Wis., 1975.

55 **Potts family.**
Mary Ann Furnace account books, 1826-89.
28 v.
In Historical Society of Pennsylvania collections.
Ledgers (1833-38), daybooks, blast book (1826-37), settlement book, memorandum book of other furnaces (1827-34), pig iron book, cordwood book, provision book, receipt books, and other records and accounts. Almost all of the records are for the 1830's, when the furnace was owned by Joseph D. Potts; later records consist of a discharge book (1885-89)
Presented by George Thompson, agent for the executors of John Savage, 1950.

56 Potts, Joseph D.
Student notes, 1849. - 1 v.
In Massachusetts Institute of Technology, Institute Archives and Special Collections (Cambridge, MA).
The volume is entitled "Notes on Chemistry," and is signed Joseph D. Potts, December 4, 1849. The volume also includes sections entitled "Notes on Analyses," "Quantitative Analyses," and "Notes on the Blowpipe."

57 Potts, William John.
Papers, 1864-1875 and n.d. - 1 folder.
In Historical Society of Pennsylvania (Philadelphia).
Analytical chemist. - Chemistry examination on lead, two letters, and data on analyses done.

58 **Potts, William McCleery,** *b.* 1856.
Records of the Isabella Furnace, 1880-1921.
ca. 18,000 letters and 80 v.
In Historical Society of Pennsylvania collections.
Correspondence and business records of the proprietor of the Isabella Furnace, one of the last of the charcoal burning furnaces in the U. S.
Presented by William Wikoff Smith, 1944.

59 **Powers & Weightman.**
Records, 1838-99.
2 ft.
In Harvard Business School, Baker Library.
General accounts of a company which made chemicals.
Described in List of business manuscripts in Baker Library, compiled by Robert W. Lovett. 2d ed. (1951)
Gift of Merck & Co., 1940.

Prange, Louis H   1884-1957.   P60
Papers, 1952-57. 2 boxes.
In State Historical Society of Wisconsin, Area Research Center at Milwaukee.
Dairy farmer and State legislator, of Sheboygan, Wis. Correspondence and press clippings, relating to Prange's service in the Wisconsin Senate and his political campaigns. Includes discussions of highway safety, expansion of State traffic patrol, raising the drinking age, daylight savings time, right to work and full train crew laws, proprietary drugs, chiropractic medicine, Wisconsin Radio Council, and artificial food coloring in canned foods; and clippings relating to Prange's proposed anti-subversive activities law.
Gift of Henry C. Prange, Plymouth, Wis., 1960.

Pratt, Zadock, 1790-1871.   P61
Pratt-Ingersoll family papers, 1819-98. 5 ft.
In Boston University Library.
Farmer, merchant, and U. S. Representative from Prattsville, N. Y. Correspondence, clippings, and other papers of Pratt and his immediate family. Includes scrapbooks containing records of Pratt's political career in New York State and in Congress; account books (1819-39) and other records of his tannery business; printed copies of Pratt's many addresses; and records of his son George Pratt of the 20th New York Infantry. Correspondents include James Buchanan and Jay Gould.
Inventory available in the library.
Information on literary rights available in the library.
Gift of Ralph Ingersoll, 1965.

Preisler, Paul William, 1902-1971.   P62
Papers, 1902-71. 16 ft.
In University of Missouri-St. Louis, Library.
Biochemist, lawyer, and labor leader, of St. Louis, Mo. Correspondence, minutes, legal records, business papers, political campaign material, news clippings, photos, movies, memorabilia, and other material, relating to Preisler's life and career. Includes material on his youth, education, and legal practice, briefs of cases filed or defended by him, such as civil liberties issues, legislative redistricting, nonpartisan candidacies, and test cases of tax issues, his political campaigns (1937-71), American Civil Liberties Union, American Federation of Teachers, Local 420, St. Louis, and City Central Committee of the Socialist Party of St. Louis.
Information on literary rights available in the repository.
Gift.

Prenderghast.   P63
Notebook, 1817. - 35 pp.
In University of Michigan, William L. Clements Library (Ann Arbor).
Philadelphia pharmacist's recipe and account book.

P64 Priestley, Joseph, 1733-1804.
Papers, 1766-1803. – 2 boxes and 2 reels of microfilm.
In American Philosophical Society (Philadelphia, PA).
Educator, scientist, theologian and discoverer of oxygen. – Manuscripts and photostats of manuscripts in the Municipal Library of Warrington, England dealing with science and chemistry among other topics.
Described more fully in a forthcoming guide to manuscript collections at the Society.

P65 Priestley, Joseph, 1733-1804.
Papers, 1777-1948.
ca. 125 items.
In Pennsylvania State University Library.
English clergyman and chemist. Correspondence, two drafts of Priestley's autobiography, a play, sermon, will, and account book. Four letters by Priestley include one to Josiah Wedgwood. Other correspondence (ca. 1800-1948) relates to the condition and disposition of Priestley's house at Northumberland, Pa.
Unpublished guide in the library. Also described in The headlight on books at Penn State, v. 18, no. 1, p. 4-5.
Gift of the estate of Gilbert C. Pond, and purchase, 1936-61.

P66 Priestley, Joseph, 1733-1804.
Priestley family collection, 1717-1935. ca. 2 ft.
In Dickinson College Library (Carlisle, Pa.)
English clergyman and chemist. Twelve letters by Priestley; papers, engravings, and memorabilia concerning him; letters of his son and grandchildren, mostly between Birmingham, Eng., and Northumberland, Pa.; and other family papers, including those of Joseph Priestley Button and Hilaire Belloc.
Card catalog in the library.
Gift of Mrs. Temple Fay, daughter of Joseph Priestley Button, 1964.

P67 Priestley, Joseph, 1733-1804.
Letters, 1766-1803.
In Library of Congress, Manuscript Division (Washington, DC).
Microfilm of originals in Dr. Daniel Williams' library, London.

P68 Priestley, Joseph, 1733-1804.
Papers, 1766-1803. – ca. 100 items.
In University of Pennsylvania Libraries, E.F. Smith Collection (Philadelphia).
Noted English chemist and discoverer of oxygen who fled to Pennsylvania from persecution in England. – Approximately 100 letters and documents, some photostats; letters are to Jefferson, Franklin, and John Wilkinson among others.

P69 Principio Company.
Records, 1724-1903. ca. 3 ft.
In Historical Society of Delaware collections (Wilmington)
Cashbooks, ledgers, daybooks, waste books, provision and smith books, sales books, memorandum book, and coal and time book pertaining to the Northeast Forge, Kingsbury Furnace, Lancashire Furnace, and Accokeek Furnace. Includes a book of stock certificates (1878-1903) for the McDaniel and Harvey Company and a book of letters (1866-83) to Joseph Shields Wilson regarding the use of his patents.
Described in Guide to the Manuscript Collections of the Historical Society of Pennsylvania (1949), entry no. 1555.
Information on literary rights available in the repository.
Deposited by Joseph S. Wilson; on permanent loan from the Historical Society of Pennsylvania.
This entry replaces MS 61-371.

Professional Circle, Philadelphia, Pa. **P70**
Records, 1951-74. 1 box.
In Philadelphia Jewish Archives Center (Pa.)
Correspondence, history, minutes, constitution, ledger, membership lists, and newsletters, of a fraternal and beneficial organization of Philadelphia physicians, dentists, and pharmacists.
Unpublished container list in the repository.
Gift of Dr. Louis Helfand.

Proskauer family. **P71**
Papers, ca. 1814-1970. – Ca. 31 items.
In Leo Baeck Institute (New York, NY).
Papers of Gottschalk Rosenberg, b. 1814 in Holland (at that time East Prussia). Includes birth certificate and other documents relating to family history. Papers of Rosenberg's grandson, Bernhard Proskauer, a chemist (1851-1915), of Berlin. Proskauer's papers include school reports, university notebooks, university reports, commissions as professor at Institute for Infectious Illnesses in Berlin; order books; letters; certificates from medical institutions at which Proskauer studied; awards and medals conferred on Proskauer; photos; album containing funeral oration and other posthumous honors. Papers of Proskauer's son, Arthur, a doctor (1880-1960), including letter of resignation from Beth Israel Hospital in New York, 1958; letters to his wife; and photographs. In German.
Unpublished finding aid available.

Pugh, Evan, 1828-1864. **P72**
Papers, ca. 1850-1863. – 9 ft.
In Pennsylvania State University Libraries (University Park).
Agricultural chemist; professor. – Pugh studied agricultural chemistry in Germany at the University of Gottingen, and in England under Lawes. He taught agricultural chemistry, as well as other scientific subjects, at Penn State. The collection contains correspondence, diaries and journals, experimental records and observations logs, and lecture notes.

R1 Rabbitt, James Aloysius, 1877-1969.
Papers, 1895-1969. 20 ft.
In Hoover Institution on War, Revolution and Peace, Stanford University (Calif.)
Consulting engineer. Correspondence, lectures, reports, surveys, patents, news clippings, sketches, and photos, concerning economic, scientific, and technological developments in the mining and metallurgical industries of China, Japan, and the Far East, with emphasis on nickel alloys, Chinese labor, and dockyard projects. Includes material on the governments of China and Japan, the cultivation of rice in Japan, and a proposed book ms. entitled Yankee Engineer in Asia.
Unpublished preliminary inventory in the repository.
Gift of Mrs. Rabbitt, 1971.

R2 Rabinowitch, Eugene I., 1901-1973.
Papers. - 10 linear ft.
In University of Chicago Library.
Senior Chemist and Section Chief, Metallurgical Laboratory, Manhattan Project (1944-46). - Correspondence with international scientific figures about nuclear energy and its social and political implications. Records of Pugwash Conferences (1954-71) to promote international control on nuclear energy. Correspondence relating to The Bulletin of the Atomic Scientists, which he helped found.
Unpublished guide available.

R3 Randall (W & R), *Cortland, N. Y.*
Records, 1809-66. ca. 7 ft. (60 items)
In Cortland County Historical Society collections (Cortland, N. Y.)
Invoice books, ledgers, daybooks, merchandise lists, and other records (1809-25) of the W & R Randall general store, operated by William Randall (1782-1850) and his brother, Roswell Randall (1786-1871); records (1825-43) of W. Randall & Company; distillery records (1811-18); farm records (1855-66); legal records, including docket books, judgments, lists of bills given to justices for collection, and unrelated attorney's records of court cases; cancelled promissory notes; and other miscellaneous records.
Index to a ledger for 1813-14 and unpublished register in the repository.

R4 Raskob, John Jakob, 1879-1950.
Papers, 1900-56. - Ca. 400,000 items.
In Hagley Library (Greenville, DE).
Business executive and financier. - Correspondence, financial records, pamphlets, newspaper clippings, printed matter, and other papers, covering Raskob's business career with E. I. du Pont de Nemours & Company and General Motors Corporation, the development of his personal fortune, his political activities in the Democratic Party, and his civic and philanthropic interests. Besides the Du Pont Company, other companies represented in his papers which are of interest to historians of the chemical and chemical process industries are the New York Insecticide Company; Raskob Mining Interests, Inc.; amd the Wilmington Sugar Refining Company.
Described in A Guide to the Manuscripts in the Eleutherian Mills Historical Library by John B. Riggs (1970), pages 916-964.

R5 Reason, Walter M
Papers, 1912-14. 750 items.
In University of Michigan, Bentley Historical Library, Michigan Historical Collections (Ann Arbor)
Court reporter for 6th Circuit Court, Pontiac, Oakland County, Mich. Correspondence concerning Reason's invention of a puncture proof inner tube for automobile tires.

R6 Recombinant DNA Controversy.
Collection, 1972-. - 6.5 ft., 200 tapes, and 45 transcripts.
In Massachusetts Institute of Technology, Institute Archives and Special Collections (Cambridge).
This collection was organized under the direction of MIT professor Charles Weiner to document the controversy surrounding recombinant DNA. The collection covers the development of policies for this technique at the local, state, federal, and international levels. Includes interviews with various women who played a significant role in the controversy; with members of the Cambridge Laboratory Experimentation Review Board; and with nurses, social workers, city officials, and others. In addition, the collection contains correspondence; memoranda; minutes; conference proceedings; hearings transcripts; petitions; briefs; committee rosters; draft reports; a press file; videotapes; and other material.
Unpublished finding aid available.

R7 Reese, Charles Lee, 1862-1940.
Papers, 1880-1942. - 112 items.
In Hagley Library (Greenville, DE) (1590).
Chemist and Du Pont industrial research director. - Fifteen notebooks from years as a chemistry student at Johns Hopkins University, the Universities of Virginia and Heidelberg, 1880-85. Drafts of lectures and publications on industrial research; some correspondence related to these drafts and to professional honors; and biographical and genealogical data.
Unpublished finding aid available.

R8 Reese, Charles Lee, Sr.
Papers, 1775-1950. 8 boxes.
In Historical Society of Delaware (Wilmington).
Chemist; president of the ACS, 1932-34. - Collection includes family letters; record book of family deaths; University of Virginia annual reports, 1891, 1892, 1985; journals, including Chemical News Journal, 1884; reprints by C. L. Reese on petroleum, acids, and explosives, 1898-1924; manuscript composition on illuminating gas by C. L. Reese, 1878; personal correspondence of C. L. Reese; correspondence between Reese and Du Pont, 1902-1930; U.S. Patent Office abstracts on method of purifying burner gas, by C. L. Reese, assigned to Du Pont, 1911-1913; letters relating to Reese's term as president of the ACS, 1932-34; letters relating to the American Institute of Chemical Engineers, 1929-34; loose photographs, photograph albums, and negatives; and scrapbook.

R9 Reisch Brewing Company (Springfield, IL).
Records, 1862-1961. - 21 boxes & 219 volumes.
In Illinois State Historical Library (Springfield).
Company founded in 1849 by Franz Sales Reisch (1809-1875), a German immigrant, and operated by Reisch family until its closing in 1966. - Brewery records include correspondence; reports giving results of chemical analysis of yeast, hops and beer samples; a subject file; audits and other financial reports; letterpress books; day books; cash books; bottled beer sales books draft beer account ledgers; and Internal Revenue books.

R10 Remsen, Ira, 1846-1927.
Correspondence, 1879-1933. 940 items.
In Johns Hopkins University Library.
Professor of chemistry and president of Johns Hopkins University. Includes much material on saccharin and on benzoates and benzoic acid.

R11 Rensselaer Polytechnic Institute School of Science.
Papers, 1913-1981. - 12 linear ft.
In Rensselaer Polytechnic Institute Archives and Dept. of Special Collections (Troy, NY).
Correspondence, memoranda, minutes, reports, lectures, academic programs, of the school. Also included is information on various departmental committees, long range plans of the school, and various departmental newsletters.
Access restricted.
Unpublished finding aid available.

R12 Repauno Chemical Company, Repauno, New Jersey.
Payroll record book, 1882-1884. - 1 volume.
In Hagley Library (Greenville, DE).
Book covers the period March 1882 to December 1884, but lacks entries for the period April through August 1883. Recorded are the names of workers, work assignments, rates of pay, hours, and total hours and pay.

R13 Resinol Chemical Company, Baltimore, Md.
Records, 1895-1950. 168 items and 105 v.
In Maryland Historical Society Library (Baltimore) (2178)
In part, photocopies.
Letter books (1896-1911), salesbooks (1896-1918), journals (1895-1943), ledgers (1895-1918), subsidiary ledgers (1898-1918), cash journals (1897-1925), and cashbooks (1895-1946), of a company manufacturing skin care products developed by Dr. Merville Hamilton Carter; together with 3 account books (1899-1901, 1929-50) of Royal Drug Company, Baltimore, Md., acquired by Resinol; and sales summaries, balance sheets, profits and loss statements, lists of creditors, advertising pamphlets, composition book, letters of endorsement, photos, and other records, of both companies.
Unpublished inventory in the repository.
Gift of the company, 1975.

R14 Rheineck, Alfred E., 1904-1971.
Papers, ca. 1940-1968. - 11.6 ft.
In North Dakota State University Archives (Fargo, ND).
Professor and Chairman, NDSU Department of Polymers and Coatings, 1958-1971. - Collection consists of his professional files dealing with organizations and private companies, such as Du Pont, NASA, and the USDA, concerning the exchange of data and other information in the areas of polymers and coatings. The research files contain information on various compounds (alkyd resins, polymerized oils, lacquer resins) documented by his own research and from other sources. In addition, there are copies of talks given, manuscripts of his numerous publications, some NDSU academic papers, and records concerning the symposia on New Coatings and New Coatings Raw Material, sponsored by the Department, 1959-1970.
Deposited by the Polymers and Coatings Department, 1972.
Unprocessed; available for research.

R15 Rhinelander family.
Business records, 1771-1829. 55 v.
In New-York Historical Society collections.
Letter books, daybooks, cashbooks, ledgers, inventories, order and memorandum books, and other account books of Frederick Rhinelander (1742/3-1805), Phillip Rhinelander (1756-1822),

and William Rhinelander (1753-1825), brewers and merchants of crockery, cutlery, glassware, and sugar, of New York City. Includes receipt book (1779-81) of John McAdam and Company, New York City; ledger (1771-79) of McAdam and Watson, New York City; and account books of the brigs Sally and Juno.
Unpublished list in the repository.
Gift, 1974.

R16 Rho Chi Society.
Records, 1908-1971. - 20 boxes.
In State Historical Society of Wisconsin (Madison).
National pharmaceutical honor society. - Collection includes administrative records and correspondence of the national office, materials relating to the annual conventions, and records pertaining to local chapters.
Gift of the American Institute of the History of Pharmacy, Madison, Wisconsin, 1970, and by David L. Cowen, New Brunswick, New Jersey, 1972.
Unpublished finding aid available.

R17 Rhoads (J. E.) & Sons.
Records, 1727-1962. 200 ft.
In Eleutherian Mills Historical Library (Greenville, Del.) (290, 923, 1080, 1156)
Correspondence (1867-1961), some relating to European leathers, minutes and notes of meetings, property records and appraisals, licenses, daybooks (1727-1896), ledgers (1851-1954), journals (1877-1954), cashbooks, wage, payroll, and banking records, purchase and receiving records, time books (1900-08), laboratory test notebooks (1913-48), sales records, data on company history, and printed commercial and advertising material, relating to tanning and the manufacture of leather belting by a firm with factory and headquarters in Wilmington, Del. Includes records of the company's predecessor, Rhoads & McComb, personal correspondence of Jonathan E. Rhoads (1830-1914) and his descendants, and records of the Rhoads family. Other persons represented include John B. Rhoads (1865-1911), George A. Rhoads (1860-1937), William E. Rhoads (1870-1945), and Joseph Edgar Rhoads (b. 1883).
In part, described in A Guide to the Manuscripts in the Eleutherian Mills Historical Library, by John B. Riggs (1970) p. 965-967.
Open to investigators under library restrictions.
Gift, 1966; and deposits, 1963-65, 1968-71.

R18 Richards, Ellen Henrietta (Swallow), 1842-1911.
Papers, 1895-1946. - 1 document box and 4 v.
In Smith College Archives, Sophia Smith Collection (Northampton, MA).
Industrial chemist; specialist in food technology, ecology, and public health; first woman graduate of and first woman professor at the Massachusetts Institute of Technology. - Collection contains correspondence and essays relating to her founding in 1873 of the Society to Encourage Studies at Home for women and girls and to her founding of the Association of Collegiate Alumnae. Also contains photos, tests, diagrams, drawings, and reprints of scientific writings.
Published guide available.

R19 Richards, Theodore William, 1868-1928.
Papers, 1889-1928. 7 ft.
In Harvard University Archives.
Professor of chemistry at Harvard University. Personal and professional correspondence, teaching mss., lecture mss., and clippings.
Access restricted until 1978.
Information on literary rights available in the repository.

R20 Ricketts, Bernard G., 1914-1976.
Papers, 1938-75. - 1 lin. ft.
In Washington State University Libraries, Manuscripts Division (Pullman).
Professor of Metallurgical Engineering at the University of Illinois and a graduate of Washington State University. - Curricular, research and student papers.
Container list at repository.

R21 Ridenour, Louis Nicot, 1911-1959.
Papers, 1946-50. 2 ft.
In University of Illinois, University Archives.
Dean of the Graduate College and professor of physics at the University of Illinois. Correspondence, reports, publications, book reviews, and other papers relating to scientific research projects, military research and development, nuclear energy, atomic and hydrogen bombs, international understanding, loyalty and security, Ridenour's consulting work, metallurgy, solid state physics, and various Government and private organizations involved in scientific research. Correspondents include Hans A. Bethe, Vannevar Bush, Karl Compton, I. Bernard Cohen, James H. Doolittle, Lee Du Bridge, Edward M. Earle, H. H. Harris, James R. Killian, David Lilienthal, Karl Lark-Horovitz, Charles C. Lauritsen, Leon Linford, Wheeler Loomis, Carl Overhage, I. I. Rabi, Frederick Seitz, George D. Stoddard, George E. Valley, T. F. Walkowicz, Warren Weaver, and Raymond Woodrow.
Received 1964.

R22 Ridley, Grahame Brooke, 1885-1970.
Papers, 1920-70. 7 ft.
In University of Wyoming, American Heritage Center (Laramie)
Forms part of the repository's Business History Research Center collection.
Engineer and inventor. Correspondence, notebooks, personal papers and photos, patents, blueprints, maps, and other material, relating to city water systems, conservation, irrigation, soil classification, roads, walnut growing equipment, cane sugar refining, fruit packing, California mining, gasoline filters, clutches,

carburetors, drying apparatus, dehydrators, windows, plant design and operation, testing airplanes and cars, and Electric Motor Chair Company (1920). Many of the letters were written by Ridley to his wife during his extensive travels, including Ireland during World War II, where he worked for Lockheed Aircraft.
Unpublished finding aid in the repository.
Gift of Mr. Ridley's daughter, Mrs. J. Kenneth Brooke, 1974.

R23 Riegelman, Sidney, 1921-1981.
Papers, 1960-1981. - 13 cartons.
In University of California, San Francisco Archives.
Professor of Pharmacy and Pharmaceutical Chemistry at UCSF. - Collection contains correspondence, manuscripts, lecture notes, laboratory notes, records, and computer printouts, relating to research in pharmaceutical chemistry and pharmacokinetics, faculty matters, and teaching.

R24 Rising, Willard Bradley, 1839-1910.
Papers, 1866-1908. - 11 boxes and 2 cartons.
In University of California, Berkeley, Bancroft Library.
Professor of Chemistry at UC Berkeley; State Analyst of California. - Correspondence, writings, speeches, photographs, laboratory notebooks, and other papers relating to chemistry and his career at UC Berkeley. Included also are various records from the Department of Chemistry and notebooks which contain both personal business as well as scientific data concerning his work with explosives, etc.
Partially processed collection; folder list available in repository.

R25 Roberts, Milnor Oakes, 1877-1965.
Papers, 1881-1965. 30 ft.
In University of Washington Library (Seattle)
Professor of mining engineering and dean of College of Mines, University of Washington. Personal and administrative correspondence, consulting files, financial records, maps, ephemera, and photos, relating to Roberts' work as dean of College of Mines and as a consultant, and to his financial affairs. Includes material relating to American Institute of Mining Engineers, American Mining Congress, Mining and Metallurgical Society of America, the periodical Mining Engineering, and Northwest Mining Association. Correspondents include Ralph Arnold, John Casper Branner, Henry Broderick, Enoch A. Bryan, George Watkin Evans, Frank Pierrepont Graves, Joshua Green, Herbert Hoover, David Starr Jordan, Thomas F. Kane, Norman F. Lovett, Charles F. Lummis, Carl E. Magnusson, Reginald H. Parsons, Thomas A. Rickard, Henry Suzzallo, Artemas Ward, Jr., George Livingston Yates, American Smelting and Refining Company, Children's Orthopedic Hospital, Seattle, and Young Men's Christian Association.
Unpublished finding aid in the repository.
Acquired from Mr. Roberts' estate, 1965.

R26 Roberts, Paul Henley, 1891-1971.
Papers, 1915-76. 3 ft.
In Nebraska State Historical Society collections (Lincoln)
Forester. Correspondence, mss. of writings, reviews of Roberts' books, newspaper clippings, photos, and other papers, relating to his career with U.S. Forest Service, much of it in Nebraska. Includes material relating to shelterbelt programs (1930's), forest and soil conservation, Emergency Rubber Project, Colorado Spruce Beetle Project, natural resources, and Robert's work on Missouri Basin Survey Commission (1952-53).
Unpublished finding aid in the repository.
Gift of Mrs. Paul H. Roberts, 1979.

R27 **Robinson, Wirt,** 1864-1929.
Diaries, 1883-1905. 29 v.
In U. S. Military Academy Library (West Point, N. Y.)
Army officer and professor of chemistry, mineralogy, and geology, at the U. S. Military Academy. Diaries containing descriptions of personal and professional military events with numerous, detailed observations of animal and plant life in the eastern United States, West Indies, and Panama. Documents, photos, and illustrations are inserted.
Gift of the U. S. Military Academy Association of Graduates.

R28 **Robison family.**
Papers, 1766-1865. 12 ft.
In Maine Historical Society Library (Portland)
Family and business correspondence, journals, record books, reports from ship captains and agents, ledgers, household accounts, bills, receipts, bankbooks, logbook (1847-50) for the schooner Sarah, legal documents, manifests, and other papers, relating chiefly to the business activities of Thomas Robison (ca. 1742-1806), a distiller and merchant in the triangle trade out of Portland, Me., and of his grandsons, Anson Smith Robison (1818-1844) and Robert Ilsley Robison (1808-1868?), also merchants in Portland. Includes material on the capture (1796) of Thomas Robison's ship Eliza by the British ship Unicorn, the legal action to regain her, and family matters.
Unpublished inventory in the library.
Information on literary rights available in the library.

R29 **Rochow, Eugene George,** b. 1909.
Papers, 1964; 1969-1979. - 5 linear ft.
In Harvard University Archives (Cambridge, MA).
Professor of chemistry. - Correspondence, ca. 1964-1970, contains letters to and from industrial and university laboratories and research agencies (Advanced Research Projects Agency, National Science Foundation), and individual researchers. Also includes

technical reports, lists of publications, and similar material. It reflects close relations with German scientists, and documents Rochow's activities in scientific societies, particularly the ACS, and as a scientific writer. The correspondence also reflects Rochow's involvement in academic life through letters of recommendation, letters to recruit teaching assistants, and rating sheets for prizes and awards. A separate series of teaching materials relates mainly to Chemistry I (later Natural Science 3) and includes examinations, laboratory instructions, and an outline of "Lecture Demonstrations." A third series contains two book manuscripts, one on descriptive chemistry, the other entitled <u>Inorganic Chemistry</u>. A final series contains two manuscripts (photocopies of typescripts) entitled <u>Germanium and Silicon</u>.

Gift of Professor Rochow, July 26, 1978.

Unpublished finding aid available.

Access restricted. Contact archivist for further information.

30 The Rockefeller Foundation.

Archives, 1909-(1912-1970). – 3300 ft.

In Rockefeller Archives Center (Pocantico Hills, North Tarrytown, NY).

Correspondence, memoranda, reports, surveys, minutes, program and policy files, administrative and financial records accumulated in the New York office. There are also records from the Foundation's office in Paris, 1917-1959, and records of the China Medical Board, 1913-1929, and the International Health Board, 1913-1927, which were part of the Foundation. Material related to the Foundation's grant-making activities in the fields of public health, medicine, medical education, natural and social sciences, arts, humanities, and agriculture, as well as to the Foundation's direct activities in public health and agriculture. Most files related to chemistry are dated between 1930 and 1960. Included are correspondence files concerning grants to support chemistry research and teaching at approximately forty institutions in the United States.

Records more than twenty years old are open for research.

Unpublished finding aids available.

31 Rockwell, Alfred Perkins, 1834-1903.

Notebook and diary, 1851, 1855. – 1 v.

In Yale University Library, Department of Manuscripts and Archives (New Haven, CT).

Student. – Manuscript notebook and diary kept while a student at Yale College (1855). The first 17 pages are a dairy for the period October-December 1851. The remainder of the volume is used for notes on chemistry and metallurgy lectures and experiments.

Rogers, William Barton, 1804-1882. R32

Papers, 1804-1919. – 16 boxes and 10 oversize items.

In Massachusetts Institute of Technology, Institute Archives and Special Collections (Cambridge).

Founder and first president of MIT; Professor of Chemistry and Natural Philosophy; State Geologist of Virginia and leader of Virginia's first Geological Survey. – The collection documents the life and professional activities of Barton and his three brothers, Henry Darwin, Robert Empie, and James Blythe, all of whom were scientists. The collection is divided into 2 groups. The manuscripts include correspondence, notes, memoranda, and drafts. The published material consists of publications on geology and related sciences by WBR, HDR, and RER; publications on geology by HDR; documents involved in the founding of MIT; miscellaneous documents found among the Rogers Papers; sources of biography of WBR; and clippings.

The strength of the collection is the extensive family and professional correspondence. WBR's correspondence, in particular, reflects his lectures in chemistry, physics, astronomy, geology, and natural history; his involvement with developing scientific societies; and his role in the founding, organizing, and administration of MIT.

Two additional collections should be consulted: the Rogers family Papers (1811-1904), consisting primarily of family correspondence; and the William Barton Rogers II Papers (1817-1919), made up of correspondence from Henry Darwin Rogers to his nephew WBR II (1833-1893).

Gift. The WBR Collection was donated, a portion at a time, by Mrs. Walter Humphreys, Miss Frances Porter, Mrs. Joseph A. Minott, Mrs. Emma Savage Rogers, and the Goodnow Library, Sudbury, MA, between 1941 and 1978. The Rogers family Papers were given to MIT in October 1976 by Mrs. Ernest C. Adams, and were transferred to the Institute Archives in 1977. The WBR II Papers were given to MIT in 1978 by Mrs. Franklin W. Hobbs.

Unpublished finding aids available.

R33 Rolfe family.
Papers, 1885-1972. 5 ft.
In University of Illinois at Urbana-Champaign Archives.
Correspondence, record books, mss. of writings, printed matter, clippings, photos, and other papers, of Charles Wesley Rolfe (1850-1934), professor of geology at the University of Illinois, and his daughters, Martha Deette Rolfe (1879-1971) and Mary Annette Rolfe (1881-1974). Includes correspondence and photos relating to Mary A. Rolfe's YMCA and Red Cross service in France during World War I; her tape-recorded recollections of the University of Illinois, alumni, and Thomas J. Burrill, Thomas Arkle Clark, and her father, Charles Wesley Rolfe; correspondence and records (1908-67) relating to farm property near Oswego, Ill.; research paper (1914-26) on salt; and papers relating to property in Champaign and Urbana (1927-60); farms in Alvin (1911-47) and Ludlow, Ill. (1907-51); University of Illinois Classes of 1900 (1954-64) and 1902 (1948-62); Congregational church (1915-64); ceramic engineering (1907-27); Champaign Neighborhood House (1902); trip to Alaska (1938); and golden wedding anniversary and estate of Charles Wesley Rolfe and his wife, Martha Kinsman (Farley) Rolfe (b. 1849).
Unpublished finding aid in the repository.
Information on literary rights available in the repository.
Acquired, 1968 and 1975.

R34 Rolston, William A., Sr.
Papers, 1911-1952. - 846 items, and 63 printed v.
In Louisiana State University, Department of Archives (Baton Rouge).
Papers of William A. Rolston, Sr. of Baton Rouge, alumnus of LSU, and Superintendent of the Francisco Sugar Company, Francisco, Province of Camaguey, Cuba.

R35 Roosevelt Family.
Papers, 1469-1962. - 46 linear feet.
In Franklin D. Roosevelt Library (Hyde Park, NY).
The collection contains correspondence, letter books, diaries, account books, daybooks, ledgers, journals, legal papers, logs, printed material, scrapbooks, newspaper clippings, and real estate and business papers of many Roosevelt family members--direct ancestors and other relatives of both Franklin D. and Eleanor Roosevelt. Of interest for the history of the chemical process industry are papers documenting the sugar refinery business in New York City (1750-90), conducted by Jacobus Roosevelt (1692-1776), Isaac Roosevelt I (1726-1794), and James Roosevelt II (1828-1900). Also contains Delano and Hall (Eleanor's maternal grandparents) family papers.
Gift of Franklin D. Roosevelt, 1942.
Open to investigators under library restrictions.
Finding aids available.

R36 Rose, Paul C.
Papers, 1901-1912. - 6 in.
In University of Wyoming Archives-American Heritage Center (Laramie).
Sequoia and Cities Service (refinery) employee. - Manuscripts, photographs, printed material pertaining to early time oil activities near Ponca City, Oklahoma.
Gift of Mr. Paul C. Rose, 1982.
Unpublished finding aid available.

R37 **Rose, William Cumming,** 1887-
Papers, 1923-66. 1 ft.
In University of Illinois, University Archives.
Professor of chemistry at the University of Illinois. Correspondence, University committee reports on food research, Krebiozen, and honorary degrees, publications list, list of doctorates supervised, mss. of published articles, reprints, photos, and other papers relating to University of Illinois affairs, the second International Congress of Biochemistry (1952), papers for professional meetings, lectures, publications, amino acids, threonine, the physiology of amino acid metabolism, nutritional and growth requirements, and Lafayette B. Mendel.
Unpublished finding aid in the repository.
Received 1966.

R38 Rosen, Ralph, 1904-1982.
Papers, 1924-1982. - 3 ft.
In University of Missouri-St. Louis Library, Western Historical Manuscript Collection and State Historical Society of Missouri Manuscripts.
Dentist, of St. Louis, Mo., and advocate of fluoridation of public drinking water. - Correspondence (1934-1981), writings (1924-1982), literature for and against fluoridation, bulletins of St. Louis Dental Society (1932-1958), newspaper clippings, photos, and oral histories, relating to the fluoridation controversy in St. Louis and Missouri, 1940s - 1980s.
Gift of Dr. Rosen, 1982.
Unpublished finding aid in the repository.

R39 **Rosengarten and Denis,** *Philadelphia.*
Records, 1818-53.
20,000 items.
In Historical Society of Pennsylvania collections.
Business papers of a firm of Philadelphia chemists, together with some family and personal correspondence in English and German.
Presented by G. D. Rosengarten, 1948.

R40 Rossini, Frederick Dominic, b. 1899.
Papers, 1948-1982. - 174 linear ft.
In University of Notre Dame Archives, (Notre Dame, IN).
Chemist and educator. - Chiefly professional correspondence and related printed matter. Includes material relating to the American Chemical Society, American Petroleum Institute, Argonne National Laboratory, International Council of Scientific Unions, International Union of Pure and Applied Chemistry, National Academy of Sciences, National Research Council, and National Science Foundation; and material relating to Rossini's activities on behalf of Catholic and cultural groups.
Gift of Frederick D. Rossini, 1971.
Access restricted.
Unpublished finding aid in the repository.

R41 Rothermel, Johannes.
Account book, 1800-1843. - 1 v.
In Merrimack Valley Textile Museum, (North Andover, MA).
Textile worker. - Account book, in German, records the purchase of textiles or the production of yarn or cloth. Also includes dye recipes.
Finding aid available.

R42 Roughton, Francis John Worsley, 1899-1972.
Papers, ca. 1920's-1960's. - Ca. 90,000 items (90 linear ft.).
In American Philosophical Society Library (Philadelphia, PA).
Biochemist. - Collection of correspondence and documents focuses on Roughton's prolific life's work on respiratory physiology. His specific work at Cambridge University from the 1920's to the 1960's, as well as interludes in the U.S. during World War II, is covered, as is the scientific milieu in which he worked. Some of the subjects or organizations in which he was involved and for which there are documents, are: Bermuda, and Naples Zoological Research Stations; British Glue and Gelatin Research Association; Biochemical Journal; Cambridge Philosophical Society; Cambridge University, Departments of Physiology (pre-1939) and Colloid Science (post-1940); Harvard Fatigue Lab.; Medical Research Council; Trinity College Cambridge. There is also material on the following individuals, among others: Joseph Barcroft, Britton Chance, John T. Edsall, J.B.S. Haldane, William Harvey, Lawrence J. Henderson, Archibald V. Hill, Frederich G. Hopkins, Hans A. Krebs, Joseph Needham, Linus Pauling, Max F. Perutz, Ernest Rutherford, P.F. Scholander, Jefferies Wyman.

R43 Roussel family.
Papers, 1835-88.
8 v.
In Historical Society of Pennsylvania collections.
Includes journals, chiefly concerning weather (1855-56), a ledger of Eugene Roussel (1855-75), ledger of Peter Perlet, clockmaker (1861-67, 1875-81), Notes et receittes de liqueurx de table (1835) by Eugene Roussel, notes on travel in North Africa and Europe (1872-73), and a volume of medical lectures and case notes (1883-88)
Presented in part by Miss M. D. Weihrnmeyer.

R44 Ruder, William Ernst, 1886-1963.
Papers, 1907-50. 7 ft.
In Union College Library (Schenectady, N.Y.)
Metallurgist, at General Electric Company, Schenectady, N.Y. Technical reports on filaments, magnetic properties, metallurgy and micrography, and magnetostriction.
Unpublished index in the repository.
Information on literary rights available in the repository.
On loan from Research and Development Center, General Electric Company, Schenectady, N.Y., 1976.

R45 Rumford Chemical Works.
Records, 1853-1951. - 30 ft.
In Rhode Island Historical Society (Providence).
The firm was founded in 1854 as Wilson, Duggan & Co. and underwent several changes of name and ownership, eventually becoming Rumford Chemical Works. It was dissolved in 1969. Its products included chemicals used in dyeing and printing, fertilizer, baking powder, yeast powder, phosphoric acid products, cream of tartar, sulphuric acid, and sodium phosphate glasses. - Collection contains administrative records, general accounts, purchasing and receiving records, production records, sales and shipping records, correspondence, and miscellaneous materials.
Gift of Dr. Karl Holst of Essex Chemical Company, 1976.
Unpublished finding aid available.

R46 Rumford Chemical Works.
Records, 1848-1949. - Ca. 256 vols.
In Rutgers University Library, Special Collections Department (New Brunswick, NJ).
Financial records of a chemical company. - Consists of general account and sales records in three interrelated groups: North Meacham & Co., 1848-1949; Geo. F. Wilson & Co., 1856-1860; and the Rumford Chemical Works, 1858-1948.
Donated in 1967 by the Essex Chemical Corporation, purchaser of the Rumford Chemical Works in 1966.
Unpublished finding aid available.

R47 Russell, Israel Cook, 1852-1906.
Papers, 1833-1904. - 20 items and 1 v.
In University of Michigan, Bentley Historical Library, Michigan Historical Collections (Ann Arbor).
Professor of Geology at the University of Michigan. - Correspondence on scientific topics and Portland Cement; biographical information on Bela Hubbard; photograph album, 1883, illustrating the U.S. Geological Survey.
Unpublished finding aid available.

R48 Russell Process Company.
Records, 1881-1901. - 1.75 linear ft.
In Yale University Library, Department of Manuscripts and Archives (New Haven, CT).
New York-based company which licensed patents for metallurgy, especially for extracting metals from ore. Collection contains correspondence, legal and financial records. Many of the papers originated in the office of the Secretary-Treasurer, Talcott H. Russell, attorney in New Haven.
Gift of Dr. Talcott H. Russell, 1966.
Unpublished finding aid available.

R49 Ruter, Charles.
Letter press book, 1888-95. 1 v.
In State Historical Society of Colorado collections (Denver)
Letters by an operator of sampling works at Pueblo and Aspen, Colo.

R50 Rutgers University. College of Agriculture and Experiment Station.
Records, 1885-1951. - 1 v. and 2 ft.
In Rutgers University Library, University Archives Collections (New Brunswick, NJ).
Record book of experiments at the State Experiment Station, 1885-1894; College and Experiment Station budgets and financial reports, 1914-1930; correspondence and legal documents relating to claim of Albert Schatz that he was co-discoverer with Selman A. Waxman of streptomycin, 1943-1951.
Unpublished finding aid available.

S1 Sacramento Valley Sugar Company (Calif.)
Records, 1905-1963. - 94 ft.
In California State University Library (Chico)
Firm engaged in general agriculture and sugar beet industry. - Annual reports, minutes, bylaws, payroll ledgers, water rights, leaseholds, maps, photos, and other material, of the company and its associated companies: Alta California Beet Sugar Company, Hamilton Land Company, Pacific Sugar Construction Company, and Sacramento River Farms. Subjects include labor records, railroad competition, eastern and oil money, land speculation, land distribution, agricultural mechanization, crop experiments, irrigation projects, creation and distribution of electric power, ethnic minorities, fossil fuels, alternative fuels, crop diseases, water transportation, roads, bridges, levees, sugar tariffs, sugar factory mechanization, town planning, Pacific Gas and Electric, and Southern Pacific Railroad. Includes material on Thomas Robert Bard, Edward L. Doheny, Elwood Mead, Capay Rancho (Mexican land grant in California) Stanford Vina Ranch (owned by Leland Stanford), Butte, Colusa, Glenn, Tehama, Sutter, and Yolo Counties, Calif., and Klamath County, Or.
Gift of Berylewood Investment Company, Somis, Calif.
Unpublished register in the repository.

S2 St. Amand, Clarence W.
Letter book, 1884-1881. - 1 v.
In South Carolina Historical Society collections (Charleston) (34-351)
Bookkeeper for Wyllie, Teacher & Gordon, phosphate rock dealers, Charleston, S.C. -Personal and business correspondence, including material on the phosphate industry at Oak Point Mines, Bolton Mine, and others.

S3 Sampson, Jesse, 1900-1965.
Papers, 1932-1963. - 0.3 ft.
In University of Illinois Archives (Urbana).
Professor of animal pathology and hygiene, veterinary physiology and pharmacology, and department head, at the University of Illinois.- Correspondence (1962-1963) relating to encyclopedia articles, course outlines (1946), reprints of published articles on ketosis in cattle and sheep, acid-base balance in cows and ewes, and disturbances of carbohydrate metabolism in animals.
Information on literary rights available in the repository.
Acquired, 1966.

S4 Sanger, Charles Robert, 1860-1912.
Papers, 1886-1911. - Ca. 1 linear ft.
In Harvard University Archives (Cambridge, MA).
Professor of chemistry. - Collection includes a scrapbook regarding arsenic wall paper; a notebook with notes on arsenic analysis; papers relating to research in St. Louis; an assistant analyst's notebook; and a general folder.

S5 Saunders family.
Papers, 1887-1958. 23 ft.
In Hamilton College Library (Clinton, N.Y.)
Personal and professional correspondence, particularly in conjunction with the Saunders' horticultural business, diaries, and research notes on peony hybridization. Family members represented include Arthur Percy Saunders (1869-1953), dean of Hamilton College, chemist, and horticulturist, and his wife, Louise Brownell Saunders (1870-1961), educator, College Hill, Clinton, N.Y. Includes ms. poetry of James Agee and Richard Eberhart. Correspondents include James Agee, Richard Eberhart, Ezra Pound, Elihu Root and family, B. F. Skinner, and Alexander Woollcott.
Unpublished finding aid in the repository.
Gift of Miss Silvia Saunders, 1976.

S6 Savery, Thomas H   1837-1910.
Papers, 1848-1946. ca. 2800 items.
In Eleutherian Mills Historical Library (Greenville, Del.) (various accessions)
In part, microfilm (negative) and photocopies made in 1962-63 from originals held by a private owner.
Manufacturer, inventor, and business executive, of Wilmington, Del. Business, personal, and family correspondence, diaries (49 vols., 1864-1910), copybooks and ledgers (1848-1912), original and published patents (2 vols., 1868-1906) issued to Savery for papermaking machinery, and other papers. Includes data (1892-1903) on the papermaking patents of Eügen Füllner and others, on Fourdrinier paper machines, and on combustion systems of various American railroads; correspondence, reports, minutes, photos, and other papers, relating to various companies, including the Delaware Specialty Company, Harper's Ferry Electric Light and Power Company, Harper's Ferry Paper Company, Parsons Engineering Company, Pusey & Jones Company, of which Savery was president (1898-1907), Shenandoah Pulp and Paper Company, Harper's Ferry, W. Va., York Haven Paper Company, and the York Haven Water and Power Company, Philadelphia; correspondence (1899-1935) of Savery's son, William H. Savery (1865-1949); diaries (1860-1927) of Savery's wife, Sarah (Pim) Savery; and a biographical sketch of Savery by his daughter, Anne P. (Savery) Thayer.
In part, described in A Guide to the Manuscripts in the Eleutherian Mills Historical Library, by John B. Riggs (1970) p. 970.
Open to investigators under restrictions accepted by the library.
Gifts and deposits, 1961-69.
The library also has microfilm of some of the originals in the collection.

S7 Sawyer, Francis A. and Jonathan.
Records. 1841-1899. - 4147 items and 2 v.
In Duke University Library (Durham, NC).
Business correspondence, bills, receipts, memorandum book, and samples of woolen materials, of a wool manufactory owned by Francis A. and Jonathan Sawyer. Subjects discussed include prices, chemicals, dyes, and the state of the market.
Card index in the library.
Purchase, 1948.

S8 Sawyer Woolen Mills, *Dover, N. H.*
Records, 1841-99. 4147 items and 2 v.
In Duke University Library (Durham, N. C.)
Business correspondence, bills, receipts, memorandum book, and samples of woolen materials, of a wool manufactory owned by Francis A. and Jonathan Sawyer. Subjects discussed include prices, chemicals, dyes, and the state of the market.
Card index in the library.
Purchase, 1948.

S9 Schaefer family.
Papers, 1849-1935.
50 items and 4 v.
In Minnesota Historical Society collections.
In part, transcripts (typewritten) of originals in the possession of Mr. and Mrs. Donald O. Winston, Los Angeles.
Correspondence, diaries, business records, and other papers. Includes diaries of Jacob and Sarah (Miller) Schaefer describing a journey from Rockford, Ill., to San Francisco in 1849 and life in Honduras in 1855; expense and prescription books (1866-73) of a drug firm owned by Charles S. Whitaker of Mankato, Minn.; MSS. telling of trips to Europe, Colombia, and Honduras (1924-35) by Francesca (Schaefer) Whitaker Winston and Mr. and Mrs. Donald O. Winston; letters (1929-32) written by a Miss Starr, describing life in Turkey and Iran; and a MS. (1933) by Mary Thayer Hale, entitled Early Minneapolis.
Descriptive inventory in the repository.
Gift of Mr. and Mrs. Donald O. Winston, 1954-49.

S10 Schanck family.
Papers, 1741-1906. 3 ft.
In Monmouth County Historical Association Library (Freehold, N.J.)
Correspondence, diaries, financial and legal records, estate papers, church records, genealogical notes, and other papers, of a Monmouth County, N.J., family, particularly Capt. John Schanck (1745-1834), Revolutionary soldier, fuller, tanner, cordwainer, farmer, and sawmill operator, DeLafayette Schanck (1781-1862), farmer and businessman, and Rev. Garret Conover Schanck (1806-1888), Reformed Dutch clergyman, farmer, and genealogist of early Dutch families in Monmouth County. Includes diary (25 v., 1826-88) and church records (1827-67) of Rev. Garret Conover Schanck, and material relating to political and national events, Monmouth County social life, customs, and agriculture, crafts of tanner, fuller, and currier, steamboats, slavery, and genealogy of Schanck/Schenck, Smock, Ryerson, and Mandeville families.
Unpublished finding aid in the repository.
Gift of Mrs. D.J. Romaine, Hackensack, N.J., 1960, and other sources.

S11 Schlossberger, Julius Eugen, 1819-1860.
Notebook, 1857. - 1 volume.
In American Philosophical Society (Philadelphia, PA).
German chemist. - Notebook of 272 pages in German on inorganic chemistry.

S12 Schlundt, Herman, 1868–
Papers, 1914-38.
1 v. and 1465 folders.
In University of Missouri Library, Western Historical Manuscripts.
Correspondence, a scrapbook, and other papers of a professor of chemistry at the University of Missouri, including articles, chemical pamphlets, athletic bulletins, financial papers, clippings, and reports of the Committee on Junior Colleges, of which Schlundt was a member. Correspondence relates to standards for junior colleges, assistantships at the University of Missouri, recommendations, health resorts, and spring waters. Includes letters from Excelsior Springs, Mo., Hot Springs, Ark., Waukesha, Wis., Idaho Springs, Colo., and Saratoga Springs, N. Y., requesting Schlundt to analyze the mineral content of water.
Gift of Anna Schlundt and the Chemistry Dept. of the University of Missouri.

S13 Schoch, Eugene Paul.
Papers, 1859-1967. - 31 boxes
In University of Texas at Austin, Eugene C. Barker Texas History Center.
Professor of Chemistry and Chemical Engineering, 1897-1938; Director, Bureau of Industrial Chemistry, 1928, UT-Austin. - Bulk of the collection is research materials, reflecting Schoch's interest in the production of acetylene, the production of reactive hydrocarbons from methane, and the effect of electric current on natural gas. Included are notes, notebooks, notecards, blueprints and drawings, and correspondence. Materials pertaining to the university itself include masters' theses and student reports, 1924-1944, faculty meeting minutes, 1945, and miscellaneous information about the honorary society Sigma Xi. Some printed material in the form of pamphlets, booklets, and magazine articles.
Unpublished finding aid available.

S14 School records, 1715, 1770-1963.
Ca. 3 ft.
In New Haven Colony Historical Society Library (CT).
Records of public and private schools and school associations in and around New Haven, Connecticut, along with students' notebooks In particular, 2 chemistry lab notebooks of Elizabeth M. Blakeslee, ca. 1890, consisting of 275 pages.
Unpublished finding aid available.

S15 Schrack (C.) and Company, Philadelphia, Pa.
Records, 1808-1938. ca. 200,000 items.
In Eleutherian Mills Historical Library (Greenville, Del.) (various accessions)
Correspondence, accounts, bills and receipts, stock books, formulae books (1844-1912), orders, shipping records, banking records, and other business records of a paint, varnish, and color manufacturing firm. Persons represented include Christian Schrack (ca. 1790-1854), founder of the firm, who began business as a carriage builder; his partner, Joseph Stulb (d. 1898); Stulb's sons, Edwin H. Stulb (1850-1920) and Joseph Stulb, Jr.; his grandsons, Joseph Reichert Stulb (b. 1883) and Edwin H. Stulb, Jr.; and Townsend Willits.
In part, described in A Guide to the Manuscripts in the Eleutherian Mills Historical Library, by John B. Riggs (1970) p. 970-971.
Gift, 1966, and purchases, 1965-68.

S16 Schuette, Henry August, b. 1885.
Papers, 1917-1960. – 8 boxes.
In University of Wisconsin Archives (Madison).
Professor of Chemistry. – General correspondence (3 boxes); letters of recommendation for former students written by Professor Schuette (1 box); subject files (4 boxes). Collection reflects Schuette's active interest in the field of foods, particularly honey and fatty oils. Also contains correspondence with the American Chemical Society, of which he became President in 1940.
Received in September, 1968.
Unpublished finding aid available.

S17 Schultz, Jack, 1904-1971.
Papers, 1920-71. – Ca. 25,000 items (27.5 linear ft.).
In American Philosophical Society Library (Philadelphia, PA).
Geneticist, biochemist. – Correspondence (18.5 boxes), manuscripts (lectures and articles), research grant material, research data, and some personal notes from his graduate school days. He obtained his AB, AM, and Ph.D. (1929) from Columbia University, where he was the last graduate student to get his doctorate under T.H. Morgan (for a recollection of his days in Morgan's "fly-room" see, Shultz to G.W. Beadle, 7-31-70); he also worked with Morgan at the California Institute of Technology (1929-36, 1941-42). Schultz's career centered on the study of the nature and function of the gene; chemical genetics of Drosophila; cytochemical and nutritional techniques; cytochemistry of growth; and the pattern of human chromosomes.
As a Rockefeller Foundation Fellow in 1937-39, he worked closely under Torbjorn Caspersson at the Karolinska Institutet in Stockholm, Sweden (there are 11 folders of correspondence with Caspersson, 1937-71). The remainder of his career was spent at the Institute for Cancer Research, Philadelphia, Pennsylvania, where he was Senior member and head of Department of Genetics and Cytochemistry (1943-57), and Chairman of the Division of Biology (1957-69). Much of his tenure at the Institute was spent as an administrator, rather than on original research; he had much success in choosing, encouraging, and stimulating a brilliant research staff. He did not publish prolifically, much of his work being known from his lectures and informal discussions (there are 7 boxes of his lectures and articles). There is significant material relating to his participation in professional organizations: American Society of Naturalists (President, 1968); Genetics Society of America (President, 1963); National Research Council; and the National Science Foundation.
Presented by Mrs. Helen Redfield Schultz, 1983.

S18 Schulz, Helmut William, b. 1912.
Manuscript, Sept. 15, 1940. – 28 cm.
In University of California, Berkeley, Bancroft Library.
Chemical engineer, organic chemist. – "Separation of the Uranium Isotopes by Centrifugation." Photocopy of an unpublished report (typescript) submitted to Harold C. Urey in 1940, outlining a possible method for the enrichment of U-235. Also included: a biographical sketch of Schulz, a photocopy of the letter to Urey attached to the original report and a letter, Jan. 29, 1974, from Schulz describing the report.
Description in repository.

S19 Schweitzer, Paul, 1840–
Papers, 1893-1902.
5 folders.
In University of Missouri Library, Western Historical Manuscripts Collection.
Correspondence and other papers of a member of the Chemistry Dept. of the University of Missouri, including letters from Armour and Co. relating to the manufacture of oleomargarine and commercial fertilizer, and a resolution of the university faculty concerning Schweitzer's resignation.

S20 Scott, Robert C.
Papers, 1933-1962. - 12 linear ft.
In University of Washington Archives (Seattle).
Research chemist, engineer, activist. - Reports, minutes, publications, and other materials regarding Scott's activities as a member of miscellaneous cooperative associations, 1949-59. Also, student papers from his engineering course at the University of Washington, 1933-36; and ca. 12 boxes consisting of correspondence, reports, photographs, and subject files related to his employment at Adhesive Products Company; his affairs with the American Chemical Society, Puget Sound Section; and the Washington State Farm Chemurgic Committee.
Accessioned 1965.
Unpublished finding aid available.

S21 Seaborg, Glenn Theodore, 1912-
Papers, 1961-63. ca. 1 ft. and 15 reels of microfilm.
In John F. Kennedy Library (Waltham, Mass.)
Educator. Correspondence, appointment books, telephone logs, and speech files, relating to Seaborg's service as chairman of the U.S. Atomic Energy Commission.
Access restricted.

S22 Seaton family.
Business records, 1788-1954.
7 ft. (1729 items and 158 v.)
In University of Kentucky Library.
Correspondence, diaries, business records, journals, scrapbooks, clippings, pamphlets, maps, autograph albums, and photo albums of William Biggs Seaton (1855-1927) and his family, representing their interests in the development of the iron industry in eastern Kentucky and southern Ohio, river and rail transportation, and the city of Ashland, Ky. The material is related to the library's Means family papers.
Indexed in manuscript catalog.
Acquired from Mrs. Isabel Seaton Humphrey and John Means Seaton, 1956.

S23 Seattle-King County Dept. of Public Health.
Records, 1951-52.
1 reel of microfilm (negative)
In State Historical Society of Wisconsin collections.
Correspondence and data collected by the Dept. of Public Health in regard to the proposed fluoridation of Seattle's water supply.

S24 Seiberling, Frank A   1859-1955.
Papers, 1897-1955.   200 ft.
In Ohio Historical Society collections (Columbus) (347)
Founder and president of Goodyear Tire and Rubber Company and Seiberling Rubber Company, of Akron, Ohio. Business and personal correspondence, financial records, legal documents, corporate and association records, and files on community activities.
Gift of the heirs of Frank A. and Gertrude F. P. Seiberling, 1970.

S25 Seibert, Florence Barbara, b. 1897.
Papers, 1920-1970's. - Ca. 5000 items (4 linear ft.).
In American Philosophical Society Library (Philadelphia, PA).
Biochemist. - Seibert spent her professional life researching the chemistry and immunology of tuberculosis and cancer, as well as doing pioneering work on pyrogens. The correspondence, reports, and other material document her work at Yale University, under Lafayette Mendel, and at the University of Chicago, under H. Gideon Wells, as well as at the Henry Phipps Institute at the University of Pennsylvania (1932-59). There are cancer research folders concerning her later work at the Mound Park Hospital Foundation and the Bay Pines V.A. Center, in St. Petersburg, Florida. There are, as well, substantial amounts on Goucher College (her alma mater); Lilly Research Labs.; Merck, Sharpe & Dohme; and Parke, Davis & Co.
Unpublished finding aid available.

S26 Seidell, Atherton, b. 1878.
Papers, 1899-1938. - 2 boxes.
In National Library of Medicine (Bethesda, MD).
Chemist. - Includes reprints (1899-1937), minutes of meetings of the Chemical Society of Washington (1904-06), and research notes. Seidell introduced microfilming into the Army Medical Library in the 1930s. The research notes were probably for a future edition of Seidell's book, Solubilities of Inorganic and Organic Compounds.

S27 Semet-Solvay Company.
Records, 1897-1942.
54 items and 17 boxes.
In Cornell University Library, Collection of Regional History and University Archives (430)
Correspondence, statements of investments and ownerships, descriptions of construction costs, materials and operations, reports, specifications and instructions, plans for proposed plants, designs, blueprints and photos. relating to the Semet-Solvay Co., Solvay Process Co.,Tennessee Products Corp., Indiana Harbor Co. Plants, Carbonifera de Sabinas Rosita, Mexico, and other by-products coke oven plants. Includes material on other industrial processes and correspondence of European industrial scientists.

Severinghaus, Elmer Louis, 1894-1980.
Correspondence, 1920-45. - Ca. 100 items.
In American Philosophical Society Library (Philadelphia, PA).
Biochemist. - Maps, letters to Severinghaus concerning diabetes, endocrinology, nutrition, and information about grants and funding support. Other correspondents, mostly represented by one letter, include Walter B. Cannon, Alexis Carrel, George W. Corner, Hugh S. Cumming, Morris Fishbein, George A. Harrop, Alex Hrdlicka, Alfred C. Kinsey, Jacques Loeb, Karl A. Menninger, Gregory Pincus, P.A. Shaffer, Isaac Starr Jr., Norman C. Wetzel, Russell M. Wilder.
Unpublished finding aid available.
Accessioned, 1976.

Shaffer, Philip Anderson, 1881-1960.
Papers, 1910-1958. - 7.7 ft.
In Washington University School of Medicine Archives (St. Louis, MO).
Biochemist and university professor. - Office correspondence from the Department of Biochemistry, Washington University School of Medicine, and diaries.
Gift of the Dean's Office, Washington University School of Medicine.
Information on literary rights available in the library.
Unpublished finding aid available.

Shamel, Charles Harmonas, b. 1867.
Papers, 1874-1949. - 16.3 ft.
In University of Illinois Archives (Urbana).
Patent and mining lawyer. - Correspondence, diaries, autobiography, notes, photos, and other papers relating to University of Illinois affairs, Shamel's career, the Board of Trustees, faculty appointments, chemistry, evolution, agriculture, military drill, student rebellions, social affairs, lectures, politics, religious meetings, farmers' institutes, alumni activities, legal and patent affairs, and travel. Correspondents include Grace Howe, Katharine Kennard, Robert Orr, Arch Shamel, and Clarence Shamel.
Received 1965.
Unpublished finding aid available.

Shaw, Albert Duane, 1841-1901. S31
Papers, 1891-96.
1 v.
In Cornell University Library, Collection of Regional History and University Archives (1909)
Consular official and business executive. Letter book, together with a few loose papers. The letters, mainly to business associates, are primarily concerned with Shaw's unsuccessful efforts to raise funds in England for building an electric railway from Queenston, Ont., to Niagara Falls Park, with the founding of the Canadian Niagara Power Co., of which Shaw was president, and with various prospective ventures, including the possible formation of a company to utilize an electro-chemical invention by Dr. Kellner. Other letters pertain to Shaw's activities as president of the Watertown, N. Y., Y. M. C. A., his speech at the Founder's Day exercises of Cornell University (1893), the eightieth birthday of Henry W. Sage, of Ithaca, N. Y., and to a painting of William Henry Miller, Ithaca architect.

Sheehan, John Clark, b. 1915. S32
Laboratory records, 1956-1980. - 63.75 linear ft.
In Massachusetts Institute of Technology, Institute Archives and Special Collections (Cambridge, MA).
Laboratory notebooks, graphs of spectral analyses, and chemical samples documenting the development of organic compounds. Included is documentation of research on synthesis of penicillin.
Unpublished finding aid and index available.

Shekerjian, Haig, 1886-1966. S33
Papers, 1905-66. 4 boxes.
In Hoover Institution on War, Revolution and Peace, Stanford University (Calif.)
U. S. Army officer. Correspondence, speeches, notes, medals, and photos, relating to Shekerjian's career as military attaché in Greece during World War I, with the Allied Armies of the Orient, and as Chemical Officer, First Army, Commanding General of the Chemical Warfare Replacement Center, and Commanding General of Camp Silbert, Alaska.
Unpublished inventory in the library.
Gift of Mrs. Helen Shekerjian, 1972.

Shelby Iron Company, *Shelby, Ala.* S34
Records, 1862-1923.
ca. 494,000 items.
In Alabama University Library.
In part, transcripts.
Correspondence; 64 volumes of minutes (1862-90) of directors and stockholders; manufacturing records comprising charcoal reports, daily stable reports, daily cost of ore mining, time books, and payrolls in each department; commissary records (1868-); grist mill toll books; furnace record books; and other records. The material is most complete for the period 1862-90. Of particular interest are records of difficulties with the Confederate government over conscription of workers and securing supplies to maintain production. Includes records of the company's subsidiary formed in 1890 to handle its real estate business, the Shelby Manufacturing and Improvement Company.
Unpublished guide in the library.
Open to investigators under library restrictions.
Information on literary rights available in the library.
Gift of Woodward Iron Company, 1948.

S35 Shenk, John Wesley, 1875-1959.
Papers, 1900-35. 13 ft. (ca. 18,000 items)
In Stanford University Libraries (Calif.)
California Supreme Court justice. Correspondence, printed material, photos, and other family documents, relating to Judge Shenk's family's chemical firm, land in Imperial Valley, Calif., and mining properties, and charitable, civic, and fraternal organizations.
Unpublished guide in the library.
Information on literary rights available in the library.
Gift of Judge Shenk's sons, John W. Shenk III and Samuel C. Shenk, San Francisco, Calif., 1972.

S36 Shepard, Charles Upham, 1804-1886.
Papers, 1804-86. ca. 3 ft.
In Amherst College Library (Mass.)
Professor of chemistry and natural history at Amherst College. Correspondence relating to mineral collecting, chemistry notes from study with Benjamin Silliman at Yale University, and material concerning the sugar inquiry (ca. 1833). Correspondents include Lewis Feuchtwanger, Andrew Fern Holmes, and J. N. Penny.
Unpublished guide in the library.
Gift of the Pratt Geology Museum, Amherst College, 1968.

S37 Shepard, Charles Upham, 1804-1886.
Papers, 1830; 1842-1894; n.d. - 0.98 linear ft.
In Smithsonian Institution Archives (Washington, DC).
Mineralogist; authority on meteorites; professor. - Shepard's professional associations included being an assistant to Benjamin Silliman at Yale; Director of Brewster Scientific Institute and Lecturer in Natural History at Yale; and Professor of Chemistry at South Carolina Medical College. Collection consists primarily of correspondence with mineralogists concerning Shepard's research on meteorites. Also included are notebooks, scrapbooks, catalogues, and lectures of Shepard; specimen lists; and newspaper clippings. A small amount of correspondence of Shepard's son Charles U. Shepard, Jr. is present, including correspondence with Frank Wigglesworth Clark.
Unpublished finding aid available.

S38 Sherwood, Thomas Kilgore, b. 1903.
Papers, 1945-1973. - 0.35 cu. ft.
In Massachusetts Institute of Technology, Institute Archives and Special Collections (Cambridge, MA).

Chemical engineer; educator. - The collection consists of working notes for, correspondence on, and drafts and reprints of articles written on topics in chemical engineering in conjunction with the following M.I.T. professors and/or graduates: Francis S. Chambers, Jr., George K. Cheng, Patrick E. Fowles, Thomas Harper Goodgame, Bernard M. Goodwin, Kenneth Fraser Gordon, Chang Dae Han, Samuel W. Ing, Jr., Frank J. Jenny, Conrad Joannes, Jean Paul Lienroth, Jr., W. Henry Linton, Jr., Yi Hua Ma, Robert O. Maak, Daniel S. Maisel, Michaelson O. Phillips, James McKee Ryan, Warren M. Towle, and Olev Trass.

S39 Shoemaker family.
Papers, 1824-1950. 58 boxes.
In Maryland Historical Society Library (Baltimore) (1968, 1973)
Correspondence, bills, business papers, inventories, insurance policies, account books, wills, family history and genealogical material, and other papers, of a Baltimore County, Md., family. Persons represented include Samuel Moor Shoemaker I (1821-1884), his wife, Augusta Shoemaker (d. 1907), Samuel Moor Shoemaker II (1861-1933), Samuel Moor Shoemaker III (1893-1963), Episcopal clergyman and author, and his wife, Helen Smith Shoemaker. Includes letters (1865) describing the South during the Civil War; letters and papers of a religious nature pertaining to Samuel G. Hayer; letters and papers pertaining to Samuel Eccleston Harper; business papers relating to Adams Express Company, Adams White Lead Company, Deder Electric Company, Dinsmore & Kyle Company, Northern Central Railroad, and Walker-Gordon Laboratory; and papers relating to family property, the farm at the family home, Burnside, in Stevenson, Md., cattle production there, and the Shoemakers' interest in road construction and improvement in Maryland. Samuel Moor Shoemaker I's correspondents include U.S. Grant, Rutherford B. Hayes, and Enoch Pratt.
Gift of Helen Smith Shoemaker, Stevenson, Md., 1973.

S40 Shriver (A. K.) & Sons Tannery, Union Mills, Md.
Records, 1858-80. ca. 1 ft.
In Maryland Historical Society Library (Baltimore) (2035.9)
Forms part of the repository's Shriver family collection.
Correspondence, bills of sale, receipts, accounts, and other records, of a tannery operated by Andrew Keiser Shriver (1802-1884). Includes material relating to sale, condition, and price of hides and bark, and the state of the market and conditions affecting it. Correspondents include merchants in Baltimore, Md., Derby, Conn., and Philadelphia, Pa.
Unpublished inventory in the repository.
Gift of Union Mills Homestead Foundation, Westminster, Md.

41 Sill, Harley A., fl. 1930-1964.
Papers, 1901-1964. - Ca. 1,445 pieces.
In The Huntington Library, (San Marino, CA).
Mining engineer. - Sill, from Los Angeles, was a consultant who visited and reported on various types of mines throughout the western hemisphere, particularly California and the American West. Collection includes reports on ores and mines, and technical data, chiefly from California, Arizona, Nevada, Canada and Mexico. Also includes some reports on mines in other parts of the U.S. and in South and Central America. Some maps.
Gift of Mrs. Harley A. Sill, October, 1973.
Unpublished finding aid available.

42 Silliman, Benjamin, 1779-1864.
Correspondence, 1785-1867.
ca. 700 items.
In Historical Society of Pennsylvania collections.
Chemist and inventor. Correspondence on scientific matters in general as well as on chemistry and Silliman's discoveries in that field. Includes letters pertaining to the historical paintings of John Trumbull, and to social and domestic affairs.

43 Silliman, Benjamin, 1779-1864.
Correspondence, 1808-59. - 26 items.
In American Philosophical Society Library (Philadelphia, PA).
Chemist, naturalist. - Maps, miscellaneous collection of letters concerning a variety of topics: the American Geological Society, blowpipes, chemistry, the Lowell Institute, geology, natural history, etc. The correspondents are: Johan Jacob Berzelius, Alexandre Bronginart, Heinrich Georg Bronn, Parker Cleaveland, Ashbel Green, Louis McLane, William W. Mather, Andrew Orr.

44 Silliman, Benjamin, 1816-1885.
Correspondence, 1875-84.
134 items.
In University of Arizona Library.
Chemist. Letters from George A. Treadwell and others, relating to mining ventures in Arizona, together with 2 post cards, a drawing, and 2 newspaper clippings. Includes 2 letters to George A. Treadwell.
Index to correspondents.

45 Silliman family.
Papers, 1717-1911. 21 ft.
In Yale University Library (New Haven, Conn.)
Family and scientific correspondence, diaries, mss. of articles and lectures, journals, account books, and other papers, of the Silliman and allied Church, Dana, Fish, Gilman, Hubbard, and Trumbull families. Papers of Benjamin Silliman (1779-1864) and his son, Benjamin Silliman, Jr. (1816-1885), both chemists, professors at Yale, and editors of the American Journal of Science, form the bulk of the collection. Includes diary (17 v., 1840-64) of the elder Silliman, and his correspondence with his wife, Harriet (Trumbull) Silliman (1783-1850) and son; correspondence concerning the Revolution between Silliman's father and mother, Gold Selleck Silliman (1732-1790) and Mary (Fish) Noyes Silliman Dickinson (1736-1818); material relating to John Trumbull (1756-1843) and the Trumbull Gallery at Yale; and extensive correspondence between Silliman, Sr., and leading scientists and statesmen. Subjects include political and social conditions in Connecticut and New England, the development of American science in the 19th century, and the beginnings of scientific instruction at Yale and in the U.S. Correspondents include Louis Agassiz, Frederick A. P. Barnard, Jöns Jakob Berzelius, Alexandre Brogniart, George Jarvis Brush, John C. Calhoun, Josiah P. Cooke, James Fenimore Cooper, James Dwight Dana, Charles Darwin, Josiah Willard Gibbs, Charles Goodyear, Robert Hare, Washington Irving, Andrew Jackson, Sir Charles Lyell, Gideon Mantell, Josiah Meigs, Samuel F. B. Morse, Edwards Pierrepont, Gerrit Smith, Jared Sparks, Moses Stuart, John Torrey, and Josiah Dwight Whitney.
Unpublished register and partial catalog in the library.
Information on literary rights available in the library.
Gifts of the Silliman family and purchases.

Sizer, Frank L., 1856-1942. S46
Papers, 1878-1909. - 5 ft.
In Montana Historical Society collections (Helena)
In part, microfilm made in 1980 of letterpress book (1905) owned by Don Coburn, Helena, Mont.
The repository also has microfilm of letter book for 1882-1884.
Mining superintendent and engineer.- Correspondence (1882-1908), diary (1904), reports, legal documents, financial records, school notebooks, scrapbooks, and other papers, relating to Sizer's personal life and work as mining superintendent and engineer. Much of the correspondence documents in detail the operations of mines Sizer managed. Mines and companies represented include Rosario Mine, Chihuahua, Mexico, Whitlatch Mining Company, Unionville, Mont., and Spring Hill Mining Company, Unionville, Mont.
Gift of Macie Conrad, Great Falls, Mont., 1955. Formerly part of the Helena Banking Group collection.
Unpublished finding aid in the repository.

S47 **Skinner (William) & Son,** *Holyoke, Mass.*
Records, 1874–1950. 89 items.
In Connecticut Valley Historical Museum (Springfield, Mass.)
Account books, cashbooks, daybooks, dye house books, inventory books, ledgers, journals, store ledgers, payrolls, and miscellaneous timebooks and receipts, of a company that manufactured silk and rayon fabrics. Includes journals, meeting records, register and cash receipts of the Mount Holyoke Hotel Company (1909–40).
General index on cards in the repository.
Gift of the company, 1961.

S48 Skinner, William Woolford, b. 1874.
Daily journal, 1942-1944. - 4 v.
In National Agricultural Library (Beltsville, MD).
Chemist; Chief, Bureau of Agricultural and Industrial Chemistry, 1942-44. - Official manuscript diary of Skinner while Chief of the Bureau of Agricultural and Industrial Chemistry.
Unpublished finding aid available.

S49 **The Louis A. Slotin Memorial Fund** collection, 1946–62.
200 items.
In University of Chicago Library.
Mainly letters of Samuel King Allison (b. 1900), professor of physics at the University of Chicago, soliciting money for a Louis A. Slotin Memorial Fund to be used to finance lectures in the sciences at the University of Chicago; replies to the letters; and reports on the size and progress of the fund. Includes correspondence pertaining to Slotin's death in 1946 which resulted from an accident with radioactive materials, and to the disposition of his personal library; together with other material concerning his work as a physicist and physical chemist at Oak Ridge, Tenn., and Los Alamos, N.M.
Unpublished guide in the library.
Gift of Mr. Allison, 1963.

S50 Small, Lyndon F. (Lyndon Frederick), 1897-1957.
Correspondence, 1929-1955. - 4 boxes.
In National Library of Medicine (Bethesda, Md.)
Chemist with Drug Addiction Laboratory, University of Virginia and National Institutes of Health.-Correspondence, chiefly from the 1930s, including correspondence with Anti-Opium Bureau of the League of Nations. Commercial firms represented include Abbott Laboratories, American Instrument Company, E.R. Squibb, Mallinckrodt Chemical Works, and Merck. Correspondents include H.J. Anslinger, Alfred Burger, Nathan B. Eddy, Arthur B. Lamb, and William Charles White.

S51 Smith, Albert William, 1862-1927.
Papers, 1885-1923. - Ca. 50 items.
In Case Western Reserve University Archives (Cleveland, OH).
Professor and head of the Chemistry Department at Case Institute of Technology and consultant for the Dow Chemical Company. - Smith was actively engaged in research on electrolytic cells for the production of chlorine. Collection includes correspondence, transcripts of writings and speeches, student notebooks, reprints, and copies of patents granted to Smith for a method of making metallic sodium, a method of extracting gold, a process of making chloroform, and a process of extracting bromine. Also includes records related to the Midland Chemical Company, including Smith's correspondence with the company, financial vouchers, factory reports, and Minutes of the Board of Directors, 1904-1913.
Unpublished finding aid available.

S52 Smith, Alexander, 1865-1922.
Papers, 1911-1919. - 5 boxes.
In Columbia University, Rare Books and Manuscripts Library (New York City).
Physical-inorganic chemist; teacher; author of textbooks. - Professor of Chemistry at Columbia and head of the Columbia Chemistry Department from 1911 to 1919 when he retired because of illness. The collection contains his correspondence and financial records.
Unpublished finding aid available.

S53 Smith, Donald E   collector.
Records, 1884-1956.   5 reels of microfilm.
In West Virginia University Library (Morgantown) (1944)
Microfilm made in 1967 from originals loaned by Mr. Smith.
Correspondence, account books, and tax reports of Liverpool Salt and Coal Company, American Calcium Chloride Works, Liverpool Salt Company, Jackson Coal and Mining Company, and Jackson Valley Bell Farm of Hartford; and account books for G. Y. Roots and Company, Cincinnati, Ohio.

S54 Smith, (E. F.) Collection.
Correspondence, 1928-75. - 7 boxes.
In University of Pennsylvania, Van Pelt Library, E. F. Smith Collection (Philadelphia).
One of world's largest collections of published works and some manuscript materials in the history of chemistry, chemical engineering and the chemical industry. Built around the personal library of Edgar Fahs Smith (1854-1928), Professor of Chemistry and Provost at the University of Pennsylvania. - This collection represents the archives of the E. F. Smith Collection, i.e., the

administrative and public service correspondence generated by the staff of the E. F. Smith Collection. It also contains printed material, financial documents, diaries and notes. Much of the correspondence is made up of reference requests from the public and contacts with book dealers.
Unpublished finding aid available.

S55 Smith, Edgar Fahs, 1854-1928.
Papers, ca. 1880-1928. - Ca. 150 linear ft.
In University of Pennsylvania Libraries, E. F. Smith Memorial Collection (Philadelphia).
Professor of chemistry and provost at the University of Pennsylvania; bibliophile; historian of chemistry; president of the ACS. - The papers complement Smith's personal library, the nucleus of the present E. F. Smith Memorial Collection, and contain correspondence; scrapbooks of newsclippings and cards sent by wellwishers, mostly related to events in Smith's career; reports and papers by Smith's students; reprints; historical material collected by Smith relating to the Chemical Society of Philadelphia, the University of Pennsylvania, and various 19th century chemists; addresses and speeches made by Smith; Smith's honorary degrees; biographical sketches of Smith; photographs; books and other printed material; Smith diaries; news items relating to Smith's death; and assorted memorabilia.
Gift of Mrs. Margie Alice Gruel Smith.
Unprocessed; available for research.
Unpublished inventory available.

S56 Smith, George Frederick, b. 1891.
Papers, 1922-1972. - 0.6 ft.
In University of Illinois Archives (Urbana).
Professor of Chemistry at the University of Illinois.- Reprints of articles on analytical, industrial and food chemistry, quantitative analysis, general inorganic and perchlorate chemistry, chemistry of the 1,10-phenanthrolines and related compounds; biographical data; tape-recorded interview about the Chemistry Department, the G. Frederick Smith Chemical Company, Columbus, Ohio, and other aspects of Smith's interests in chemistry; photos; and other papers.
Information on literary rights available in the repository.
Unpublished finding aid available.

S57 Smith, George McPhail.
Papers, 1922-1923; - 0.5 linear ft.
In University of Washington Archives (Seattle).
Professor of Chemistry. - Lecture notebooks compiled while a faculty member at the University of Washington.
Accessioned 1974.

S58 **Smithfield plantation records**, 1900-1970.
35 ft. and 191 v.
In Louisiana State University, Dept. of Archives and Manuscripts (Baton Rouge).
Complex of sugar plantations, factories, and stores, Port Allen, La. - Correspondence, bills and receipts, cane yield reports, tax returns, stock orders, ledgers, journals, and production reports from the plantation, its factory, and stores, and from its associates, Milliken and Farwell Store, Little Texas Sugar Factory, and Westover Planting Company and Store.
Finding aid in the repository.

S59 Smithson, James, ca. 1796-1829.
Papers, 1796-1951. - 2.95 linear ft.
In Smithsonian Institution Archives (Washington, DC).
Chemist and mineralogist; founder of Smithsonian Institution. - Collection contains the few surviving original Smithson manuscripts, 1796-1878; documents related to securing the Smithson bequest; research and correspondence about Smithson's life and lineage, 1881-1951; material regarding removal of Smithson's remains to America, 1903-1905; photographs; and publications.
Unpublished finding aid available.

S60 Smithsonian Institution. Department of Mineral Sciences.
Records, 1936; 1938; 1948-1977; n.d. - 10.8 linear ft.
In Smithsonian Institution Archives (Washington, DC).
Created in 1963 as part of a reorganization in the National Museum of American History. - Record unit consists of correspondence documenting the operation of the Department of Mineral Sciences, 1963-1977, and its predecessor, the Division of Mineralogy and Petrology of the Department of Geology, 1948-1963. Correspondence concerns identification and acquisition of specimens; participation in professional societies and mineral exhibitions; and other topics. Correspondents include geologists and mineralogists; mineral collectors; mining companies; colleges, universities, and government agencies; and the general public.
Restricted.
Unpublished finding aid available.

S61 Smithsonian Institution. Division of Medical Sciences.
Records, ca. 1890-1977. - 18 linear ft.
In Smithsonian Institution Archives (Washington, DC).
The Smithsonian's interest in the medical sciences dates back to the Toner Lecture series established by Joseph Toner in 1872. A number of reorganizations have produced the Division of Medical Sciences, Department of Science and Technology, in the Museum of History and Technology. Records document Division activities, chiefly for the years, 1917-1975. Included are correspondence concerning topics in history of medicine and pharmacy; memoranda; annual reports; and administrative files. Special topics include advancements in public health and rehabilitative medicine.
Restricted.
Unpublished finding aid available.

S62 Smithsonian Institution. Division of Meteorites.
Records, 1963-1970; n.d. - 4 linear ft.
In Smithsonian Institution Archives (Washington, DC).
Division was established in 1963 as part of a reorganization in the National Museum of Natural History. Record unit consists of incoming and outgoing correspondence, concerning identification and acquisition of specimens; publication of scientific manuscripts; exhibits; participation in professional societies; and divisional administration. Correspondents include geologists and mineralogists, meteorite collectors and dealers, Smithsonian staff and administrators, government agencies, and the general public.
Unpublished finding aid available.

S63 Smithsonian Institution. Division of Mineralogy and Petrology.
Records, 1932-1963. - 4.3 linear ft.
In Smithsonian Institution Archives (Washington, DC).
Division established in 1911. Name prior to 1911 was Division of Mineralogy. Record unit consists primarily of Edward P. Henderson's official correspondence files as Associate Curator of the Division. A smaller amount of files were created by Roy S. Clarke, Jr., who was appointed Chemist of the Division in 1957. Correspondence is both incoming and outgoing and deals exclusively with the national collection of meteorites under the care of the Division. Correspondents include geologists and mineralogists, meteorite collectors and dealers, Smithsonian staff and administrators, government agencies, research foundations, and the general public.
Unpublished finding aid available.

S64 Smithsonian Institution. Division of Physical Sciences.
Records, 1956-1976. - 3 linear ft.
In Smithsonian Institution Archives (Washington, DC).
Division was established in the Department of Science and Technology in 1957 to be responsible for collections in the history of astronomy, chemistry, astrophysics, geology, meteorology, and classical physics. Records consist of public inquiries concerning general scientific instruments; memoranda, layout plans, photographs, and scripts for exhibits; correspondence with foreign and domestic science museums, colleges and universities, professional scientific societies, and manufacturers and collectors of scientific instruments; administrative records; and other document types.
Restricted.

S65 Smithsonian Institution. Office of the Smithsonian Chemist.
Records, 1880-1883. - 0.33 linear ft.
The Office of Smithsonian Chemist was officially created ca. 1876. Prior to that time, the Institution maintained informal relations with a number of chemists, who had use of the Smithsonian's facilities to carry out their own work as well as some done for the Institution. Most of this work dealt with composition of minerals or with queries from government departments. Frederick W. Taylor served as the Smithsonian's chemist from 1877 until 1884, when he resigned. The Office was abolished, and the United States Geological Survey performed chemical work in the Smithsonian as needed. The records consist of one letterpress book recording Taylor's reports of chemical analysis to various inquirers.

S66 Smithsonian Institution. Section of Agriculture, Division of Agriculture and Mining.
Records, ca. 1923-1973. - 5.9 linear ft.
In Smithsonian Institution Archives (Washington, DC).
The Section of Agriculture has roots dating back to 1921, and has undergone several administrative reorganizations. Records document the activities of the

Section and its predecessors, and include correspondence, memoranda, exhibits scripts, and reports. Annual reports covering the Sections of Chemical Industries, Chemical Technology, Foods, Manufactures, and Organic Chemistry are also contained in the records. Annual reports for the Division of Agriculture and Mining, 1969-1973, also contain material concerning the Section of Mining.
Restricted.

67 **Society of Separatists of Zoar.**
Records, 1816-1942. 23 ft.
In Ohio Historical Society collections (Columbus) (110)
Correspondence and personal, official, legal, business, and religious papers. Includes records relating to various economic enterprises of the society including a brewery, tannery, and hotel; material relating to the Zoar Post Office, Zoar Foundation, and other Zoar agencies and persons after the society's dissolution; and papers of its successive leaders: Joseph M. Bimeler, Jacob Sylvan, Christian Wiebel, Jacob Ackerman, Simon Beuter, and Lewis Zimmerman.
Unpublished box inventory in the repository.
Gift of the Zoar Historical Society through the Zoar Foundation, 1940-50.

68 **Sondericker, Jerome.**
Notebook, ca. 1890. - 1 v.
In Massachusetts Institute of Technology, Institute Archives and Special Collections (Cambridge, MA).
Mechanical engineer; educator. - The collection consists of an undated notebook containing the "Report on the Uniform Method for Testing Cement by the Committee of American Society of Chemical Engineers," translation of a German work by A. Wohler, and drawings and notes on a torsion machine.

9 **Sorby, Henry, 1826-1908.**
Diaries, 1859-1908. - Microfilm.
In American Philosophical Society (Philadelphia, PA).
English geologist and mineralogist. - Microfilm copy of original diaries held at the Sheffield University Library in England.

70 **Special Fabrics Company, Saylesville, Rhode Island.**
Records, 1931-1950. - 3 linear ft.
In Rhode Island Historical Society (Providence).
Began as technical committee of Sayles Finishing Plants bleachery. Formally organized in 1921. Products were tracing cloth and paper, blueprint paper, tag cloth, fabric-backed tape, waterproofed cloth, and book binding cloth. Sold in 1971 to Holliston Mills, Inc. - Collection consists of fabric sample books and minutes of the Special Fabrics Committee meetings and technical discussions.
Gift of Sayles Finishing Plants, December 25, 1971.
Unpublished finding aid available.

Spencer, Kenneth Aldred, 1902-1961. S71
Papers, 1918-1961. - 30 linear feet.
In University of Kansas Libraries (Lawrence).
Business executive and industrialist of Kansas City, Missouri. - Correspondence, diaries, appointment books, scrapbooks, personal memorabilia, and speeches relating to Spencer Chemical Company and the Pittsburg and Midway Mining Company (Pittsburg, Kansas).
Gift, Helen Spencer, 1966
Unpublished inventory and calendar in the repository.

Spillman, William Jasper, 1863-1931. S72
Notebooks, 1885-1886. - 11 v.
In University of Missouri-Columbia Library, Western Historical Manuscripts Collection and State Historical Society of Missouri Manuscripts.
Student at University of Missouri. - Notebooks contain lecture notes taken by Spillman (graduated 1886). In particular, 2 volumes of applied chemistry notes, from a course taught by Dr. Paul Schweitzer.
Received July 31, 1946.
Unpublished finding aid available.

Spogen, Dominic, fl. 1886-1895. S73
Reminiscence, n.d. - 0.1 linear ft.
In Montana Historical Society (Helena).
Smelterman and coking foreman for the Anaconda Copper Mining Company. - Collection contains a photocopy of a 24-page typescript reminiscence written by Spogen, describing his experiences in building and operating the Anaconda smelter and working in coal and coke production at Belt, Montana (ca. 1886-1895). The reminiscence is entitled "History of Marcus Daly Starting the Reduction Works at Anaconda, Montana."
Loaned for copying by D. Spogen of Belt, Montana in July, 1962.

S74 Spokane Brewing and Malting Company.
Records, 1887-1932. ca. 89 ft. (ca. 70,000 items)
In Washington State University Library (Pullman)
Correspondence ledgers, minutes, financial records, inventories, brewing records, and other papers of the Spokane Brewing and Malting Company (established 1900) and its predecessors: Henco Brewery (established 1886), New York Brewery (established 1886), and Galland-Burke Brewing and Malting Company (established 1891).
Unpublished container list in the library.
Information on literary rights available in the library.
Gift of Julius Galland, Jr., 1937.

S75 **Stabler-Leadbeater Apothecary Shop**, *Alexandria, Va.*
Records, 1813-ca. 1880.
8 v. and 1 box.
In New York Historical Society collections.
Account books and business papers.

S76 Stanley, Wendell Meredith, 1904-1971.
Papers, ca. 1930-1971. - 1 box, 31 cartons.
In University of California, Berkeley, Bancroft Library.
Professor of biochemistry and molecular biology; director of the Virus Laboratory, UC Berkeley; Nobel laureate. - Correspondence, subject files, manuscripts of writings and speeches, reprints, course and departmental materials, photographs and negatives.
Finding aids available.

S77 Stearns, Frank W. (Frank Waterman), 1856-1939.
Papers, 1683-1956. - 3 ft.
In New England Historic Genealogical Society collections (Boston, Mass.).
Department store executive, of Boston, Mass.
Correspondence, memos, and clippings, relating chiefly to Amherst College (of which Stearns was a trustee) and the hiring and firing of president Alexander Meiklejohn; material on R.H. Stearns Company department store in Boston, relating mainly to the building of a new store in 1909; and papers of and relating to ancestors, collected by Stearns's son, Foster Waterman Stearns (1881-1956), U.S. representative from New Hampshire. Collected papers include papers of William Smith Clark (1826-1886), professor of chemistry, botany, and zoology at Amherst, Union Army officer, president of Massachusetts Agricultural College, and father-in-law and grandfather, respectively, of Frank Waterman Stearns and Foster Waterman Stearns; one letter by William Richards (1793-1847), missionary and adviser to the Hawaiian royal court; and papers of the Sprague and Frost families with details of frontier life in New York, Pennsylvania, and Ohio, 1820s-1830s.
Bequest of Foster Waterman Stearns, 1956.
Inventory in the repository.

Stewart, Francis Edward, 1853-1941. S78
Papers, 1866-1938. - 19 boxes.
In State Historical Society of Wisconsin (Madison).
Physician, corporate pharmacist, and author. - Stewart worked to establish pharmaceutical standards, change patent and trademark laws, and increase cooperation between physicians and pharmacists. The papers reflect those activities, and include correspondence, articles, and personal papers.
Gift of Mrs. Robert P. Fischelis, Ada, Ohio, 1981.
Unpublished finding aid available.

Stine, Charles M. A., 1882-1954. S79
Papers, 1925-1946. - 84 items.
In Hagley Library (Greenville, DE) (1706).
Research chemist and executive of the Du Pont Company. - Articles, addresses, and biographical items with a few items of correspondence mostly dealing with chemical engineering and the chemical industry; scientific research and education and the relationship of chemical technology to society.
Unpublished finding aid available.

Storer, Francis H., 1832-1914. S80
Papers, 1876-1916.
In Harvard University Archives (Cambridge, MA).
Professor of Agricultural Chemistry; Dean of Bussey Institute. - One folder of scattered incoming correspondence and one folder of outgoing correspondence to John A. Hemshaw. Biographical material and reprints.

Storie, Raymond Earl, 1894- S81
Papers, 1917-74. 20 ft.
In University of Wyoming, American Heritage Center (Laramie)
Forms part of the repository's Conservation History and Research Center collection.
Soil scientist, University of California at Berkeley. Soil surveys of California and elsewhere, writings and reports on soil surveys, articles, notebooks, manuals on identifying and categorizing soils, maps, biographical material, certificates, clippings, and photos of geological formations. Includes material on Storie's Index for Rating the Agricultural Value of Soils (1933); evaluating land for irrigation, grazing, and timber; and University of California Soil Survey Program, 1939-60, which Storie directed.
Unpublished finding aid in the repository.
Gift of Mr. Storie, 1976-77.

S82 Stout, Jacob
Stout-Alston papers, ca. 1750–1905. 4 ft.
In Friends Historical Library of Swarthmore College.
Correspondence, medicinal "receipts," account books, bills, records of crops, price list, tax statement, receipts, administrative papers, agreements, bonds, briefs and testimony, deeds, indentures, promissory notes, wills, and other business and legal papers, of Jacob Stout and Jonathan Alston, 18th century Quaker tanners and merchants of Smyrna, Dover, and Leipsic, Del. Includes an anonymous journal, discharges from military duty, and a list (1773–87) of birth dates for the Allston (Alston) family.
Checklist in the library.
Acquired in 1927.

S83 Stout, Wilbur Elihu, 1876–1961.
Papers, ca. 1930–55. 6 ft.
In Ohio Historical Society collections (Columbus) (408)
Geologist. Histories and reference material on charcoal and blast furnaces in Ohio and ms. of a book on the iron industry in Ohio.
Information on literary rights available in the repository.
Gift of Walter Stout, Westerville, Ohio, 1974.

S84 Strait, Louis A., 1907–1975.
Papers, 1943–1970. – 6 cartons.
In University of California, San Francisco Archives.
Professor of Biophysics and Pharmaceutical Chemistry at UCSF. – Collection contains correspondence, committee minutes, and other material, relating to teaching, the Spectrographic Laboratory, and the Academic Senate.

S85 Straits Sugar Company.
Correspondence, 1898–1900. 1 v.
In Franklin Institute Library (Philadelphia)
Correspondence of John Turner, manager in the Straits, with J. Arnold in London, the company secretary, and letters to Turner from John Bruce, in charge at Gedong, and G. W. Stothard, in charge at Rubana. Subjects include the opening and planting of new estates, and the various aspects of sugar growing, refining, and marketing in Malaya.
Purchase.

S86 Stritch family.
Papers, 1875–1927. 110 items.
In Tennessee State Library and Archives (Nashville)
Correspondence, notes, bills, receipts, notices, school reports, invitations, and programs, primarily of Samuel Cardinal Stritch (1887–1958) during his years as a student at the North American College in Rome, of his father, Garrett Stritch (1841–1896), of the Cardinal's grandfather, Thomas Stritch, living in Ireland, and of several uncles and brothers. Subjects include anticlerical riots, the Cardinal's school work and classmates in Rome, Methodists in Rome, and travel in Italy; together with material on Garrett Stritch's work with the Sycamore Manufacturing Company, makers of sporting and blasting powder. Other persons represented include Rev. John Farrelly, Eugene Castner Lewis, and Rev. John Morris.
Unpublished register in the repository.
Information on literary rights available in the repository.
Permanent deposit by Miss Catherine Stritch, 1968.

Strunz Soap Company, Pittsburgh, PA. S87
Records, 1861–1917. – 2 ft.
In Archives of Industrial Society (Pittsburgh, PA).
Letter book (1884–1913), general ledgers, sundry accounts ledger, journal, inventory book (1868–99), and customer index of a company producing industrial soaps, poultry feed, glycerine, and fertilizer for the local Pittsburgh market.
Gift of E. J. McGrael, president of the company, 1964.
Unpublished finding aid available.

Stuart, Alexander Hugh Holmes, 1846–1867. S88
Student notebook, 1867. – 116 p., 8 x 6 1/4 in. holograph.
In Virginia Historical Society (Richmond).
Bound volume. – Contains notes on chemistry lectures delivered by Socrates Maupin at the University of Virginia.
Gift of Archibald G. Robertson, Richmond, Va., March 20, 1963.

Stubbs, William C. S89
Papers, 1896–1904; 1924. – 22 items.
In Louisiana State University, Department of Archives (Baton Rouge).
Professor of Agriculture at LSU; Director of the Louisiana Sugar Experiment Station. – Correspondence of William C. Stubbs. Subjects include sugar planters; agriculture curriculum; agricultural experiment stations; geological survey of Louisiana, 1902; and the St. Louis Exposition, 1904.

Stubbs, William Carter, 1846–1924. S90
Papers, 1867–1925. 228 items.
In University of North Carolina Library, Southern Historical Collection (703)
Chemist, agriculturist, educator, and director of the Louisiana State Sugar Experiment Station, 1885–95, and of the U. S. Experiment Station in Honolulu, 1900. Chiefly business papers, including speeches, writings, reports, and descriptions of different varieties of tobacco, peas, cotton, and potatoes.
Unpublished description in the library.
Purchased, before 1940.

S91 Stubbs, William Carter, 1846-1924.
Papers, 1825-1936. - 11,743 items.
In College of William and Mary, Earl Gregg Swem Library (Williamsburg, VA).
Professor of Chemistry at the Agricultural and Mechanical College in Auburn, Alabama, and Director of the Louisiana Sugar Experiment Station. - Correspondence and genealogical papers of Stubbs, his wife, Elizabeth, and their families. In addition, the collection contains 2 notebooks relating to Stubbs' chemical activities: a bound volume containing his notes on soil experiments, and another volume containing his notes on chemical analysis.
Unpublished finding aid available.

S92 Student Chemistry Notebooks.
Notebooks, 1833-1893. - Ca. 15 items.
In University of Virginia Libraries (Charlottesville).
Notes taken in the chemistry lectures of John Patten Emmet (1796-1842), Robert Empie Rogies (1813-1894), Socrates Maupin (1808-1871), and John William Mallet (1832-1912). Collection also contains a notebook of readings in chemistry, 1823, compiled by John Patten Emmet, and the 1868-69 chemistry notebook of Walter Reed.

S93 Student Notebooks, 1870-1931.
89 v.; 47 items.
In Cornell University Archives (Ithaca, NY).
Student lecture notebooks, especially those of Albert Huntington Hooker, Jr., in chemistry, geology, and other subjects (1914-20); and notebooks of Gilbert Holmes Crawford, Jr. in mechanical and electrical engineering (laboratory reports, 1908-10).

S94 Stumm, Werner, 1924-
Papers, 1956-70. ca. 3 ft.
In Harvard University Archives (Cambridge, Mass.)
Teacher of applied chemistry at Harvard University. Personal and professional correspondence, mss. of writings, proposal and reports relating to research, processed material relating to professional meetings, and teaching material.
Access restricted.
Information on literary rights available in the repository.
Acquired from the Division of Engineering and Applied Physics, Harvard University, 1973.

S95 Sugar Beet Industry, New York State.
Oral history interviews, 1965-66. - 6 items (288 pp. typescript).
In Cornell University Archives (Ithaca, NY).
Interviews conducted by Gould P. Colman, concerning the planning, establishment, and operation of a sugar beet industry in a thirteen-county area of New York State from 1961 to 1966. Among subjects discussed are the acquirement of background knowledge about raising sugar beets and producing beet sugar; production practices; and sources of both reliable and unreliable information concerning beet production. Interviewees were businessmen, professors, economists, and sugar beet experts.
Restricted in part.

S96 Sullivan, John F.
Papers, 1912-1932. - 10 boxes, 1 pkg.
In New York Public Library, Rare Books and Manuscripts Division (New York).
Consulting engineer; New York City planner. - Correspondence deals primarily with the construction of the U.S. nitrate plants (Sullivan was in charge of construction at the Muscle Shoals, Alabama plant in 1919). Other subjects include: the Narrows Tunnel; West Side improvement; and the construction of bridges and tunnels in New York City. Reports and notebooks dealing with the Muscle Shoals project are included. Also blueprints, maps of New York, hydrographic charts, newspaper clippings, printed matter, photographs, and lantern slides.
Received February 2, 1937.
Unpublished finding aid available.

S97 Sumner, James Batcheller, 1887-1955.
Papers, 1908-58. 8 ft.
In Cornell University Libraries, Collection of Regional History and University Archives (Ithaca, N. Y.) (29/2/671)
Professor of biochemistry and director of the Laboratory of Enzyme Chemistry, Cornell University. Correspondence (1953-54) relating to the organization of biochemical research in Brazil and Sumner's proposed lecturing there, notes, including his chemistry notebook (1908) from Harvard University, articles by Sumner and others, graphs, photos, and ca. 300 glass slides made in the course of his research, printed matter, and other papers. Includes material relating to the Nobel Prize for scientists, correspondence, biographical sketches, notes, and clippings used by Prof. Leonard A. Maynard in the preparation of his memoir of Sumner published in 1958.
Gift of Mrs. Sumner, 1963; biographical material gift of Prof. Maynard, 1968.

S98 Sun Oil Company.
Marcus Hook Refinery records, 1903-54. 16 ft.
In Eleutherian Mills Historical Library (Greenville, Del.) (382 and 993)
Cashbooks, ledgers, receipts and disbursement records, coal record books, inventories, material and labor cost records, customer and gasoline sales records, and other accounts and business records (1903-29); together with papers (ca. 1937-54) of George H. Voelker, equipment engineer, containing purchasing and testing information on the first commercial hot gas turbine unit, jointly put in operation by the Brown Boveri Corporation of New York and the Sun Oil Company, and including correspondence with Allis-Chalmers Manufacturing Company, Adolph F. Meyer, and others. Other persons represented include Joseph N. Pew (1848-1912), founder of the company, and other members of the Pew family.
Restricted in part.
In part, described in A Guide to the Manuscripts in the Eleutherian Mills Historical Library, by John B. Riggs (1970) p. 973.
Gifts, 1962 and 1967.

S99 Survey of Sources for the History of Biochemistry and Molecular Biology.
Archives, 1975-79. - 18 linear ft.
In American Philosophical Society Library (Philadelphia, PA).
Maps, project correspondence with individuals and groups, autobiographical questionnaires of biochemists, information on sources in China, and files relating to the project's search for sources. In addition, there is information concerning the computer system created to produce the guide. See David Bearman and John T. Edsall, eds., Archival Sources For the History of Biochemistry and Molecular Biology (APS, 1980).
Unpublished finding aid available.
Presented by the Survey of Sources, 1979.

S100 Sutro, Adolph Heinrich Joseph, 1830-1898.
Papers, 1853-1931. - Ca. 2,786 items.
In The Huntington Library, (San Marino, CA)
Mining engineer; Mayor of San Francisco. - Sutro established the Sutro Metallurgical Works at Dayton, Nevada, in 1861, utilizing the invention he and John Randohr had made for a metal extracting process. He had constructed a ten-stamp mill in 1862, when he conceived the idea of a tunnel to relieve the problems of floods, high temperatures, and noxious gases in the mines of the Comstock Lode. Collection includes letters, manuscripts, documents, and maps. Subjects are San Francisco, land development, street railways, politics, the Sutro Library, and mining in Nevada.
Purchased from George D. Lyman, March, 1943, and W.P. Wreden, November, 1961.
Unpublished finding aid available.

S101 Swain, Robert Eckles, 1875-1961.
Papers, 1916-1938. - 1 linear ft.
In Stanford University Archives (Stanford, CA).
Professor of Chemistry and Acting University President, Stanford. - Correspondence as department chairman and pertaining to the honor system in the department. Draft, typescripts, and notes for his address at the Stanford memorial service for David Starr Jordan (1932).

S102 Swift, Harold H., 1885-1962.
Papers. - 107 linear ft.
In University of Chicago Library.
University Trustee (1914-62) and Chairman, Board of Trustees (1922-48); Chairman of the Board, Swift and Company. Files relating to Chemistry Department (1913-37), including correspondence with John U. Nef, Sr. and Julius Stieglitz; material concerning Kent Chemical Laboratory, Jones Laboratory, and the Research Institutes (1945-50).
Partially restricted.
Unpublished guide available.

S103 Szent-Gyorgyi, Albert, b. 1893.
Oral history interview, 1967. - 4 reels; 7 hours.
In National Library of Medicine (Bethesda, MD).
Chemist; Nobel laureate. - Interview conducted by Harlan B. Phillips in Woods Hole, Massachusetts, June 17-22 and September 10-23, 1967. Autobiographical memoir, including descriptions of his early life and education, his research, and his philosophy. Dr. Szent-Gyorgyi received the Nobel Prize in 1937 for his work on the isolation of ascorbic acid (vitamin C). Among the subjects discussed is Dr. Szent-Gyorgyi's association with the National Institutes of Health.
Transcript, with index and appendix (190 p.).
Information on restrictions available in the Library.

S104 Szent-Gyorgyi, Albert, b. 1893.
Papers, 1958-1970. - 39 items.
In Library of Congress, Manuscript Division (Washington, DC).
Biochemist and chemist. - Correspondence and drafts of Szent-Gyorgyi's works, including typewritten drafts of "The Crazy Ape" (1970) and "The Sick Ape."
Gift of Albert Szent-Gyorgyi, 1964-1970.

T1 Talbot, Arthur Newell, 1857-1942.
Papers, 1877-1942. - 2.5 ft.
In University of Illinois Archives (Urbana).
Professor of Engineering, Mathematics, and Municipal and Sanitary Engineering at the University of Illinois. - Correspondence of Talbot with engineers, the alumni, equipment manufacturers, and public officials concerning railway engineering, sewage treatment and water purification plants; notes and specifications for public works; papers (1878-1881) relating to university societies, and alumni reunion programs; copies of examination questions (1885-1894) for mathematics, drawing, and railway and sanitary engineering; mss. of talks or reports on the Wabash River pollution (1894), sanitary science and health in town and country (1899-1909), paving bricks (1905), Chicago city hall investigation (1910), engineering college graduates, and early University of Illinois alumni and faculty; and clippings about Talbot's service in receiving Army construction projects (1918).
Unpublished finding aid available.

T2 Talbot Mills, North Billerica, Massachusetts.
Records, 1851-1957. - 200 linear ft.
In Merrimack Valley Textile Museum, (North Andover, MA).
Company formed in 1839 as C.P. Talbot & Co. by Charles P. and Thomas Talbot (brothers) to manufacture dyewood. Eventually expanded to include chemical and dye manufacturing, as well as a woolen mill. - Collection contains labor records from the 1850's, including the payroll for the dye and chemical works and for the woolen mill. Also includes a file on water power from this early period. Bulk of the papers cover the period from 1890's to the 1930's. Includes executive correspondence files and manufacturing cost statistics. Gift of the company, 1960-1961.
Unpublished finding aid available.

T3 Tanner, Herbert Battles, 1859-1933.
Papers, 1739-1933.
1 v. and 26 boxes.
In State Historical Society of Wisconsin collections.
Physician and public official of Kaukauna, Wis. Correspondence, official reports, and miscellaneous papers of Tanner, together with a box of documents from the office of a notary public near Mexico City dealing with lawsuits (1739-1865); registers of electors in Kaukauna in 1896; and ledger (1894-97) of a firm of Kaukauna druggists. Tanner's papers concern his work as State supervisor of inspectors of illuminating oils, 1894-1900; the management of the Rio Tamasopo Sugar Company in San Luis Potosi, Mexico, a company in which Tanner and a brother were financially interested; his operation of a Kaukauna drug store; his activities as Kaukauna's first health officer, secretary of the Fox River Valley Medical Association, physician (with emphasis on his services as local representative of the Chicago and Northwestern Railroad and other corporations), mayor of Kaukauna, 1888-96, an investor in the Kaukauna Electric Light Company, and active member of the South Kaukauna Congregational Church; his candidacy for Congress in 1900; and his work on local history, particularly the career of Captain Hendrick Aupaumut, and on his own family genealogy. Includes monthly reports of subordinate district oil inspectors (1895-97), together with correspondence concerning appointment of these inspectors, as well as correspondence concerning the gubernatorial campaigns of 1894 through 1900 and the relations between Tanner's office and the Standard Oil Company.

T4 Tatum, Arthur Lawrie, 1884-1955.
Papers.
3 ft.
In State Historical Society of Wisconsin collections.
Professor and pharmacologist. Correspondence, diary, speeches delivered to various scholarly groups, articles, financial records, class lectures, records of experiments performed, data from cases observed in connection with Tatum's research, and an itinerary and map relating to his trip to Guatemala for field observation of tropical diseases (1948). The early correspondence is chiefly with poundmasters who furnished him with research data, and with various institutions of learning.
Inventory in repository.
Gift of Mrs. Tatum, 1957.

T5 Tatum, Edward Lawrie, 1909-1975.
Papers, 1936-1975. - 25 ft.
In Rockefeller University Archives, Rockefeller Archive Center (Pocantico Hills, North Tarrytown, N.Y.).
Biochemical geneticist, Rockefeller University. Correspondence (1946-1975); mss. of published papers, lectures, and conferences; laboratory notes and data (1937-1975); and photographic material, primarily technical. Includes material relating to Nobel prize, scientific committees, neurospora, and population control.
Gift of Barbara Tatum and Margaret Easter.
Finding aid in the repository.

T6 Taussig, Charles William, 1896-1948.
Papers, 1928-48. 80 ft.
In Franklin D. Roosevelt Library (Hyde Park, N. Y.)
Businessman and Government advisor and official. Correspondence, minutes of conferences and committee meetings, reports, statistical data, memoranda (including that of conversations), file of speeches and articles, outlines, printed congressional bills and other material, and newspaper clippings relating to Taussig's career as chairman of the Board and president of the American Molasses Company and the Sucrest Corporation. The papers also document Taussig's work as a member of President Franklin D. Roosevelt's "brain trust"; as technical advisor to the World Economic Conference; as Chairman, National Advisory Committee of the National Youth Administration; as U. S. Chairman, Anglo-American Caribbean Commission; as advisor to U. S. Delegation, U. N. Conference, 1945; and as advisor to the Secretary of State. Includes information on world trade, especially the sugar market, economic, social, and political conditions in the Caribbean area and the U. S. policy of trusteeship for dependent areas. Correspondents include Adolf A. Berle, Spruille Braden, Lawrence Cramer, Jerome Frank, Cordell Hull, Harold Ickes, Charles Judd, Fiorello La Guardia, Raymond Moley, Eleanor Roosevelt, Franklin D. Roosevelt, Sir Frank Stockdale, Rexford Guy Tugwell, Sumner Welles, and Aubrey Williams.
Unpublished shelf list in the library.
Information on literary rights available in the library.
Gift of Ruth Adler Taussig, 1951.

T7 Taylor, Charles Fayette, b. 1894.
Papers, 1917-1973. - Ca. 40.5 linear ft.
In Massachusetts Institute of Technology, Institute Archives and Special Collections (Cambridge).
Professor of Engineering. - Taylor's research interests were the internal combustion engine, detonation, combustion, fuel composition, friction, air capacity, piston/valve ratios, high speed Diesel engines, and engine thermodynamics. He taught courses at MIT and supervised student work in aeronautical, chemical, and mechanical engineering. In addition, he did consulting work for government and industry. The papers fall into 3 broad categories: correspondence, teaching materials, and technical reprints. Subjects covered in the correspondence file include companies such as General Electric, Westinghouse, and Thiokol Chemical. The teaching material consists of lecture notes, reading lists, problem sets, quizzes and class rosters, and thesis material. The technical reprint file is the largest part of the collection and is an extensive compilation of early research literature on the internal combustion engine. The Taylor Papers also include a collection of glass slides which Taylor used for lectures and book illustrations.
Gift of C. Fayette Taylor, 1964.
Unpublished finding aid available.

T8 Taylor, Edward R., 1844-1919.
Papers, 1812-1925. - 8.4 cu. ft.
In Cornell University Libraries, Department of Manuscripts and University Archives (Ithaca, NY).
Chemical manufacturer; head of Taylor Chemical Company; inventor of the electrothermal process for the production of carbon disulfide. - Includes ledgers, cash books, check stubs, diaries, letter books, accounts, proceedings of the American Association for the Advancement of Science (1891-1908, various locations), song books, statements, invoices, bills, receipts, and printed matter, some related to the Taylor Chemical Company.
Unpublished finding aid available.

T9 Taylor, Frederic William, 1860-1944.
Papers, 1897-1944.
5 ft. (ca. 400 items)
In University of California at Los Angeles Library.
Agriculturist. Correspondence, clippings, certificates, awards, photos., printed matter, and other papers concerning Taylor's activities as Director General of Agriculture, El Salvador, and developer of rubber production in the U. S. Includes 14 scrapbooks containing letters, reports, and printed matter compiled by Taylor while superintendent of horticulture and director of concessions at the Buffalo Exposition (1901) Titles of the volumes include McKinley's assassination; Annals, horticulture; Concessions reports; Foods and their accessories; and Reports of the Agricultural and Horticultural Divisions.
Gift of Mrs. Taylor, 1944.

T10 Taylor, John, 1808-1887.
Family papers, 1838-1978. - 35 ft.
In University of Utah Libraries, Special Collections Department (Salt Lake City).
Businessman; third President of the Mormon Church. - Collection includes papers of John Taylor, his son John W. Taylor, and grandsons Samuel W. and Raymond W. Taylor. John Taylor was instrumental in attempting to first establish a sugar refining process in Utah in the 1850s. Includes correspondence between Samuel and Raymond Taylor about this attempt.
Gift of Samuel and Raymond Taylor, 1969-1970 and 1978.
Unpublished finding aid available.

T11 Taylor-Wharton Iron & Steel Co., High Bridge, New Jersey and Easton, Pennsylvania.
Records and papers, 1742-1950. - Ca. 400 items.
In Hagley Library (Greenville, DE) (acc. #1292).
The company's origin dates from 1742 when the Union Furnace was built in Hunterdon Co., NJ by William Allen (1704-1780) and Joseph Turner (1701-1783), both of Philadelphia. The firm was important during the 19th century in producing equipment for railroads. In 1892 it introduced the manufacture of manganese steel in the United States.
Records include: accounts of the original owners; original deed; day books; organizational records of the company at various stages; auditors' salary rolls; order and shipping records; sales and cost journals; annual reports; and a financial record book. In addition, there are items pertaining to the 175th and 200th anniversary celebrations of the company, including news items, genealogical data, correspondence, imprints, and advertising material; and miscellaneous editions of newspapers, all with articles concerning the Taylor-Wharton Iron & Steel Co.
Described more fully in the Supplement to A Guide to Manuscripts in the Eleutherian Mills Historical Library (1978).

T12 Terrell, James Wharey, 1829-1908.
Papers, 1813-1908.
2113 items and 8 v.
In Duke University Library.
Businessman, of Webster, N. C. Correspondence, account sheets, bills, receipts, deeds, contracts, summonses, and other papers, relating to Terrell's work as business manager for William Holland Thomas, conditions among the Cherokee Indians in North Carolina, the tanning business, merchandising, and wagon making.
Card index in the library.
Acquired, 1950.

T13 Thomas family.
Papers, 1883-1978. ca. 8 ft. (ca. 8000 items)
In Winthrop College Library (Rock Hill, S.C.)
In part, transcripts (typewritten) of Louise E. Thomas' diaries.
Correspondence, diaries, genealogies, speeches, financial and estate records, brochures, newspaper clippings, memorabilia, and photos, chiefly 1920-45, of a Rock Hill, S.C., family. Includes material on Winthrop Training School and Winthrop College, where Roy Zachariah Thomas, Sr., was professor of chemistry; Thomas Tours, an educational tour business which he founded; life during the Depression; ROTC training at Clemson University during World War II; training in the U.S. Army Air Corps during the war; and effect of the U.S. atomic bombing of Japan.
Unpublished finding aid in the repository.
Described in Winthrop College Archives and Special Collections: A Guide to the Manuscript and Oral History Collections (1978).
Gift of Louise Thomas Miller.

T14 Thompson, Charles Manfred, 1877-1963.
Papers, 1900-63. 2 ft.
In University of Illinois, University Archives.
Professor of economics and dean of the College of Commerce and Business Administration of the University of Illinois. Correspondence, autobiography, articles on economics and historical topics, reports, printed matter, and photos, relating to Thompson's career. Includes material on the investigation (1911-14) of the route traveled by Abraham Lincoln and his family from Kentucky through Indiana to Illinois, Red Grange, the Illinois Post-War Planning Commission, and the synthetic rubber program (1942).
Unpublished finding aid in the repository.
Received 1963-64, 1966.

T15 Thompson family.
Papers, 1813-44. ca. 350 items, 1 box, and 2 v.
In Maryland Historical Society Library (823, 990, 1442)
Correspondence, bills, receipts, day book, other business papers, and miscellaneous personal papers, of a Baltimore merchant and commission merchant family, principally Hugh Thompson (1760-1826), James Thompson (d. 1836), and James Thompson (fl. 1833-44). Includes correspondence with London and Dublin merchants, lists (1833-35) of policies on ships and cargoes, and commission orders and sales of chemicals and drugs (1837-44). Correspondents include Baring Brothers, John Gordon, Hall Harrison (1774-1830), George Law, Robert Oliver, David Steuart, and William S. Thompson.
The vol. listing policies on ships and cargoes is indexed.
Gifts of W. Bryant Tyrrell from the estate of Frank C. Kirkwood, 1946, and Jeannette B. Dubbin, 1953.

T16 Thompson family.
Papers, 1607-1903.
ca. 6000 items and 125 v.
In Historical Society of Pennsylvania collections.
Correspondence and other papers of Jonah Thompson and his descendants, Philadelphia Quakers, humanitarians, merchants, ironmasters, and landowners. Includes Jonah Thompson's manuscript books on the manufacture of iron, nails, machinery, etc. (1783-1829), and journals written abroad (1812); George Thompson's papers (1769-1876) relating to land holdings at Fort Pitt, legal controversies, administration of estates, Thompson's business interests, etc. (1831-76), and to Eastern Penitentiary (1839-50); accounts of John Thompson with Robert Morris; John J. Thompson's papers (1803-1903) relating to the iron industry, manufacture of steam engines, machinery, and other interests; papers of estates in which the Thompsons were interested; Joseph Trotter's papers (1725-1868) on literature, ethics, and commerce; Newbold family papers (1755-1838), letter and daybooks, material on Mexico; Abel James' account with Lawrence Growdon (1765-68); papers of John Harper (1776-79), Thomas Pleasants (1776-93), Henry Drinker (1771-83), Joseph Galloway (1782-92); accounts, legal papers, bonds, deeds, surveys, Quakeriana, monthly Quaker meeting reports; copies of deeds (1676-89), Sir William Johnson letter to Teedyuscung (1760), New England colonists' petition against the Sugar act (1763), petition for a road from Philadelphia to Chester (1764); North Carolina land warrants, surveys, leases (1796-1838); papers of the Deaf and Dumb Institution, Philadelphia (1813-17); Thaddeus Stevens letters (1813), William C. Poultney journal (1805), journal of a trip to Europe on the Montyuma (1827), James Gallagher journal (1831), Hanover Furnace records (1793-1838), Bridgewater, N. J. copper mine records (1831), and broadsides on fire extinguishers (1848)
Purchase.

T17 Thompson, James, fl. 1833-1844.
Account books, 1833-1844. - 2 v.
In Maryland Historical Society Library (Baltimore).
Baltimore commission merchant who dealt in chemicals and drugs. - Volume I lists policies on ships and cargoes, 1833-1835, with 3,143 entries, fully indexed in front; Volume II is a day book for purchases and sales of chemicals and drugs, 1837-1844. Many items connected with Philadelphia and New York firms.
Gift of W. Bryant Tyrrell, September, 1946.

T18 Todd, Clare Chrisman, 1880-1954.
Papers, 1895-1939. 33 ft.
In Washington State University Library (Pullman)
Professor of chemistry and Dean of the College of Sciences and Arts, Washington State University, Pullman. Correspondence, reports, notes, and other papers, relating to the Dept. of Chemistry and College of Sciences and Arts of the University of Washington, Pullman, the U. S. Office of Education's survey of land-grant colleges and universities (1929), and research in chemistry.
Unpublished container list in the library.
Information on literary rights available in the library.

T19 Torrey, Henry A., 1871-1910.
Notebook, undated. - 1 volume.
In Harvard University Archives, (Cambridge, MA).
Assistant Professor of Chemistry - Research notebook.

T20 Tower family.
Papers, ca. 1820-1850. - 3 ft.
In University of Michigan, William L. Clements Library (Ann Arbor).
The Charlemagne Tower family operated a brewing industry in Waterville, Oneida County, New York, throughout the early nineteenth century. Collection contains family and business correspondence.

T21 Traube, Isidor, 1860-1943.
Autobiography, 1930, 1942. - Ca. 150 pages (photocopy).
In American Philosophical Society Library (Philadelphia, PA).
German chemist. - Maps, and a brief autobiographical sketch by Traube written in 1930 (11 pages), a longer version of 1942 (101 pages), and a bibliography (28 pages). This manuscript also includes notes on Traube's life by John T. Edsall.
Presented by John T. Edsall, 1982.

T22 Traverse City, Michigan. C. A. Bugbee Drug Company.
Formula book, 1881. - 1 v.
In University of Michigan, Bentley Historical Library (Ann Arbor).
Formula book of Charles A. Bugbee, proprietor of the C. A. Bugbee Drug Company, Ltd. of Traverse City, Michigan.

T23 Travis, Eugene Porter.
Papers, 1880s - 1948. - 1 cu. ft.
In Chewelah Historical Society (WA).
Miscellaneous papers of Travis, a pharmacist at Chelawah and Northport, Washington; also includes papers of his father J.J. Travis, a physician and pharmacist, including a diary kept at Chewelah in 1889, and a notebook of drug remedies.

T24 Tredegar Company, *Richmond*.
Records, 1836-1957. 507,505 items and 1118 v.
In Virginia State Library, Archives Division.
In part, photocopies (negative) of originals owned by the Tredegar Company and Mr. D. G. Haase.
Correspondence; minutes of the directors and stockholders; accounts; ledgers; cashbooks, receipts, and disbursements; sales data; journal voucher books; stock certificate record books; monthly statements; War Dept. contracts (1907-41); reports to Government agencies; maps, plats, and surveys; suit papers for the Chesapeake and Ohio Railway v. Tredegar Company (1936); material regarding the merger of the Tredegar Company and the Albemarle Paper Company (1956-57); and other papers of the Tredegar Company, an iron works. Includes papers of the Anderson family and records of minor family enterprises including journals and ledgers of Joseph Reid Anderson, founder of the company; private correspondence of Archer Anderson, his son; private notebooks of E. R. Anderson; papers regarding the estates of Archer Anderson, Sr., St. George M. Anderson, and Joseph Reid Anderson; and records of the Anderson Association, Jessup-Purchase Association, J. R. Anderson & Company, and the A. Anderson (Special) Bacon Syndicate Purchase.
Accession analyses and detailed lists in the library.
Gift of the Tredegar Company, 1952, and its successor, the Albemarle Paper Manufacturing Company, 1958.

T25 Tregaskis, Richard, 1837-1914.
Correspondence, 1865-1907. 2 v.
In Detroit Public Library, Burton Historical Collection.
Accountant and inventor. Scrapbooks of correspondence relating to recommendations of Tregaskis as an accountant and to the sale of the process he invented for soldering aluminum.

T26 Trinity College. Student Essays.
Manuscript essays, 1858-1905.
In Trinity College Archives (Hartford, CT).
Manuscript essays (mostly handwritten) on a variety of prescribed topics in applied chemistry prepared by undergraduate students. Essays range in date from 1858 to 1905. Selected topics covered are: Water: Chemical Constitution, Properties and Uses (1858); Carbonic Acid (1859); Fire Damp (1863); Chlorine and its Compounds (1864); Potassium (1877); Explosives (1881); The Periodic Law (1894); Acetylene (1902); Radium (1905). Other topics include combustion; ozone; the atomic constitution of matter; phlogiston; photography; the spectrum of light; the electro-magnetic telegraph; the Voltaic Pile; the steam engine. Several essays on each of the above topics were prepared at various dates during the latter half of the nineteenth century.

T27 Trommer Extract of Malt Company, Fremont, Ohio.
Records, 1874-1890. - 21 ft.
In Rutherford B. Hayes Presidential Center (Fremont, OH).
Correspondence, invoices, and account books, of a medicinal extract manufacturing company formed by Ralph and Stephen Buckland, and Drs. Gustavus A. Gessner, and John B. and Robert H. Rice.

T28 Tucker, Henry St. George, 1853-1932.
Student notebooks, 1871-1876. - 4 v. Various sizes. Holograph.
In Virginia Historical Society (Richmond).
Kept while a student at Washington and Lee University, Lexington, Virginia, and include notes made from lectures on chemistry (by John Lyle Campbell), English grammar (by Edward Southey Joynes), Latin (by Carter Johns Harris) and Latin (by Charles Alfred Graves).
Gift of John Randolph Tucker, Richmond, Virginia, November 30, 1949.

T29 Tucker, Willis Gaylord, 1849-1922.
Portfolio of letters addressed to Willis Gaylord Tucker from relatives, friends, colleagues, 1862-1917. - 63 items.
In New York Academy of Medicine Library.
Professor of chemistry and toxicology, and registrar, Albany Medical College (N.Y.). - Correspondents include John Call Dalton, Elisha Harris, Abraham Jacobi, John Ordronaux, Lawson Tait, and Theodore Roosevelt.

T30 Turner, Francis John, b. 1904.
Papers, 1950-1980. - 6 cartons.
In University of California, Berkeley, Bancroft Library.
Professor of Geophysics, UC Berkeley. - Research notes, correspondence, photographs, grant proposals, writings, reviews, reprints regarding Turner's work with crystals of minerals at high temperatures and pressures.
Partially processed collection; folder list available in repository.

T31 Twitchell, Ernest, 1863-1929.
Notebooks, 1895-1917. - 0.2 ft.
In Emery Industries, Inc., Research Library (Cincinnati, OH).
Chemist; Perkin Medal winner in 1917; - Pocket notebooks with notations on formulations, analyses, etc. relating to his search for a method to determine the resin in the fatty acids of soaps, and methods for the hydrolysis of oils and fats for the production of free fatty acids and glycerin.

T32 Tyson family.
Papers, 1775-1889. 2 boxes.
In Maryland Historical Society Library (Baltimore) (2107)
Chiefly correspondence of Isaac Tyson, Jr. (1792-1861), his wife, Hannah A. Wood Tyson, and their sons, R. W. Tyson, James Tyson (1828-1900), and Jesse Tyson (1826-1906), with frequent references to Tyson Mining Company, Baltimore, Md., engaged in chrome mining. Includes letters to Isaac Tyson, Jr., from his son, R. W. Tyson, from Haverford School, Haverford, Pa., and school letters from his other sons; letters to James Tyson on boxing in Sykesville, Md.; letters from Jesse Tyson on a business trip to London, Eng., Glasgow, Scot., and Frankfurt, Ger., on the chrome industry in Europe and French and English attitudes toward the U.S.; and letters relating to activities of the Society of Friends, of which the Tysons were members (including the Tyson's attitudes on the Civil War), development of the chrome industry in the U.S., and other family matters.
Gift of Miss Hannah Ann Tyson Watters, Baltimore, Md., 1975.

U1 Ullman family.
Papers, 1857-1965. - 0.6 linear ft.
In Case Western Reserve University Archives (Cleveland, OH).
German Jewish family, especially Morris Ullman (1835-1908), who immigrated to United States in 1849 and eventually established Ullman, Einstein & Co., one of the largest liquor distilleries in the United States. - Personal and general correspondence; financial papers; legal papers; distillery formula books relating to the Ullman, Einstein & Co.; miscellaneous Ullman family receipts, certificates, newspaper clippings, and scrapbooks. In particular, the collection contains papers of Ullman, Einstein & Co., including correspondence, receipts, articles of incorporation and formulas, 1865-1917.
Donated by Rufus M. Ullman, 1975.
Open to the serious scholar.
Unpublished finding aid available.

U2 Ulster Iron Works.
Notebook, 1816-1866. - 474 pp.
In University of Michigan, William L. Clements Library (Ann Arbor).
Iron works whose home office was in Saugerties, New York. - Bound manuscript notebook contains technical descriptions of processes, machinery, and products; recipes, labor and costs. Covers iron works at several locations in America and England.

U3 Underhill, Frank Pell, 1887-1932.
Papers, 1899-1932. - 4.75 linear ft.
In Yale University Library, Department of Manuscripts and Archives (New Haven, CT).
Pharmacologist. - Underhill received his degree from Yale in 1900; was an assistant in physiological chemistry from 1900 to 1903; was Instructor, Assistant Professor, and Professor of Pathological Chemistry between 1903 and 1918; Professor of Experimental Medicine from 1918 until 1921; and Professor of Pharmacology and Toxicology from 1921 until 1932. The collection contains research material, including data on poison gas collected during World War I by the Chemical Warfare Services and for his book, The Lethal War Gases (1919). Included also are experimental data on food and diet and correspondence relating to his teaching career at Yale University, as well as on his publications.
Papers were given to Yale Medical Library by the estate of Frank P. Underhill. Transferred to Manuscripts and Archives in 1952.
Unpublished finding aid available.

U4 United States. Assay Office, Helena, Mont.
Records, 1900-07. .1 ft.
In Montana Historical Society collections (Helena)
General correspondence (1901-07), financial records (1900-05), reports (1901-05), and miscellaneous papers.

U5 United States. Bureau of the Mint, Branch Assay Office, Seattle, Washington.
Records, 1898-1955. - 99 cu. ft.
In National Archives and Records Service, Archives Branch (Seattle, WA).
Correspondence, assaying records, registers of deposits of gold and silver, and miscellaneous financial records. (RG 104).

U6 United States. Hygienic Laboratory.
Letterpress books of the Division of Chemistry, 1905-15. - 4 v.; 30 cm.
In National Library of Medicine (Bethesda, MD).
The first chief of the Division of Chemistry was Joseph H. Kastle; he held that position from 1905 to 1909. Vol. 1: Aug. 22, 1905-June 1, 1908; vol. 2: June 5, 1908-July 1, 1909; vol. 3: June 23, 1909-Aug. 8, 1911; vol. 4: Aug. 1, 1911-Jan. 7, 1915.

U7 United States. Special Committee to Study the Rubber Situation.
Records, 1942. - 8 ft.
In Franklin D. Roosevelt Library (Hyde Park, NY).
Correspondence, stenographic notes and verbatim transcripts of committee hearings and proceedings, reports on rubber production, technical studies on synthetic processes, and copies of congressional hearings. The committee, in its final report, recommended continued and increased use of the Polish and butylene glycol processes for rubber production. Members of the committee are Bernard Baruch, Karl Compton, Samuell Lubell, and James B. Conant, chairman.
Open to researchers under restrictions accepted by the library.

U8 United States Chemical Warfare Association, Boston Section.
Records, 1924-1926. - 0.3 cu. ft.
In Massachusetts Institute of Technology, Institute Archives and Special Collections (Cambridge, MA).
This organization was formed in 1924 to promote the use of chemical warfare in national defense. The records include correspondence regarding the purpose and establishment of the organization, minutes of meetings, information on membership, treasurer's reports, and several publications.

U9 United States Pharmacopoeial Convention, 1900-
Records, 1819-1969. - 154 boxes.
In State Historical Society of Wisconsin (Madison).
Collection includes correspondence, minutes, proceedings, reports, circulars, and financial records relating to the revision of the Pharmacopoeia of the United States and the annual conventions.
Gift of the American Institute of the History of Pharmacy, Madison, Wisconsin, and the United States Pharmacopoeial Convention via Mrs. Carol G. Wingo, Bethesda, Maryland, 1970, 1973.
Unpublished finding aid available.

U10 United States Rubber Company.
Records of the Boston office, 1876-1900.
3997 items.
In Harvard Business School, Baker Library.
Minutes of executive committee and directors' meetings, lists of shares of stock, memoranda of agreements, production records, statements of financial condition, and correspondence.
Gift of H. Stuart Hotchkiss, 1935.

U11 University of California, Berkeley. College of Chemistry.
Records, 1874-1955. - 10 boxes, 1 v.
In University of California, Berkeley Archives.
Correspondence, notes, subject files, and other material of faculty members, including Edmond O'Neill, G. N. Lewis, William Bray, Wendell Latimer, John I. Winkler, and Willard B. Rising. Minutes of faculty meetings (1909-1916).
Finding aid available.

U12 University of California, Berkeley, history: oral history interviews, 1954-70. 38 items.
In University of California, Berkeley, Bancroft Library.
Transcripts of tape-recorded interviews with persons prominent in the history of the University of California, Berkeley, from 1900-70, particularly in the development of the departments of physics, chemistry, education, art, agriculture, philosophy, medicine, public health, environmental design, dramatic arts, and English; the president's office and administration and finances; athletics; student housing and organizations; extension education; and working relations with the State Legislature. Interviewees include Raymond Thayer Birge, Allen C. Blaisdell, Ralph Works Chaney, James H. Corley, Ira Brown Cross, Mary Blossom Davidson, William R. Dennes, Ruth Donnelly, Carroll "Ky" Ebright, Clinton W. Evans, John Gregg, Brutus Hamilton, William Charles Hays, Joel Henry Hildebrand, Claude B. Hutchinson, Marguerite Kulp Johnston, Benjamin H. Lehman, Ferdinand D. Lessing, Victor F. Lenzen, Ralph P. Merritt, Karl F. Meyer, Lucy Sprague Mitchell, Joseph R. Mixer, Eugen Neuhaus, John Francis Neylan, Stephen C. Pepper, Robert Langley Porter, Peter J. Shields, Ida Wittschen Sproul, Frank C. Stephens, Katherine A. Towle, Walter Treadway, Robert M. Underhill, Henry C. Waring, Baldwin M. Woods, Jean C. Witter, and William Wilson Wurster.
Catalog and card index in the university's Regional Oral History Office, which conducted the interviews, retaining the original tape recordings and depositing the transcripts in the Bancroft Library. Each transcript also is indexed.
Partially restricted.
Information on literary rights available in the library.

This is a composite collection of individual interviews; new material is added regularly.
Photo-offset duplicates have been made. The Library of the University of California, Los Angeles, has a copy of the transcripts. Information on the location of other copies is available from the Regional Oral History Office.

U13 University of Chicago. Physical Sciences Division.
Papers, 1937-1949. - 13.5 linear ft.
In University of Chicago Library.
Includes Departments of Chemistry, Physics, Astronomy, and Meteorology; Institute for the Study of Metals; and Institute for Nuclear Studies. Departmental budgets, research reports, correspondence, salaries, fellowships, and scholarships.
Partially restricted.
Unpublished finding aid available.

U14 University of Illinois. Department of Chemistry and Chemical Engineering.
Records, ca. 1868-1978. - 14.6 linear ft.
Collection includes Departmental History, 1868-1961 (printed circular); The Study of Chemistry at the University of Illinois, 1907, a printed booklet for potential students; William A. Noyes inauguration addresses, 1907; lecture and symposium announcements and proceedings; departmental publications; programs and announcements; programs and announcements of the Illinois ACS Section; programs of the High School Teachers Summer Institute; announcements and correspondence of the Chemistry Show; departmental reprints; reprints of the Inorganic Division; Organic Division publications; Organic Seminar abstracts; Inorganic Seminar abstracts; Chemical Engineering Division publications; Biochemical Division studies; Analytical Division studies; course materials; chemical engineering research reports; chemical engineering class of 1930 alumni letters; Illinois Chemist; chemical engineering class of 1933 newsletters; Alumni Newsletter; departmental letterbooks; rare earth research journals; and Illinois Chemist file.

U15 University of Michigan. Chemical Laboratory
Papers, 1868-1917. - 23 items.
In University of Michigan, Bentley Historical Library, Michigan Historical Collections (Ann Arbor).
Records of the chemical laboratory of the University of Michigan. - Includes communications with the university administration concerning funding, purchases, faculty appointments, and building plans; and reports, including history and development of the chemical laboratory, 1909-1920.
Unpublished finding aid available.

U16 University of Michigan. College of Pharmacy.
Papers, 1873-1962.
In University of Michigan, Bentley Historical Library, Michigan Historical Collections (Ann Arbor).
Records of the University of Michigan College of Pharmacy. - Includes minutes of the College and correspondence files of Deans Howard Lewis, Charles H. Stocking, and Thomas D. Rowe concerning the teaching of pharmacy, the administration of the College, and the activities of the American Pharmacy Association.
Unpublished finding aid available.

U17 University of Michigan. Department of Chemistry.
Papers, 1856-1921. - 16 items and 2 v. (oversize).
In University of Michigan, Bentley Historical Library, Michigan Historical Collections (Ann Arbor).
Records of the University of Michigan Department of Chemistry. - Collection contains communications to Board of Regents requesting funding or recommending appointments; report, 1904, on "The Need of a New Chemistry Building"; and records, 1885-1909, of experiments performed in quantitative chemistry, including some experiments of Moses Gomberg.
Unpublished finding aid available.

U18 University of Michigan. Department of Mineralogy and Petrography.
Records, 1915-1933. - 50 items.
In University of Michigan, Bentley Library (Ann Arbor).
Correspondence and records of the Department. Correspondents include Edward H. Kraus.

U19 University of Missouri. Department of Agricultural Chemistry.
Records, 1910-1930. - 4 boxes, and oversize material.
In University of Missouri-Columbia Library, Western Historical Manuscripts Collection and State Historical Society of Missouri Manuscripts.
Correspondence of L. D. Haigh, P. F. Trowbridge, and A. G. Hogan (1884-1961). Subjects include animal feed analyses and fertilizer inspections. Also includes material on nitrogen in beef extracts; feed stuff legislation; Association of Official Agricultural Chemists; and departmental annual reports and project reports, 1916-1927. Oversize materials include photographs and blueprints.
Transferred to the Library, December 7, 1984.
Unprocessed as of 1985. Available for research.
Unpublished finding aid available.

U20 University of Missouri. Department of Veterinary Physiology and Pharmacology.
Records, 1962-1967. - 63 folders.
In University of Missouri-Columbia Library, Western Historical Manuscripts Collection and State Historical Society of Missouri Manuscripts.
Correspondence with university departments; veterinary medicine schools; private and federal research organizations; U.S. Air Force aeromedical research departments; commercial drug, chemical and medical supply companies; professional veterinary medicine associations and journals; prospective students; independent researchers. Other correspondence concerns department business and procedures, equipment, staff, graduate program, applications for graduate study, conventions, recommendations, the expanding department, the American Society of Veterinary Physiologists and Pharmacologists, and the department chairman's correspondence (faculty hiring, etc.). Other series are: curriculum; the graduate program and budget; copies of exams and laboratory schedules; and departmental information, including lists of department members, requirements for a Ph.D. in physiology, publications by department members, and research interests and activities of department members.
Received July 25, 1967, from the Department of Veterinary Physiology and Pharmacology.
Unpublished finding aid available.

U21 University of Notre Dame Radiation Laboratory.
Records, 1946 - ca.1966. - 44 linear ft.
In University of Notre Dame Archives, (Notre Dame, IN).
Laboratory established in 1946 to study the effects of high energy radiation through the work of faculty members, research chemists, and doctoral and postdoctoral students. It is part of the Atomic Energy Commission's system of research centers. - Collection includes research notes, logs, and reports; correspondence, personnel records, budget material, and other administrative records; documents relating to numerous scientific meetings and conferences; and reprints of articles.
The collection also includes some personal papers of Milton Burton, Notre Dame chemistry professor and Director of the radiation laboratory from its inception in 1946 until his retirement in 1971. Burton's papers include correspondence, research records, lectures, and documents relating to his travels and participation in scientific meetings. Because of his close association with the laboratory, it is difficult to clearly separate Burton's personal papers from the laboratory's records.
Collection restricted; researchers should apply to the University Archivist.

U22 University of Pennsylvania. Department of Chemistry.
Records, 1920-1959. - 21 boxes.
In University of Pennsylvania Archives (Philadelphia).
Correspondence of the Chemistry Department at the University of Pennsylvania. -- Subjects include applications for assistantships and instructorships; Executive Committee actions; Engineering Alumni Society; student placement; summer courses; and other topics.
Received March 11, 1981.
Inventory at the repository.

U23 University of Texas. Department of Pharmacology and Toxicology.
Scrapbook, 1964. - 8 items and 2 v.
In University of Texas Medical Branch, Moody Medical Library (Galveston).
Scrapbook, "Historical Development of the Department of Pharmacology and Toxicology," prepared by Hans Ash. Also, letters between Ash and Dr. R. M. Calder concerning writing of the history.

U24 University of Washington. Chemical Engineering Department.
Publications, 1956-73. - 1 ft.
In University of Washington Archives (Seattle).
Publications and bulletins issued by the Department.
Unpublished inventory available.

U25 University of Washington. Chemical Engineering Department.
Records, 1950-1981. - 14 boxes.
In University of Washington Archives (Seattle).
General correspondence, with the ACS, the AIChE, Dow and Monsanto Chemical Companies, and others; intra-university correspondence; minutes of faculty meetings; annual and miscellaneous reports; subject files; and subgroups of files, on the UW Engineering College Executive Committee, the UW Forest Products Institute, the UW Nuclear Engineering Committee, the Promotions and Tenure Committee, and others.
Promotions and Tenure Committee material is restricted.
Unpublished finding aid available.

U26 University of Washington. Nuclear Engineering Department.
Records, 1959-70. - 4 cu. ft.
In University of Washington Archives (Seattle).
Administrative records containing reports, curriculum records, publications, grant and contract records, and correspondence.

U27 University of Washington. Pharmacy College.
Records, 1936-71. - 1 cu. ft.
In University of Washington Archives (Seattle).
Correspondence, printed matter, and reports. The name was changed to Pharmacy School in 1971.
Unpublished inventory available.

U28 University of Washington. Pharmacy School.
Records, 1971-77. - 0.5 cu. ft.
In University of Washington Archives (Seattle).
Administrative reports, conference and convention files, and curriculum records.

U29 University of Wisconsin. School of Pharmacy Library.
Reference files, 1850 -. - Ca. 150 ft.
In University of Wisconsin, School of Pharmacy Library (Madison).
In part, photocopies (negative).
Letters, laboratory records, minute books of organizations, student notebooks, account books, prescriptions, pictures, pamphlets, circulars, reprints, broadsides, and other printed matter, relating to pharmacy and dating chiefly from 1875 to 1930. Most of the material was originally collected by Edward Kremers, professor of pharmacy, and is now being maintained and actively developed jointly by the University of Wisconsin School of Pharmacy and the American Institute of the History of Pharmacy.
Described in Bulletin of the Medical Library Association, v. 47 (1950), no. 2, p. 141-142.
Gift and deposit, 1890-.

U30 Upjohn Company, Kalamazoo, Michigan.
Papers, 1887-1934. - 9 items.
In University of Michigan, Bentley Library, Michigan Historical Collections (Ann Arbor).
Pharmaceutical company. - Historical sketch of the company, typewritten copy of minutes of three Board of Directors meetings, photostats of patents, articles of incorporation, and other papers.

U31 Upjohn family.
Papers, ca. 1795-1974. - 5 ft., 1 v. and 7 items.
In University of Michigan, Bentley Historical Library, Michigan Historical Collections (Ann Arbor).
Correspondence, sermon notes, daybooks, accounts, timber records, genealogical and biographical material, printed matter, and other papers (collected by Robert U. Redpath and E. Gifford Upjohn) of a pioneer family of upstate New York and Hastings and Kalamazoo, Michigan, and the related Clough, Kirby, and Mills families. Includes journal (1830) of a trip to America and on the Erie Canal; historical records of Upjohn Pharmaceutical Company, Kalamazoo, Mich.; minutes (1833) of Greenbush Debating Society (NY); papers concerning work of Clough family members as missionaries in southern India; and material concerning family activities, medical practice, and daily life. Persons represented include John Everett Clough (1836-1910), Gratia (Clough) Upjohn (1874-1954), Henry Uriah Upjohn (1843-1887), Lawrence Northcote Upjohn (1873-1967), Millie Kirby Upjohn (1852-1920), Uriah Upjohn (1808-1896), William Upjohn (1770-1847), and William Upjohn (1807-1887).
Unpublished finding aids in the repository.

U32 Urey, Harold Clayton, 1893-1981.
Papers, 1934-81. - 61 linear feet.
In University of California, San Diego, Libraries.
Chemist; Nobel laureate. - Urey is best known for the discovery of deuterium, for which he received the 1934 Nobel Prize in chemistry. He was a member of the Columbia University faculty, 1929-45, and participated in research conducted there for the Manhattan Project. He continued his career at the University of Chicago, 1945-58, and at the University of California, San Diego, 1958-70. The bulk of the collection concerns Urey's work after 1945. The collection includes correspondence, manuscripts of lectures and speeches, committee records, laboratory notebooks, and research notes. Also included are files concerning the Emergency Committee of Atomic Scientists (1946-49); a small amount of material on the Rosenberg and Sobell cases; and several inches of correspondence documenting Urey's work as a science advisor to President-elect John F. Kennedy. There is also extensive documentation of Urey's developing interests in and research on rocketry and space science and exploration during the late 1950s and 1960s.

V1 Vanderbilt, Cornelius, 1873-1942.
Papers, 1776-1953. - 3 linear ft.
In Yale University Library, Department of Manuscripts and Archives (New Haven, CT).
Capitalist; Yale alumnus (A.B., 1895, Ph.B., 1898, M.E., 1899); officer in the U.S. Engineers, 1917-1918. - The bulk of the papers is from 1901 to 1928. Collection contains chiefly correspondence, memoranda, a notebook, and memorabilia relating to Vanderbilt's service with the U.S. Engineers. Also included are technical notebooks on mathematics, chemistry, and engineering (1897-1926); checkbooks; guestbooks; a nurse's record of Vanderbilt's case of typhoid fever (1902-1903); and inventories of the Vanderbilt Library in New York City and of his yacht.
Gift of Mrs. Cornelius Vanderbilt and the Vanderbilt Estate, 1944-1945; and Donald Wing, 1968.
Unpublished finding aid available.

V2 **Van Horn, Frank Robertson**, 1872-1933.
Papers, 1893-1932. 1 ft.
In Case Western Reserve University Archives (Cleveland, Ohio)
Professor of mineralogy and geology at Case Institute of Technology and director of athletics. Correspondence, photos, reprints of Van Horn's writings in mineralogy and geology, and notes taken (1893-97) during his doctoral studies at Heidelberg University.
List of items available in the repository.

V3 Van Slyke, Donald Dexter, 1883-1971.
Oral history interview, 1969. - 3 reels; 5 hours.
In National Library of Medicine (Bethesda, MD).
Chemist. - Interview conducted by Peter D. Olch, in Upton, Long Island, May 27-28, 1969. Autobiographical memoir, including descriptions of the institutions, research, and individuals important in Dr. Van Slyke's career. Dr. Van Slyke was Director of the Chemical Laboratory of the Rockefeller Institute Hospital, 1913-48, and from 1949 to 1971 was associated with the Medical Department of Brookhaven National Laboratory. Among the persons mentioned in the interview are Emil Fischer, Simon Flexner, and Albert Baird Hastings.
Transcript, with index and appendix (130 p.). Includes a copy of autobiographical essay prepared in 1952 (pp. 131-144).
Information on restrictions available in the Library.

V4 Van Zoeren, Gerrit John, 1884-1980.
Scrapbook, 1835-1955. - 1 v. (contains 150 items).
In Hope College Archives (Holland, MI).
Industrialist and Hope College alumnus. - Scrapbook, entitled "How Chemical Specialties Grew," documents the founding and development of a chemical company in Zeeland, Michigan, until its absorption by Miles Laboratories, Elkhart, Indiana. Includes manuscripts, clippings, and photographs.

V5 Van Zyl, Gerrit, 1894-1967.
Papers, 1928-1967. - 190 items.
In Hope College Archives (Holland, MI).
Professor of Chemistry at Hope College. - Correspondence, including letters recommending Van Zyl for the Manufacturing Chemists' Association College Chemistry Teacher's Award; proposals to research corporations; surveys of Hope College science graduates who obtained degrees; slides used with Van Zyl's speeches; clippings; articles; citations; and student paper, "Dr. Van Zyl and the Growth of the Hope Chemistry Department," by Kevin Echart, 1973.

V6 Varian Associates.
Records, ca. 1948-1972. - 6 cartons.
In University of California, Berkeley, Bancroft Library.
Manufacturer of electronic laboratory instrumentation. - Corporate records and reports, including financial statements, minutes of meetings of the board of directors, studies, and news releases concerning electronics research and production.
Finding aids available.

V7 Varney, Thomas, d. 1900.
Papers of Maria and Thomas Varney, 1849-1878. 1 box.
In California Historical Society Library (San Francisco)
Diaries (1849, 1852-1854, 1857-1878) of Varney, inventor and businessman, and his wife, Maria (d. 1900). The 1849 diary relates to Varney's voyage to California. Three of the journals contain his experiments with gunpowder, cartridges, engines, and a burner for use as a lamp, notes of progress in a property suit, drafts of letters to newspapers, and letter concerning a fraudulent patent. Mrs. Varney's diary is chiefly a scrapbook of newspaper clippings of letters to the editor written by her.

Vassar, Matthew, 1792-1868.
Papers, 1826-68. 3 ft.
In Vassar College Library (Poughkeepsie, N. Y.)
Poughkeepsie, N. Y., brewer, merchant, and founder of Vassar College. Correspondence, diaries, account books, and business records. Includes letters from Sarah J. Hale, editor.

Vermeule, Adrian, Jr.
Papers, ca. 1885. - 1 manuscript box.
In Rutgers University Library, University Archives Collections (New Brunswick, NJ).
Student; Rutgers alumnus. - Student notebooks and lecture notes, including lectures in geology by George H. Cook, and lectures in chemistry.
Unpublished finding aid available.

Vernon-Benshoff Company, Pittsburgh, Pa.
Records, 1911-72. 1 ft. and 4 cassette recordings.
In University of Pittsburgh Libraries, Archives of Industrial Society (Pa.)
Correspondence, technical papers, contracts and legal papers, accounting records, tape recordings and transcripts of company history, publications, memorabilia, and photos, of a company which developed the dental base Vernonite; and writings by Harold Vernon.
Unpublished finding aid in the repository.
Gift of Harold Mark Vernon.

Vickery, Hubert Bradford, 1893-1978.
Papers, 1769-1978. - 32 linear ft., 1 microfilm reel, 1 framed item.
In Yale University Library, Department of Manuscripts and Archives (New Haven, CT).
Biochemist. - Vickery received degrees from Dalhousie (1915 and 1918) and from Yale (1922). He served at the Connecticut Agricultural Experiment Station as Assistant Biochemist, Biochemist in Charge, and Emeritus Biochemist between 1922 and 1978; was a lecturer at Yale from 1924 to 1963; and was an observer at the Bikini Island atomic bomb tests. Collection contains correspondence, manuscripts, reprints, and research material relating to Vickery's work at the Connecticut Agricultural Experiment Station. Also included are files Vickery created from his Bikini Island observations, and genealogical materials and legal and financial papers of the Bradford family.
Gift of Jeannette Opsahl, 1979.
Unpublished finding aid available.

Villars, Donald Statler, b. 1900.
Papers, ca. 1935-1963. - 200 cartons.
In University of California, Berkeley, Bancroft Library.
Physical chemist. - Correspondence, reports, reprints of articles, notebooks, subject files, teaching and research films, photographs, research notes and other material relating to Villars' work in physical chemistry. Includes computer programs and material regarding the application of statistics to chemical experimentation.
Partially processed collection; folder list available in repository.

Vinograd, Jerome, 1913-1976.
Papers, 1931-1976. - 25 boxes.
In California Institute of Technology Library (Pasadena).
Chemist and molecular biologist, California Institute of Technology. - Correspondence, mss. of writings, lecture notes, conference files, proposals and grants, and personal papers. Includes material relating to DNA. Bulk of the collection is dated 1960-1976. Correspondents include William R. Bauer, Piet Borst, Lionel Crawford, Francis H. Crick, Henryk Eisenberg, Lawrence I. Grossman, Bruce Hudson, and Jacob Lebowitz.
Finding aid in the repository.

Virginia Brewery.
Records, 1866-1904. - 1 linear ft.
In Montana Historical Society (Helena).
The Thorn-Smith Brewery, located in Virginia City, Montana, was purchased by Henry Gilbert of Berks Co., Pennsylvania in 1864. Christopher Richter, a German cooper and brewer, joined the firm as a full partner in 1867. Closed when Montana's prohibition laws went into effect (ca. 1916-1918), it re-opened in 1934 and remained open until its charter expired in 1974. Its name was changed from Virginia Brewery to Gilbert Brewery in 1872. - Collection includes records of Virginia Brewery and Gilbert Brewery: invoices, brewer's book, several daybooks (one containing German entries), 2 account ledgers, and a beer account book.
Unpublished finding aid available.

Voegtlin, Carl, 1879-1960.
Autobiographical notes, 1958. - 29 p.
In National Library of Medicine (Bethesda, MD).
Biochemist. - Copy of typescript. Includes journal obituary notice.

V16 **Vogel, Hermann Wilhelm,** 1834–1898.
    Papers, ca. 1880–98.
    1 box.
    In Columbia University Libraries.
    German photochemist and photographer. Correspondence, autobiography, MSS. of articles and studies, notebooks, proofsheets, reprints, and photos.

V17 **Voorhees, Louis Augustus.**
    Papers, 1848–1943. ca. 1 ft.
    In Rutgers University Library (374)

Correspondence, bills, and other papers of Voorhees as chief chemist of the New Jersey State Agricultural Experiment Station and as a Masonic official in New Brunswick, N. J.; Civil War camp drawings and University of Pennsylvania medical lecture tickets (1848–79) belonging to his father, Dr. Charles H. Voorhees; and papers (1898–1943) concerning the Van Voorhees Family Association.
    Gift of Mr. Voorhees.

W1 **W.P. Fuller and Company.**
Letter books, 1896-1921. - 5 v.
In California Historical Society Library (San Francisco)
Pioneer West Coast firm manufacturing and selling paints, oil, and glass, founded by William P. Fuller as successor of Fuller & Heather of San Francisco and Sacramento, Calif., 1857-1868, and of Whittier, Fuller & Company of San Francisco, 1868-1894. - Letters sent to the company from its subsidiaries, Pioneer White Lead Works and Pioneer Color Works, and from branch officers in California and Portland, Or., primarily regarding orders for supplies and paints, reports, and other routine business matters.

W2 **Wadleigh, Frank A., 1857-1933.**
Papers, 1917-1930. 10,000 items.
In Colorado Historical Society (Denver).
Denver businessman and founder of Oil Shale Company. - Collection contains correspondence, business papers, reports, and scrapbooks of newspaper clippings pertaining to oil shale research and development in the United States.

W3 **Waite, Frederick Clayton, 1870-1956.**
Papers, ca. 1940-50. ca. 5 ft.
In Case Western Reserve University Archives (Cleveland, Ohio)
Professor of medicine. Correspondence, documents, notes, memoranda, and other papers. Includes material relating to the history of medical and dental education, with working papers for Waite's histories of Western Reserve College, the Western Reserve University Schools of Medicine and Dentistry, and of the teaching of biology and chemistry (1826-1940) at the university.

W4 **Waldo, Edward Hardenbergh, 1866-1950.**
Papers, 1817-1943. - 1.3 ft.
In University of Illinois Archives (Urbana).
Professor of Electrical Engineering at the University of Illinois. - Correspondence, genealogy, class notes and lectures, scientific data on electrochemistry, differential and partial differential equations, motors, transformers, and wiring systems, and other personal and family papers. Includes 2 scrapbooks (1890-1905) relating to Waldo's engineering positions with the Crocker-Wheeler Company, General Electric, and the University of Pennsylvania; "A Method of Transformer Design" (ca. 1915-1920) and "Some Notes on Transformer Design" (1929); and a journal of D. Allen.
Information on literary rights available in the repository.
Acquired, 1963.
Unpublished finding aid available.

W5 **Walker, Elbridge Gerry.**
Papers, 1801-1903. 1155 items.
In Duke University Library.
Druggist and court clerk, of Scottsville and Glasgow, Ky. Correspondence and business papers dealing primarily with the drug business and containing information on the practice of medicine in ante bellum Tennessee and Kentucky. Includes papers relating to Walker's later activities as clerk of the Circuit Court of Allen Co., Ky., and personal letters of various members of the Walker family of Kentucky.
Card index in the library.
Acquired, 1959.

W6 **Walter (George) Brewing Company, Appleton, Wis.**
Records, 1880-1975. 3 ft.
In State Historical Society of Wisconsin, Area Research Center at Green Bay.
In part, photocopies of newspaper clippings.
Correspondence (1907-73), minutes, annual and other reports, administrative, financial, sales, land, and legal records, union contracts (1946-67), building plans, publicity file, and other records, of a brewing company liquidated in 1974.
Unpublished finding aid in the repository.
Gifts of Roland J. Marx, Appleton, Wis., 1975-77.

W7 **Walton, James Henri, 1878-1947.**
Letters, 1917-1946. - 4 boxes.
In University of Wisconsin Archives (Madison).
Professor of Chemistry. - Correspondence, both to and from Professor Walton, dealing with letters of recommendation requested of him, solicitations from inventors seeking his professional advice, and correspondence from colleagues.
Received in January, 1964 and October, 1967.
Unpublished finding aid available.

W8 **Wardall, William J.**
Papers, 1931-1943. - 8.9 ft.
In University of Illinois Archives (Urbana).
Businessman. - Correspondence, minutes, reports, proceedings, briefs, exhibits, press releases, clippings, and other papers, relating to Wardall's service as trustee in the reorganization of McKesson & Robbins Drug Company and the Associated Telephone Utilities Company and the United Telephone and Electric Company. Includes material on the imposter Philip Musica, alias F. Donald Coster, elected President of McKesson and Robbins.
Unpublished finding aid available.

W9 Warner, Irving, 1882-1964.
Papers, 1794-1964. – 3,330 items.
In Hagley Library (Greenville, DE) (acc. #1518).
Consulting engineer; lime expert; executive of the Warner Co. of Wilmington, DE and Philadelphia. – Company incorporated in 1885 as the Charles Warner Co., reincorporated in 1929 as the Warner Co. Has concentrated on sand, gravel, lime, and mixed concrete. Papers include information on company history, administrative and technical files, personal correspondence and financial records, biographical sketches and notes, and genealogical data. Administrative papers contain some financial records of the company. Largest group is the technical file, numbering 1,300 items, and including correspondence, reports, patents, and monographs. Topics covered include Cedar Hollow pulverized limestone, changing properties and efficiency of the rotary lime kiln, grading calculations for the sand and gravel industry, and others.
Described more fully in the Supplement to A Guide to Manuscripts in the Eleutherian Mills Historical Library (1978).

W10 Washburn, Edward Roger, 1899-1967.
Papers, 1930-66. ca. 2 ft.
In University of Nebraska Archives (Lincoln)
Professor of chemistry at the University of Nebraska. Correspondence, notes, and articles, relating to chemistry and other scientific interests.
Preliminary inventory in the repository.

W11 **Washington Association of Scientists.**
Records, 1947-57. 31 folders.
In University of Chicago Library.
Corrsepondence, constitution, bylaws, minutes of meetings and elections, financial records, mailing lists, policy statements, records of activities, and memoranda of an organization devoted to public education and agitation for civilian and international control of atomic energy, collective security, and increased scientific research.
Unpublished guide in the library.
Information on literary rights available in the library.
Deposited by the Federation of Atomic Scientists, 1959.

W12 Watkins, Jabez Bunting, 1845-1921.
Business papers, 1864-1946. – 627 linear feet.
In University of Kansas Libraries (Lawrence).
Businessman of Lawrence, Kansas. – Correspondence, financial records, and other materials relating to Watkins' various business interests including the Calcasieu Sugar Company, and the Lake Charles Sugar Company which he established in Louisiana.
Unpublished finding aid in the repository.
Gift of James C. Malin, 1942; Dick Williams, 1961.

W13 Watson, James Dewey, 1928-
Papers, 1945-54, 1968. 2 ft.
In Harvard University Archives (Cambridge, Mass.)
Teacher of molecular biology at Harvard University. Personal correspondence (1945-53), research and lecture notes, and mss. of writings, including drafts of The Double Helix (1968), based on Watson's research in genetics and findings on DNA.
Access restricted.
Information on literary rights available in the repository.
Acquired from Mr. Watson, 1974.

W14 Waynick, Capus Miller, b. 1889.
Papers, 1775-(1932-1966) - 1980. – Ca. 14,000 items.
In East Carolina University Library, East Carolina Manuscripts Collection (Greenville, NC).
North Carolina journalist, administrator, and diplomat. – Papers include 83 page typescript, "Early History of the Vick Chemical Company and Personal Reminiscences of H. S. Richardson."
Unpublished finding aids in the library. Also described in "East Carolina Manuscripts Collection, Bulletin No. 9."
Gift of Mr. Waynick, High Point, NC, 1980.

W15 Weakly, Harry Elmer, 1899-1974.
Papers, 1920-74. 3 ft.
In Nebraska State Historical Society collections (Lincoln)
Research soil scientist, of Lincoln and North Platte, Neb. Chiefly mss. of articles and research reports on soil conservation, climate, irrigation, drought, trees, and experimental work in agriculture; together with correspondence (1931-73) relating to projects and research; notes, bird sightings, and other ornithological records (1931-67); and material relating to Weakly's interest in the Boy Scouts.
Unpublished finding aid in the repository.
Gift of Ward Weakly, 1978.

W16 Weber, Harold C.
Papers, 1952-1956. – 0.25 linear ft.
In Massachusetts Institute of Technology, Institute Archives and Special Collections (Cambridge, MA).
Chemical engineer; educator. – Printed material relaing to battery additive AD-X2.

W17 Webster, John White, 1793–1850.
Papers, 1826–35. 1 box.
In Harvard University Archives.
Professor of chemistry and mineralogy at Harvard University. Correspondence and other papers.
Information on literary rights available in the repository.

W18 Weeks, Carl, 1876–1957.
Papers, 1922–57. ca. 3 ft.
In University of Iowa Libraries (Iowa City)
Cosmetics manufacturer, of Des Moines, Iowa. Business correspondence, notes, merchandising and promotional material, speeches, and advertisements, relating to Weeks' activities as a drug and cosmetics manufacturer.
Unpublished index in the library.
Information on literary rights available in the library.
Gift of Carl Weeks, 1959.

W19 Weiss, John M., 1885–1963.
Papers, 1916–62. 8 linear feet.
In University of Pennsylvania Libraries, E. F. Smith Collection (Philadelphia).
Chemical engineer. - Weiss had offices in New York from the early 1920's to the end of his career. Collection includes progress reports issued by Weiss & Downs, Inc. and by J. M. Weiss and Company; letters; memos; working notes; reports of site visits and phone conversations; plant drawings; reprints; photographs; and a series of 25 laboratory notebooks. All of the material is rich in detailed technical information. Weiss was particularly interested in plant design for the production of maleic anhydride, phthalic anhydride, and phenol. These interests are reflected in his papers. Collection also includes Weiss's 40+ patents.
Unpublished finding aid available.

W20 West family.
Papers, 1697–1880. ca. 2 ft. (ca. 1400 items)
In University of Michigan, William L. Clements Library (Ann Arbor)
Correspondence and other papers relating to the West family of Boston, Mass., and Charlestown, N. H. The majority of the papers are of Benjamin West (1776–1829), chiefly relating to the operation of his sugar refining firm at Boston, to Boston municipal government and the estate of his uncle, Benjamin West (1746–1817), of Charlestown, N. H. Includes papers of West's father, Samuel West (1738–1808), a Unitarian clergyman, concerning his transfer from a parish at Needham, Mass., to the Hollis Street Church, Boston. Correspondents include Samuel Bradlee, Richards Child, Caleb Davis, Joshua Davis, Benjamin West (1776–1829), Enoch Hammond West, and Samuel West.
Purchase, 1965.

W21 West Virginia Pulp and Paper Company.
Records, 1884–1953.
275 ft.
In Cornell University Library, Collection of Regional History and University Archives (1781)
Correspondence, account books, deeds, contracts, leases, mortgages, records of suits and other legal papers, Government wartime directives, newspaper clippings and other printed matter, maps, statistics, and miscellaneous papers pertaining to all operations of this company. Intracompany correspondence and other papers relate to activities of various departments of the home and branch offices regarding executive and personnel administration, accounting, finance, insurance, taxation, stock transfers, issues and redemptions, advertising, public relations, industrial and labor relations, operation of mills, research, purchasing of equipment and property, wartime activities, conservation, postwar planning, housing, drought relief, and stream pollution. Includes account books of predecessor firms.
Access restricted.

W22 West Virginia Pulp and Paper Company.
Records, 1884–1953. (MS 62–4074)
—— —— Addition, 1925–54. ca. 6 ft.
In Cornell University Libraries, Dept. of Manuscripts and University Archives (Ithaca, N.Y.) (1781)
Business correspondence, office records, deeds and other legal papers, temporary stock certificates, annual reports, newspaper clippings, and other materials pertaining to stock issues and redemptions, finance, property purchases and sales, employee relations, and taxation.
Gift of the company, 1959.

W23 West Virginia University Library collection of historical materials (MS 62–1606)
—— Addition, 1773–1962. 31 items, 10 v., 2 folders, 12 reels of tape recordings, and 12 reels of microfilm.
In West Virginia University Library (507, 1127, 1140, 1144, 1148, 1150, 1193, 1210, 1212, 1243, 1247, 1250, 1252, 1255, 1280, 1337, 1346, 1356, 1371, 1374, 1403, 1428, 1443, 1452, 1496, 1515, 1539, 1564, 1581, 1598, 1611, 1612, 1616, 1647)
In part, transcript (typewritten), photocopies and microfilm made from originals chiefly in private hands.
Historical and biographical sketches, notes, reminiscences, and miscellaneous papers, chiefly relating to West Virginia local history. Names, places, and subjects represented include Nicolo Calabrese, Joseph William Cooper (1823–1898), William Henderson French (1812–1872), John Henderson (Disciples of Christ minister), William Lowther (1742–1814), Kenneth Kyle McCormick, William Presley Lewis Neale, Robert Richardson (1806–1876), Tayloe, Vanbibber, and Wells families; the coal industry; East Florida; iron furnaces in the Alleghany-Botetourt-Roanoke Co., area of Virginia; the labor movement in West Virginia; the oil industry; salt manufacturing in Mason Co., W. Va.; the Ohio Valley; Revolutionary soldiers of Monongalia Co., W. Va., West Virginia National Guard; Doddridge, Hardy, Lewis, Marion, Mason, Monongalia, Pendleton, and Tyler Counties, W. Va.; Fort Lewis, Helvetia, Logan, Mannington, Morgantown, Pierpoint-Avery Community, Ripley, Simpson Chapel and Community, Sistersville, and Wardensville, W. Va.
A composite collection consisting of groups of papers or single items listed and described in Guide to manuscripts and archives in the West Virginia collection, West Virginia University Library, no. 2 (1958–1962).
Acquired from various sources, 1959–62.

W24 **West Virginia. University.**
Business records collected by the University, 1795-1940.
ca. 135 items, ca. 20 v., 2 boxes, 1 bundle, 2 folders, and 6 rolls of microfilm.
In West Virginia University Library (63, 119, 180, 312, 372, 415, 416, 480, 535, 550, 609, 650, 654, 678, 679, 688, 705, 713, 788, 801, 833, 838, 846, 854, 864, 894, 982, 1001, 1029, 1051)
In part, transcripts (typewritten)
Ledgers, account books, daybooks, journal, minutes, casebook, deeds, reports, survey book, photos., clippings, a few letters, and related printed matter. Persons and institutions represented are James Byrnside, who operated a general store at Peterstown; the Chesapeake and Ohio Railway Company; A. B. Clark, who kept a general store; E. R., Henry, and R. H. Demain, contractors, of Morgantown; the Donnally and Steele Kanawha Salt Works; a general store, near Fairview; William Hall, of Egypt; Henry Hammett, cabinet and coffin maker of Parkersburg; Joseph Harman, who ran a general store at Petersburg; Jordan Harrison, of Skimino, York Co., Va.; Samuel Harrison, who had an inn and a cooperage business; the Talbott House of St. George, owned by W. B. Jenkins & Brother; Richard and Samuel McClure, general merchants, of Wheeling; the Mannington Times; Mountain Lake and Salt Sulphur Springs Turnpike Company, and Salt Sulphur Springs Road; Robert T. Moore, retail merchant of Wellsburg; the Morgantown Bridge Company; Julius G. Muhleman, who had a manufacturing firm in Buck Hill Bottom, Ohio; various steamboat companies operating on the Ohio river; G. W. Orr & Son, who owned and operated the Henry D. Fortney Grain & Feed Mill, at Independence; John C. Palmer, attorney of Wellsburg; W. J. Cox and A. M. Jackson, physicians at Knob Fork and Porter Falls, Wetzel County; the Red Sulphur Hotel, owned by Hunter & Company; Sebastian Rehm, a physician of Mason Co., W. Va. and Pomeroy, Ohio; a tailor, in Morgantown; William Thomas, who owned a general store and grist mill at Blacksville; the United States Leather company, of Marlinton; J. W. Vandiver, a dealer in agricultural implements at Burlington; and the West Virginia Pulp and Paper Company. Includes a narrative of the Bates and Harrison families; observations on Indians; 23 issues of several newspapers published in Marion County; and a survey book of Phillip Lybrook, of Giles County, Va.
Indexed in part.
Described separately in Guide to manuscripts and archives in the West Virginia collection, by Charles Shetler (1958)

W25 **Western Reserve University. Department of Chemistry.**
Records, 1942-1967. - 1 linear ft.
In Case Western Reserve University Archives (Cleveland, OH).
Correspondence, newsletters, and brochures; relating to curriculum, faculty, students, and the planning and construction of Willis Science Center.

W26 **Wetherill, Charles M., 1825-1871.**
Papers, 1836-70. - ca. 90 items.
In University of Pennsylvania Libraries, E.F. Smith Collection (Philadelphia).
Chemist at the Smithsonian Institution and the U.S. Department of Agriculture. - Letters written to him by such well known 19th century scientists as Alexander D. Bache, Wolcott Gibbs and Joseph Leidy; also a group of manuscript lecture notes, scientific papers and diplomas and notes of Wetherill's.
Inventory at the repository.

W27 **Wetherill, George D. (and Co.), Philadelphia, Pennsylvania.**
Ledgers, 3 v.: 1829-35; 1871-77; 1874-86.
In Hagley Library (Greenville, DE) (1652).
Wholesale dealer in paints, drugs and spices. - Vol. 1: Invoices, 1829-31; also copies of letters sent, 1829-35 (ca. 300 pp); Vol. 2: Inventories of chemicals - quantities, locations, prices and other details, 1871-77. (290 pp., w/index); Vol. 3: Inventories of all stock and machinery, listed by each of the company's property locations, 1874-86 (382 pp.).
Unpublished finding aid available.

W28 **Whann family.**
Business records, 1832-84. 2 items and 3 v.
In Eleutherian Mills Historical Library (Greenville, Del.) (331)
Daybook (1832-84) of Thomas Whann, fertilizer manufacturer, of Strickersville, Pa.; ledger (1839-54) of Joseph Whann, shoemaker and dealer in hides and skins; ledger (1862-65) of John Whann, fertilizer manufacturer, of Wilmington, Del.; articles of dissolution (1862-65) of Walton, Whann & Company, dealers in farm implements and fertilizer, Wilmington, Del.; and articles of agreement (1867) between John Whann and James Morrow to manufacture and sell cloth under the name of John Whann & Company, Wilmington, Del.

W29 **Wheatley, Charles Moore, 1822-1882.**
Papers, 1840-82. ca. 500 items.
In American Philosophical Society Library (Philadelphia, Pa.)
Mine owner and scientist. Correspondence and other papers relating to mines, shells, paleontology, and mineralogy, sketch of a mine near Phoenixville, Pa., list of furnace products sent to the Centennial Exhibition, Philadelphia, 1876, and other papers.
Unpublished finding aid in the repository.
Information on literary rights available in the repository.
Gift of Mrs. Alfred Bendiner, 1969.

W30 **Whipple, William, 1880-1962.**
Papers, 1794, 1844-1958. - 164 items, 5 mss. v., and 162 printed v.
In Louisiana State University, Department of Archives (Baton Rouge).
Professor of Steam Engineering (sugar technologist). - Papers consist of correspondence, reports, and record books for the Cinclare Central (sugar) Factory of West Baton Rouge Parish. Also periodicals and published volumes of Louisiana State University, Princeton University, United States Naval Academy, Delta Psi Fraternity; copies of The Southern Review; and copies of sugar bulletins and journals.

W31 **Whitby, George S.**, 1887-1972.
Papers, 1908-1971. – 14 boxes, 31 loose items, 1 file.
In University of Akron, American History Research Center (Akron, OH).
Professor of rubber chemistry at the University of Akron; Director of the Office of Rubber Research, 1942-1953; 1954 Charles Goodyear Medalist (ACS Rubber Division). – The collection is related mainly to the scientific and business aspects of Whitby's life and contains little on his work prior to the 1920s. The majority of the papers deal with Whitby's work in the 1940s, especially during World War II. Series include correspondence; speeches, lectures, and papers; laboratory notes and notebooks; student theses; journal articles and clippings; business papers; speech notes; personal data; photographs; and a small number of Mrs. Whitby's personal papers.
Unpublished finding aid available.

W32 **White, Hugh**, 1798-1870.
Papers, 1750-1933. 19 ft.
In Cornell University Library, Collection of Regional History and University Archives (610, 654, 898, 944, 1315)
Railroad builder, cement manufacturer, and U. S. Representative from New York. Correspondence, diaries, account books and other business papers, tax receipts, deeds, inventories, contracts, surveys, and other land records, maps, photos, church records, genealogical data, and other papers. Most of the papers relate to land in northern New York State and include material on William Constable, Hezekiah Beers Pierpont, and William Constable Pierrepont. Many papers relate to the cement business at Chittenango and at Greenkills, near Kingston, N. Y. Most of the papers relating to White's political interests fall between 1840 and 1867 and include material on the Whig Party, patronage, and national as well as local issues. Letters of various members of the Axtell, Duston, Lawrence, Mansfield, Mills, Niles, Porter, Van Wagenen, White, Young, and other related families cover most of the 19th century and deal with a great variety of subjects, illustrating the spread of a family across the continent and reflecting American life and history of the period. Correspondents include James Buchanan, John Chambers, Gilbert L. Cross, Gale and Seaton, L. P. Hickok, Washington Hunt, T. Batter King, William J. McAlpine, William L. Marcy, George E. Parsons, William H. Seward, John W. Taylor, R. J. Walker, Benjamin Wright, and William E. Young.
Described in Reports of the curator and archivist, Cornell University, Collection of Regional History and University Archives (1946-48) p. 42-45, (1948-50) p. 74-75, and (1950-54) p. 55.

W33 **White, Canvass**, 1790-1834.
Papers, 1814-35. 2 ft.
In Cornell University Library, Collection of Regional History and University Archives (609, 822)
In part, photocopies.
Canal engineer, and cement manufacturer, of Whitestown, Oneida Co., N. Y. Correspondence, legal and business papers, and other papers relating to White's service in the War of 1812, his engineering activities, and his cement manufacturing business with his brother Hugh at Chittenango, N. Y. Includes material relating to the Camden and Amboy Railroad and Transportation Company, Chambly Canal, Champlain Canal, Chesapeake and Ohio Canal, Cohoes Company, Delaware and Raritan Canal, Delaware Breakwater, Enfield Falls Canal, Erie Canal, Groton Aqueduct, Louisville and Portland Canal, Mohawk River, Pennsylvania Canal, Union Canal, and Utica and Schenectady Railroad. Correspondents include Major Bender, Edwin A. Douglass, General Jesup, William Lehman, George Olmsted, William Read, Peter Remsen, Alfred Smith, and Joseph Trumball.
Described in Annual reports of the curator and archivist, Collection of Regional History and University Archives, Cornell University (1946-48) p. 41-42 and (1948-50) p. 74.

W34 **White (J & J) Paper Company, Canton, Mass.**
Records, ca. 1929-45. 10 ft.
In Harvard University, Graduate School of Business Administration, Baker Library (Boston, Mass.)
Correspondence, orders, and bills (sampled), of a firm which produced waterproof paper for use in building, packing of perishable products, and bleaching.
Unpublished finding aid in the repository.
Gift of Dr. Charles V. Reynolds, Jr., 1978.

W35 **White, Reuben G**
Family papers, 1819-1929. ca. 3 ft.
In University of Texas Library, Texas Archives (Austin)
Letters, legal papers, financial papers, clippings, and other papers, concerning members of the White and Peebles families, early Texas history, conditions during the Republic and the Civil War, the founding of Hempstead, White's cotton plantation, development of cotton fertilizer and insecticide, and personal affairs. Includes letters to R. M. Bozman while a student at the University of the South (1895-99), Mrs. Rachel White's correspondence with the Daughters of the Republic of Texas about protection and restoration of the Alamo, T. H. Pointer's account book (1914-17), and Sarah M. Peebles' diary and book on exercises in writing and arithmetic. Persons represented include W. R. Baker, C. E. Everett, John G. Peebles, M. A. Peebles, Richard Rogers Peebles, Henry Rosenberg, J. T. Stanfield, and Robert M. Wyatt.

W36 **Whited, Oric Ogilvie**, 1854-1912.
Papers, 1877-1923.
3 v. and 1 box.
In Minnesota Historical Society collections.
St. Paul real estate dealer. Record books, a chemistry notebook kept by Whited while a student at the Massachusetts Institute of Technology (1906), and a scrapbook of clippings about land to be sold in Wisconsin and Minnesota (1919-23).

W37 Whitehorn, John Clare, 1894-1973.
Papers, 1916-73. 37 ft.
In American Psychiatric Association Archives (Washington, D.C.)
Professor of psychiatry and director of Dept. of Psychiatry at Johns Hopkins University, Baltimore, Md. Correspondence, personal papers, mss. of writings, including books, articles, lectures, and speeches, photos, clippings, memorabilia, printed material, and other items, relating to Whitehorn's personal and professional life, including biochemical and physiological research at McLean Hospital, Waverly, Mass. (under Dr. Otto Folin), teaching at Johns Hopkins and Washington University, St. Louis, Mo., service on National Research Council (1932-57), and association with American Board of Psychiatry and Neurology, American Psychiatric Association, of which he was president, American Society for Research in Psychosomatic Problems, and other professional societies. Correspondents include Karl A. Menninger, William C. Menninger, Adolf Meyer, and Merrill Moore.
Unpublished inventory in the repository.
Access restricted.
Information on literary rights available in the repository.
Gift of Joan Whitehorn Boggs, 1973.

W38 Whitfield, James Vivian, 1894-1968.
Papers, 1945-68. 550 items.
In East Carolina University Library, East Carolina Manuscript Collection (Greenville, N. C.)
Foreign Service officer and State official, of North Carolina. Papers documenting Whitfield's campaign for conservation legislation, relating to water pollution control and abatement in North Carolina, White House Conference on Water Conservation (1959-60), investigation of phosphate mining, Federal support of pollution control, water quality standards, Holly Tree State Park, forestry, seafood conservation, and activities of the North Carolina Dept. of Water and Air Resources. Correspondents include James T. Broyhill, T. Wade Bruton, Charles F. Carroll, Harold D. Cooley, Sam J. Ervin, Jr., L. H. Fountain, James A. Graham, Luther H. Hodges, B. Everett Jordan, Horace R. Kornegay, Alton A. Lennon, Dan K. Moore, Robert W. Scott, Stewart L. Udall, Roy A. Taylor, and Basil Lee Whitner.
Unpublished finding aids in the library. Also described in "East Carolina Manuscript Collection, Bulletin no. 3."
Gift of Mrs. Whitfield, Wallace, N. C., 1970.

W39 Whitmore, William Vincent, b. 1862.
Papers, 1860-1935.
ca. 1 ft. (ca. 50 items)
In Arizona Pioneers' Historical Society collections (Tucson, Ariz.)
Physician. MSS. of Whitmore's works including Anesthesia and anesthetics, Married women as teachers, and Some phases of the development of the University of Arizona; biographies of Howard Billman, Joel Ives Butler, James W. Coleman, Henry Edward Crepin, Byron Cummings, George Emery Goodfellow, John Charles Handy, Narcisco Hereu Matas, Homer Leroy Shantz, Guillermo R. Servin, George Davis Troutman, Leonard Wood, and other Arizonans; and reprints of articles on Early medical conditions in Arizona.
Unpublished calendar in the repository.
Gift of the Whitmore family.

Widtsoe family. **W40**
Papers, ca. 1850-1966. 7 ft.
In Utah State Historical Society collections (Salt Lake City)
Mss. of books, articles, reviews, and speeches, of John Andreas Widtsoe (1872-1952), agricultural chemist, president of the Agricultural College of Utah and University of Utah, and Mormon churchman, relating to both religious and secular activities; papers of his wife, Leah Eudora (Dunford) Widtsoe (1874-1965), relating chiefly to women and the Mormon church, nutrition, health, and church activities; correspondence, journals, and biographical information, of Anna (Widtsoe) Wallace (b. 1899), their daughter, and her son, John Widtsoe Wallace (b. 1926); and other family and general correspondence.
Register published by the repository in 1976.
Information on literary rights available in the repository.
Gift of G. Homer Durham, Eudora Widtsoe Durham, and Anna Widtsoe Wallace, 1966.

Wightman, Joseph Milner. **W41**
Papers, 1839-1880. - 95 items.
In Duke University Library (Durham, North Carolina).
Instrument maker of Boston, Suffolk County, Massachusetts. - The collection consists of 93 letters written to Wightman; a clipping attached to a letter; and a 4x6" paper fragment. Most of the letters date from the 1840's and early 1850's and concern the ordering of scientific apparatus for universities, colleges and academies. Several letters contain scientific drawings and detailed descriptions of the kind of instrument desired. Other subjects include brazing and soft soldering processes; magic lanterns; and the role of chemistry in medical education. Correspondents included professors of chemistry, mathematics, mineralogy, and geology; clergymen; and founders and presidents of colleges.
Unpublished finding aid available.

Wilber, Francis Augustus, 1851-1891. **W42**
Papers, 1872-1916. 6 v. and 2 folders.
In Rutgers University Library (310 and 1794)
Chemistry professor, of New Brunswick, N. J. Correspondence and a diary relating to a U. S. Government survey (1880-81 and 1889) of New Jersey and New York State mining conditions, accounts (1879-85), and other papers, together with letters received by Wilber's wife, Laura Birge (Parker) Wilber and descriptive letters (1872) written by her sisters in Tuckerton, N. J. Includes an essay (1878) on water supply for cities and towns, a thesis (1879) on gauging the Raritan River, and other papers relating to Wilber's life as a Rutgers College student and assistant and professor of analytical chemistry.
Cataloged in the library.
Gift of Mrs. Sidney P. Noe.

W43 **Wilder, Edward B.**
Letterpress books, 1869-1905. – 5 v.
In University of Utah Libraries, Special Collections Department (Salt Lake City).
Mining agent. – Letters relating to Wilder's work in Utah and Nevada, particularly evaluating and negotiating sales of mining properties. Includes references to copper, nickel, and cobalt mines and oil shale.
Unpublished finding aid available.

W44 **Wiley, Harvey Washington, 1844-1930.**
Papers, ca. 1854-1944.
99 ft. (ca. 70,000 items).
In Library of Congress, Manuscript Division.
In part, transcripts.
Chemist, teacher, author, and lecturer. Correspondence (including letterpress volumes), diaries, legal papers, clippings, memoranda, printed matter, articles, speeches, lectures, essays, maps, blueprints, graphs, tables, memorabilia, and photos. The papers relate to three major periods: Wiley's early years as student and teacher; his service as Chief of the Bureau of Chemistry of the U. S. Dept. of Agriculture (1883-1912) during which he led the movement which culminated in passage of the Pure food and drugs act of 1906; and his retirement years, during which he lectured on the Chautauqua circuit and directed the Good housekeeping magazine's Bureau of Food, Sanitation, and Health. Includes a list of papers published by Wiley, and a bound volume of souvenirs of a memorial dinner held in 1944.
Unpublished finding aid in the Library. Also described in the Library's Quarterly journal of current acquisitions, v. 10, no. 3 (May 1953) p. 154-155.
Information on literary rights available in the Library.
Gift of Mrs. Wiley, 1952-56.

W45 **Willard, Hobert Hurd, 1881-1974.**
Papers, 1906-74. 2 ft.
In University of Michigan, Bentley Historical Library, Michigan Historical Collections (Ann Arbor)
Professor of chemistry, University of Michigan, and specialist in analytical chemistry. Correspondence, class notes, reprints of publications, and newspaper clippings, relating to Willard's professional activities.

W46 **Willcox family.**
Papers, 1724-1858.
ca. 1100 items.
In Historical Society of Pennsylvania collections.
Pennsylvania family, proprietors of papermills that became leading suppliers of paper for Provincial, Continental, and Federal currency. Correspondence and documents pertaining to the early manufacture of paper in America, together with samples of paper, bank and treasury notes of Revolutionary and post-Revolutionary times, a list of clients in Delaware County and their correspondence (1724-1835), and account books of Nathan Edwards of Black Horse Hotel, near Media (1730-84).
Presented by Joseph Willcox.

W47 **William G. Johnson Company, Uncasville, Conn.**
Records, 1830-92. 2 ft. (17 items)
In Merrimack Valley Textile Museum (North Andover, Mass.)
Sales ledgers and memoranda (1830-35) relating to William G. Johnson's operations as an import-export merchant in Buenos Aires, Argentina. Daybooks, cashbooks, ledgers, order books, and payroll and inventory records of a dyewood factory operated (1834-92) by Johnson and his heirs.

Williams family. W48
Papers, 1775-1974. – ca. 21 ft.
In Albany Institute of History and Art (N.Y.)
Transcript (typewritten) of Fred Stanley Williams' diary; location of original unknown.
Family and business correspondence and records, including Chauncey P. Williams' (1817-1894) letter books (1865-1886) from an Albany lumber business and ledger (1861-1888) and material concerning Albany Horseshoe Nail Factory and Albany Chemical and Aniline Company; Chauncey P. Williams' (1860-1936) diaries, correspondence, account books, and scrapbooks (1878-1936) of events including World War I and Albany banking and society; Fred Stanley Williams' diary of 1st American Amazon expedition; correspondence and papers of Chauncey P. Williams III, relating to Albany Academy; papers relating to Jerseyfield Patent, Herkimer and Montgomery Counties, N.Y., and New York State Kansas Committee; papers of McClure and Strong families of Albany, and Hough family, including Martha Hough Williams, wife of Chauncey P. Williams I, her mother, Ruth P. Hough, teacher, of Connecticut, and Rev. Stanley Hough, president of Oneida Collegiate Institute, N.Y. Hough family papers contain material on abolitionism and Kansas Immigrant Society (including letter from Gerrit Smith).
Gift of Chauncey P. Williams III, 1975.

Williams, Robert Ramapatnam, 1886-1965. W49
Papers, 1911-1975. – 10,500 items.
In Library of Congress, Manuscript Division.
Chemist and nutritionist. – Diaries (1938-1953), correspondence, subject files, laboratory notebooks, speeches and writings, and miscellany relating chiefly to Williams' research in nutritional chemistry, especially his work in the development of vitamin B1 (thiamin). Other topics include his relations with Merck & Company, I. G. Farben, the Research Corporation, and Carnegie Corporation; patent disputes; his travels in Southeast Asia and the Philippines; and his establishment of the Williams-Waterman Fund for the Control of Dietary Diseases. Correspondents include Edwin R. Buchman, Joseph K. Cline, Casimir Funk, M. C. Kirk, E. J. Lease, John C. Merriam, Howard Poillon, and Robert Waterman.
Gift of the Research Corporation, 1979.
Finding aid available.

Willstätter, Richard Martin, 1872-1942. W50
Papers, 1910-64. 112 items.
In Leo Baeck Institute collections (New York, N.Y.)
German chemist. 4 letters written by Willstätter to Fritz Haber and 106 letters (1910-34) by Fritz Haber to Willstätter; photocopy of Willstätter's memoirs (1940) including 2 handwritten postscripts; newspaper articles; and photo. In German.
Card-catalog in the repository.
Gift of Mrs. H. Bruch, Winnebago, Ill., 1961.

W51 Wilson, Edgar Bright, Jr., b. 1908.
Papers, 1935-1974. - 113 boxes.
In Harvard University Archives (Cambridge, MA).
Professor of chemistry. - The bulk of the collection is general correspondence. A second series contains material relating to the Woods Hole Oceanographic Institute. A third series deals with research contracts and contains correspondence, proposals, and reports.
Gift of Professor E. B. Wilson, November 6, 1975.
Access restricted. Contact archivist for further information.

W52 Wilson, John Lewis (1898-1983) and family.
Papers, 1884-1983. - 18 cu. ft.; 18 boxes.
In Minnesota Historical Society (St. Paul).
Chemist. - Collection documents activities of Wilson family, especially John Lewis Wilson. Includes Wilson's family correspondence (1915-1982) and professional correspondence, including that with Georg R. Schultze, a German chemist (1930-1982); diaries and memo books (1893-1982); and papers which document Wilson's activities in clubs, organizations (1925-1980), and the Presbyterian Church (1930-1982). Also includes material reflecting Wilson's career as a chemist at Economics Laboratory (1932-1974) and his association with Jamestown College, Jamestown, North Dakota, his alma mater (1920-1975).
Gift of Josephine Wilson, Ashmore, Illinois, 1983.
Unpublished finding aid available.

W53 **Wilson, Robert Erastus,** 1893-1964.
Papers, 1926-64. 14 ft. (ca. 3500 items)
In Library of Congress, Manuscript Division.
Chemical engineer, businessman, and Commissioner of the U. S. Atomic Energy Commission. Business correspondence, mss. of Wilson's articles and speeches, clippings, printed matter (largely reprints of his speeches), and files of material collected on various subjects of interest to Wilson. The papers deal mainly with the Standard Oil Company of Indiana, the petroleum industry in general, and atomic power.
Index and unpublished finding aid in the Library.
Gift of Mrs. Wilson, 1964.

W54 Winchell, Alexander, 1824-1891.
Papers, 1837-1891. - Ca. 26 ft.
In University of Michigan, Bentley Historical Library, Michigan Historical Collections (Ann Arbor).
Professor of Geology, Zoology, and Botany at Syracuse University, Vanderbilt University, and the University of Michigan and Chancellor of Syracuse University. - Correspondence of Winchell, members of his family, friends, and university colleagues, concerning Winchell's scholarly interests in the natural sciences, family affairs, personal business ventures, the University of Michigan, and other institutions where he taught; 41 diaries (1846-1891); notebooks of Winchell's trips with expense accounts; personal and business account books; addresses, editorials, scientific reports, lecture and other notes; and articles and essays on philosophical, religious, educational, and scientific subjects, including astronomy, biology, chemistry, geology, mathematics, and paleontology. Extensive correspondence with prominent scientists and others of the day.
Gift of Alexander N. Winchell and the Minnesota Historical Society.

W55 Winfield, John Augustus, 1904-
Papers, 1965-69. ca. 75 items.
In University of North Carolina Library, Southern Historical Collection (Chapel Hill) (3925)
Farmer and Government employee, of Beaufort Co., N. C. Correspondence, proceedings of public hearings, agency resolutions, proposed State legislation, and clippings, relating to Winfield's work with the U. S. Dept. of Agriculture in North Carolina, pertaining to ground-water problems following the establishment of phosphate mines and fertilizer plants near Aurora, N. C.
Gift of Mr. Winfield, through L. L. Hiday, 1971.

W56 Withers, Robert W   d. 1854.
Papers, 1794-1890. 54 items and 8 v.
In University of North Carolina at Chapel Hill, Library, Southern Historical Collection (3235)
In part, microfilm.
Planter and businessman, of Greene and Hale Counties, Ala. Accounts and memoranda (2 v.) pertaining to Withers' medical practice and sale of drugs, his lumber and flour mills, agricultural methods, plantation and household expenditures, costs of river transport, weather, and births of family members, slaves, and racehorses; business papers pertaining to Withers' research in an effort to construct a steam mill fed by artesian wells to power a cotton factory, and to his other business interests; and notes (1852-53) of a son at the University of Virginia on chemistry lectures of John Lawrence Smith (1818-1883). Bulk of the collection is dated 1820-53.

W57 Withers, William A.
Papers, 1844; 1883-1908; 1923. - Ca. 150 items.
In North Carolina State Office of Archives and History Collections (Raleigh).
Professor of Chemistry (1889-1924) and

Vice-President (1916-1923) of NC A&M/NC State College. - Includes letters of recommendation from Davidson and Cornell Colleges, from N.C. Agricultural Experiment Station, and from Gov. Alfred M. Scales; and letters to him from Augustus Leazar on choice of experiment station director; and from several professors about the Southern Improvement, Immigration, and Manufacturing Company in Moore Co. (1902-1906). Other correspondence refers to experiments in agricultural chemistry (1885-1897), food adulteration (1898), and miscellaneous personal topics.

W58 Wolcott, Edson Ray, 1877-1966.
Papers, 1899-1965. - 4 ft.
In American Institute of Physics, Niels Bohr Library (New York, NY).
Physicist, metallurgist, and inventor. - Documentation of Wolcott's research and personal and family memorabilia. Much of the collection consists of material Wolcott accumulated while working on an unpublished history of electricity and its applications entitled Evolution of Electricity.
Gift of Edson Wolcott, 1965.
Finding aid available.

W59 Women In Science.
Collection, 1935-40. - 18 boxes and 1 oversize v.
In Schlesinger Library (Cambridge, MA).
Correspondence, financial records, clippings, photos, and publications collected for a Radcliffe College exhibit held in 1936 in connection with Harvard University's tercentenary celebration. Correspondence concerns the exhibit, while the published articles which made up the final exhibit are by women who were conducting research in astronomy, biology, geology, chemistry, physics, and the medical sciences.
Published and unpublished guides available.

W60 Women In Science and Engineering Oral History Collection.
1976-1977. - 9 transcripts; 7 ms. boxes; 87 audio tapes.
In Massachusetts Institute of Technology, Institute Archives and Special Collections (Cambridge, MA).
The core of the collection consists of nine life history interviews conducted by Shirlee Sherkow of the M.I.T. Oral History Program. Among those interviewed were Ellen Henderson, biochemist; Christina Jansen, metallurgist and engineer; and Giuliana Tesoro, engineer and chemist. With the goal of examining the women's motivations in pursuing careers in science and engineering, the interviews explored their early home environments, school and research experiences, roles as wives and mothers, and views on special problems confronting women scientists and engineers. The taped interviews were transcribed, and use copies of the transcripts are available for research. In addition, the collection includes tapes of events, such as lectures, meetings, and panel discussions. Working papers include bibliographies, publications gathered during background research, biographical information on the women interviewed, and reprints of publications by the women interviewed.
Transferred to the Institute Archives by the M.I.T. Oral History Program in 1977.
Access to most of the collection is unrestricted. Restrictions on access, photocopying, quotation, or publication of excepts pertain to some transcripts. For specific restrictions see each transcript. Tapes of interviews conducted by Sherkow are not available for use.

W61 Wood, Frederick William, 1857-1943.
Papers, 1868-1943. - Ca. 9500 items and 190 blueprints.
In Hagley Library (Greenville, DE) (Acc. # 884).
Engineer and manufacturer of Baltimore, Md.; Superintendent of the Pennsylvania Steel Company in the 1880s and President of the Maryland Steel Company at Sparrow's Point from 1891 to 1916. - Collections includes records of the two aforementioned steel companies plus other firms with which Wood was associated later in his career. These steel firms' records include material concerning ore processing, metallurgical experiments, and other materials of interest to metallurgical historians.
Described more fully in the Supplement to A Guide to Manuscripts in the Eleutherian Mills Historical Library (1978).

W62 Wood, George Bacon, 1797-1879.
Notes, 1833-34. - 3 v.; 20 cm.
In National Library of Medicine (Bethesda, MD).
Chemist. - Notes upon lectures delivered by G. B. Wood at the College of Pharmacy. Written by William Elmer. Vol. 1: Lectures 1-21, Nov. 12, 1833-Jan. 6, 1934; Vol. 2: lectures 21 (cont.)-40, Jan. 6-Feb. 20, 1834; Vol. 3: lectures 41-45, Feb. 21-Mar. 6, 1834.

W63 Wood, Henry A
Papers, 1839-60. 98 items.
In Rosenberg Library (Galveston, Tex.)
Physician, of Galveston, Tex. Family correspondence, bills, receipts, notes, and other papers, relating to a Galveston drug firm of which Wood appears to have been the owner.
Gift of H. M. Trueheart, 1922.

W64 Wood, Henry Ellsworth, 1855-1932.
Papers, 1854-1932. - 537 items.
In The Huntington Library, (San Marino, CA).
Mining engineer. - In 1878 Wood opened an assay office and laboratory in Leadville, Colorado. In 1889 he established the same business in Denver, and in 1898 he added the Henry E. Wood Ore Testing Works, which became internationally known. Collection includes letters, manuscripts, documents (including an account book, 1873-78), and photographs. Subjects include reminiscences regarding mining in Colorado, a copy of "Colorado in 1868" (taken from a notebook of Wood's father William Cowper Wood, on the expedition to Colorado with Major John W. Powell), and business affairs.
Purchased from Joseph W. Jones, April, 1955.
Unpublished finding aid available.

W65 Woodman, Ruth Cornwall, 1895-1970.
Papers, 1914-69. 12 ft.
In University of Oregon Library (Eugene)
In part, transcripts.
Scriptwriter. Correspondence, radio and television scripts, research notebooks, and script index, relating to the Death Valley Days radio and television series, with which Mrs. Woodman was associated; research data and draft of a history of the Pacific Coast Borax Company; correspondence with agents and publishers; and 46 letters (1914-16) written by Mrs. Woodman while a student at Vassar College.
Inventory with the collection.

W66 Woodward, Robert Burns, 1917-1979.
Papers, ca. 1940-1979. - 110 linear feet.
In Harvard University Archives (Cambridge, MA).
Organic chemist; Nobel laureate, 1965. - Collection includes personal and professional correspondence, scientific notes and data, manuscripts, material relating to teaching and to various professional activities, including his receipt of the Nobel Prize. Also includes research reports and other papers relating to the Woodward Research Institute at Basel, Switzerland.
Gift of the Woodward family, December 3, 1982. Transferred from the chemical laboratory.
Unprocessed as of late 1985 and unavailable for research for several years.

Worcester, Dean Conant, 1866-1924.
Papers, 1900-1924. - 4 ft.
In University of Michigan, Bentley Historical Library, Michigan Historical Collections (Ann Arbor).
Professor of Zoology at the University of Michigan, and government official and business executive in the Philippine Islands. - Correspondence, newspaper clippings, scrapbooks, and pictures dealing with Worcester's activities in the Philippines as a member of the U.S. Philippine Commission, Philippine Secretary of the Interior, Assistant to the President of the Philippine Refining Company, Vice-President and General Manager of the Agusan Coconut Company and President of the Philippine Desiccated Coconut Company, and publications concerning the Islands. Correspondents include William H. Taft and Leonard Wood.
Gift of Mrs. Kenneth B. Day, 1957.

Wormald, William.
Student notes, ca. 1848. - 1 v. 23 cm.
In National Library of Medicine (Bethesda, MD).
Notes of chemistry lectures given at the Royal Agricultural College, Cirencester, Gloucestershire, England, 1848.

Worth family.
Papers, 1844-1951. - 693 items and 8 volumes.
In Duke University Library (Durham, NC).
Correspondence, business records, and other papers, pertaining chiefly to family matters, business affairs, opposition to the Civil War, politics in North Carolina, fertilizer manufacturing and marketing, textile industry, Zebulon Baird Vance, and patronage during the early years of Woodrow Wilson's presidency. Includes letters to Jonathan Worth (1802-1869), lawyer and Governor of North Carolina, to his son, David Gaston Worth (1831-1897), commission merchant and manufacturer, when he attended the University of North Carolina and when he was Superintendent of the salt works at Wilmington, North Carolina, during the Civil War; correspondence of David with his

wife, Julia Anna (Stickney) Worth, and his son, Charles William Worth, when attending Bingham School and the State University at Chapel Hill,; and letters of Barzillai Gardner Worth. Other correspondents include Edwin Anderson Alderman, Robert Bingham, Josephus Daniels, Hannibal Lafayette Godwin, John Wilkins Norwood, Lee Slateer Overman, James Hinton Pou, Cornelia (Phillips) Spencer, and George Taylor Winston.
Card index in the library.
Purchase, 1969-1971.

Wright, Alexander, John and Peter.
Papers, 1811-1845. - 1 box.
In Merrimack Valley Textile Museum, (North Andover, MA).
Family of textile workers. - Papers include: an account book, 1811-1818, which pre-dates the arrival of the Wrights from Scotland and contains entries on spinning; seven volumes of dyers' recipes with samples; and lists of colors of yarn used in undated samples of weaving. Most concern carpet manufacture at Medway, Massachusetts, 1826-1828, and removal to Lowell, 1829, where the Lowell Manufacturing Company employed all three men.
Finding aid available.

Wright, Samuel G., 1781-1845.
Papers, 1793-1839. - Ca. 3,750 items.
In Hagley Library (Greenville, DE) (1665).
Philadelphia merchant; ironmaster; private financier and land investor; farmer; and member of the New Jersey legislature. - Includes 19 volumes of bound business records, 1809-1853, some of which relate to his iron furnaces in New Jersey and Delaware and letters received by Samuel G. Wright, 1813-1839 (approximately 2740 items) which deal extensively with his furnace enterprises.
Unpublished finding aid available.

Wrinch, Dorothy Maud, 1895-1976.
Papers, 1932-70. ca. 52 ft.
In Smith College, Sophia Smith Collection (Northampton, Mass.)
Crystallographer, biochemist, mathematician, and physicist. Personal and professional correspondence and notebooks, relating to Wrinch's research and views on crystal structure, cyclols, peptides, mineral twins, x-ray methods, B-12, insulin, viruses, and polyhedra; travel diaries, lectures and outlines (1965) for Smith College courses, symposium and conference reports and records, models, memorabilia, articles by colleagues and contemporaries, and bibliography of Wrinch's works; and letters and writings of her daughter, Pamela Wrinch Schenkman (1927-1975), political scientist. Wrinch's correspondents include David Harker, Irving Langmuir, E. H. Neville, Linus Pauling, and the Rockefeller Foundation.
Access restricted temporarily.

Wulling, Frederick John, 1866-1947. W73
Papers, 1884-1948. 7 ft.
In University of Minnesota Library, University Archives.
Pharmacist, educator, and dean of the University of Minnesota College of Pharmacy. Correspondence with outstanding men in the fields of pharmacy and pharmaceutical education, reports, speeches, and publications relating to the American Association of Colleges of Pharmacy, American Conference of Pharmaceutical Faculties, American Institute of the History of Pharmacy, American Pharmaceutical Association, Minneapolis Civic and Commerce Association, Minneapolis Division of National Safety Council, Minneapolis Society of Fine Arts, Minnesota State Pharmaceutical Association, National Association of Boards of Pharmacy, National Association of Retail Druggists, and United States Pharmaceutical Board. Correspondents include William C. Alpers, Howard Armbruster, Wilhelm Bodeman, Charles E. Caspari, W. W. Charters, John B. Christgau, Andrew G. DuMez, Reginald Dyer, Robert G. Eccles, A. Foxton Ferguson (British folksong artist), Hugo Kantrowitz, Evander F. Kelly, Edward Kremers, Charles H. La Wall, Joseph Price Remington, Lucius E. Sayre, F. A. Upsher-Smith, and George Urdang.
Unpublished description in the library.
Open to investigators under library restrictions.
Information on literary rights available in the library.

Wurtz, Henry, ca. 1828-1910. W74
Papers, 1861-1885. - 1.71 linear ft.
In National Museum of American History Archives Center (Washington, DC).
Chemist and editor. - Collection documents Wurtz's professional activities and his contributions to the theory and practice of chemistry. He studied at the College of New Jersey (later Princeton). Interest in scientific pursuits was awakened by studies under Joseph Henry and John Torrey. Between 1854 and 1856 he was State Chemist and Geologist of the New Jersey Geological Survey. In 1858 he was appointed Professor of Chemistry and Pharmacy at the National Medical College of Washington, D.C. (later George Washington University). During this time he served as chemical examiner in the U.S. Patent Office. In 1861 he opened a private laboratory for general consulting work in New York. During the years 1877-1887 he was engaged in developing processes for increasing the yields of paraffin oils and other by-products of the distillation of coal. He devoted the remaining years of his life to his private consulting practice and took out numerous patents relating to his research.
Unpublished finding aid available.

Y1 **Yandell family.**
Papers, 1831-1884. - 0.75 linear ft.
In University of Louisville, Health Sciences Library, Historical Collections Department (KY).
Lunsford Pitts Yandell (1805-1878), chemistry teacher at Transylvania University and founder of Louisville Medical Institute (1837); son David W. Yandell (1826-1898), surgeon, teacher at University of Louisville, President of American Medical Association in 1872; son Lunsford P. Yandell, Jr., teacher of clinical medicine at University of Louisville from 1869 until his death at the age of 47. - Collection includes original manuscripts by the Yandells and a number of printed pamphlets and journal extracts, as well as photocopies of articles by and about them.
Papers assembled from various locations of the Health Sciences Library in 1938.

Y2 **Yoe, John Howe, 1892-1975.**
Papers, 1920-1961. - 3,110 items.
In University of Virginia Libraries (Charlottesville).
Organic chemist. - Collection contains organic chemistry notebook, ca. 1920, 75 p.; Yoe's bound record of graduate study at Princeton, 1923; correspondence, 1944-61, hand and typewritten; and monographs, 1942-45, on chemical warfare (progress reports by Yoe and students on investigations and research conducted at the University during the war).
Gift of Mr. Yoe, June 6, 1963, September 20, 1963, and January 13, 1971.

Y3 **Yoerg Brewing Company (St. Paul, Minn.)**
Records, 1858-1953. - 43 ft.
In Minnesota Historical Society collections (St. Paul)
Founded 1849, by Anthony Yoerg as Yoerg Brewery; closed 1920-1933. Reopened 1933-1953 as Yoerg Brewing Company. - Journals (1888-1913), ledgers (1858-1911, 1933-1953), daybooks (1858-1884), sales books (1882-1913), cashbooks (1887-1913), materials accounts (1875-1905), records of fermented liquors kept for internal revenue (1883-1902, 1933-1941), financial statements (1933-1953), and federal stamp tax records (1941-1953).
Gifts of George M. Brack, receiver, 1954 and 1955.
Unpublished inventory in the repository.

Y4 **Yost, Don M., 1893-1977.**
Papers, 1936-1975. - 17 boxes.
In California Institute of Technology Library (Pasadena).
Professor of inorganic chemistry, California Institute of Technology. - Chiefly correspondence; together with mss. of writings, notes, reprints, and personal papers. Most of the collection is dated 1940-1960 and relates to Yost's World War II research with National Defense Research Committee - Office of Scientific Research and Development, his postwar work in chemistry and mathematics, and the political and social views of Yost and his correspondents. Includes comments on the Oppenheimer case, atomic secrecy, Sputnik, and Yost's opinions, as an active Catholic, on science and religion. Correspondents include members of the Iron Nail Club and other colleagues.
Finding aid in the repository.

Z1 **Zeller, Edwin Adrian.**
    Papers, 1886–1946.
        1400 items.
        In Historical Society of Pennsylvania collections.
        Businessman of Germantown, Pa. Personal and business papers reflecting Zeller's interests in sugar refining, in the Presbyterian Church, and in Philadelphia reform politics. Includes reports and communications from the City Club of Philadelphia, of which Zeller was a member.
        Purchase, 1959.

Z2 **Zoeller, Edward V**   1857–1944.
    Papers, 1885–1934. ca. 500 items.
    In University of North Carolina Library, Southern Historical Collection (1780)
        Pharmacist, of Tarboro, N. C. Letters, chiefly 1891–1912, relating to pharmacy, written to Zoeller from other pharmacists and officers in pharmacy organizations in North Carolina; letters after 1890 relating to the Farmers' Oil Mills (manufacturers of cottonseed products in Tarboro) of which Zoeller was secretary-treasurer; and records (1890–1917) of the Farmers' Oil Mills. Correspondents include Franklin Wills Hancock, of Oxford, N. C., William Simpson, and Francis Preston Venable, of Chapel Hill, N. C.
        Bequest of Mr. Zoeller, 1948.

Max Tishler, Merck chemist and research administrator from 1937 to 1970, was inducted into the National Inventors Hall of Fame in 1982 for the synthesis of riboflavin and sulfaquinoxaline. Courtesy Merck & Co., Inc.

NAME INDEX

Abbott, George Alonzo, A2
Adams, Mark F., A8
Adams, Roger, A9
Alexander, Jerome, A23
Alexander, John H., A24
Andrews, Roy Chester, A66
Arceneaux, Claude Joseph, A69
Arrhenius, Svante August, A74
Ashdown, Avery Allen, A76
Atwater, Wilber Olin, A87
Audrieth, Ludwig Frederick, A89

Babb, Albert L., B2
Babcock, Stephen Moulton, B6, B7
Bache, Franklin, B9
Bachmann, Werner Emmanuel, B10
Baekeland, Leo H., B13
Bahn, Gilbert Schuyler, B14
Bailey, Alton Edward, B15
Bailey, Jacob Whitman, B16
Bancroft, Wilder Dwight, B18, B101
Bartell, Floyd Earl, B25
Bartlett, Paul D., B26
Bartow, Edward, B28
Bartow, Virginia, B29
Basore, Annie Terrell, B31
Basore, Cleburne Ammen, B31
Baxter, Gregory Paul, B37
Beck, Lewis C., B38
Beck, Lewis Caleb, B39
Bellinger, Frederick, B43, B44
Benson, Henry K., B46
Bergmann, Max, B48, B49
Black, Joseph, B53, B54, B55
Blomquist, Alfred Theodore, B60
Bodansky, Meyer, B61
Bolton, Henry Carrington, B64
Boltwood, Bertram Borden, B65
Booth, James Curtis, B67
Bray, William Crowell, B76
Brewer, William Henry, B79
Brewster, Carl Milton, B80
Brinton, Clement Starr, B82
Brown, Harry Fletcher, B87
Brown, Herbert C., B88
Browne, Charles Albert, B92, B93
Brownell, Lloyd Earl, B95
Bruson, Herman A., B96
Bunsen, Robert Wilhelm Eberhard, B100

Caldwell, George Chapman, C2, C3
Carter, Herbert Edmund, C14
Chandler, Charles Frederick, C23
Christman, Adam Arthur, C31
Clark, George Lindenberg, C34
Clark, John Dustin, C35
Clark, William Mansfield, C36
Clark, William Smith, S77
Clarke, Frank Wigglesworth, C38, C39
Clarke, Hans Thacher, C40
Cleland, Elizabeth, M46
Cloke, John B., C43
Cohn, Mildred, C45
Collier, Peter, C49
Conant, James Bryant, C52, H42
Cooke, Josiah Parsons, C58
Coolidge, Albert Sprague, C59

Cooper, Thomas, A55, C61
Cori, Carl Ferdinand, C63
Cottrell, Frederick Gardner, C65
Crafts, James Mason, M34
Craig, Lyman Creighton, C73
Cross, Paul C., C77
Crowell, Azariah Foster, C78
Culver, Stephen Berry, C81
Curie, Marie (Sklodowska), C82
Curtis, Harry Alfred, C85

Dabney, Charles William, D1
D'Alelio, G. Frank, D2
Daniels, Farrington, D4
Davis, Marguerite, D12
Davy, Sir Humphry, D15
Day, Jesse Erwin, D16
De Barr, Edwin, D20
Debye, Peter Joseph Wilhelm, D21, D22
De Milt, Clara, D25
Derick, Clarence G., D26
Deutch, John M., D28
D'Ianni, James D., D31
Donnell, John W., D34
Donohue, Jerry, D35
Drabkin, David L., D39
Drake, Quaesita Cromwell, D41
Draper, John Christopher, D42
Draper, John William, D42, D43
Du Bois, Alfred, D48
Dudley, William Lofland, D50
Duman, Jean Baptiste A., D55
Du Pont, Alfred Victor, D71
Du Pont de Nemours, Eleuthère Irenée, D72
Du Pont de Nemours, Pierre Samuel, D73
Du Pont, Ernest, D74
Du Pont, Eugene, D75
Du Pont, Eugene, Jr., D75, D76
Du Pont family, D76, D77, D78, D79
Du Pont, Francis Gurney, D80
Du Pont, Francis Victor, D81
Du Pont, Henry, D82
Du Pont, Lammot, D83
Dushman, Saul, D84
Du Vigneaud, Vincent, D85

Ebaugh, William Clarence, E3
Edsall, John Tileston, E6, E7
Ehrlich, Paul, E11
Eisenschiml, Otto, E12
Eliot, Charles William, E13
Englis, Duane T., E15
Erdmann, Otto, E18
Eyring, Henry, E22

Fajans, Kasimir, F2
Fankuchen, Isidor, F5
Ferry, John D., F10
Ferry, Ronald M., F11
Fieser, Louis P., F13
Findlay, Alexander, F14
Fischer, Emil, F16
Flory, Paul J., F20
Forbes, George Shannon, F21
Forbes, Robert Humphrey, F22

Foreman, Edward R., F24
Fox, Denis Llewellyn, F26
Franck, James, F27
Frank, Adolf, F28
Frank, Joseph Otto, F29
Frazer, John F., A55
Frazer, Persifor, A55
Frear, William, F30
Freeland, Emile C., F31
French, Charles C., F32
Fulmer, Elton, F36
Funk, Casimir, F38
Fuson, Reynold Clayton, F39

Garrett, Alfred Benjamin, G3
Garvan, Francis P., G4
Giauque, William Francis, G10
Gibbes, Lewis Reeves, G11
Gibbs, Wolcott, G12, G13
Gilchrist, Peter Spence, G18
Godfrey, Almon T., G26
Gordon, Louis, G32
Gorham, John, G33
Greenewalt, Crawford, G41
Griffln, Edward Lawrence, G42
Gustafson, Ben G., G49
Gwathmey, Allan Talbott, G55, J8

Haber, Fritz, H1, W50
Halford, Ralph Stanley, H4
Halsband, Ruth Alice, H6
Hamilton, Cliff Struthers, H7
Hammett, Louis Plack, H10
Hare, Robert, A55, H14, H15, H16, H17
Harmon, Robert Rogers, H18
Harper, Henry Winston, H19
Harris, James Courtland, H20
Harte, Robert A., H24
Harteck, Paul, H25
Hassid, William Zev, H45
Hastings, Albert Baird, H46
Hedrick, Benjamin Sherwood, H52, H53
Hellerman, Leslie, H57
Henderson, Ellen, W60
Henderson, Lawrence Joseph, H61
Herrington, Barbour Lawson, H65
Herter, Christian A., H67
Herty, Charles Holmes, H68
Hildebrand, Joel Henry, H71
Hill, Walter Nickerson, H73
Hogness, Thorfin, H77
Hopkins, Arthur John, H86
Hopkins, B. Smith, H87
Horsford, Eben Norton, D41, H88
Houdry, Eugene Jules, H89
Huffman, Eugene Harvey, H93
Hussey, Robert E., H98

Ingersoll, Arthur William, I4
Ipatieff, Vladimir N., I9

Jackson, Charles Loring, J1
Jackson, Charles Thomas, J2, J3
Jacobs, Walter Abraham, J5
Jaffa, Meyer E., J6
Jeffress, Elizabeth Talbott (Gwathmey), J8

Johnson, Norman G., J14
Johnston, James A., J15
Jones, Grinnell, J18
Jones, Richard Uriah, J20

Kahlenberg, Louis, K2
Kamen, Martin David, K4
Keating, William H., A55
Kedzie, Frank Stewart, K5
Keller, Arthur G., K9
Kilgore, Benjamin Wesley, K17
Kimberly, John, K19
King, Victor L., K22
Kleinheksel, J. Harvey, K26
Knight, Harry Granger, K27
Kohler, Elmer Peter, K29
Kotch, Alex, K31
Krauskopf, Francis Craig, K33
Krebs, August Sonnin, K34
Kunitz, Moses, K37

Ladd, Edwin Freemont, L2, L3
Laitinen, Herbert August, L5
Lamb, Arthur Becket, L6
Lamb, George G., L7
La Motte, Arthur, L8
Langley, John Williams, L10, L11
Langmuir, Irving, L12
Lavoisier, Antoine Laurent, A55, L15
Lawrence, Ernest O., L17
Le Conte, Joseph, L20
Lehninger, Albert L., L23
LeMaistre, Frederic J., L24
Leslie, Eugene Hendricks, L27
Levene, Phoebus Aaron Theodor, L28
Levison, Wallace Goold, L29
Lewis, Gilbert Newton, L31
Lewis, Howard Bishop, L32
Libby, Willard Frank, L34
Lipmann, Fritz Albert, L39
Little, Arthur Dehon, L40
Lloyd, John Uri, L41
Loeb, Jacques, L44, L45
Long, Cyril Norman Hugh, L46
Longsworth, Lewis Gibson, L48
Lord, Richard Collins, L51
Lovering, Mary Campbell, L54
Lowry, Homer Hiram, L55
Luck, James Murray, L58
Luckey, Thomas D., L59

McAdams, William Henry, M1
McChesney, Joseph Henry, M4
McCoy, Herbert N., M5
McFarland, David Ford, M9
McGill, John Thomas, M10
MacInnes, Duncan Arthur, M11
Mackintosh, James Buckton, M14
McMillan, Edwin M., M17
Mah, Richard S. H., M20
Mallmann, Paul, M21
Maron, Samuel Herbert, M24
Mason, William Pitt, M28
Mathews, Joseph Howard, M38
Maver, Mary Eugenie, M39
Mayer, Ralph, M42
Maynard, Leonard Amby, M43

Mees, Charles Edward Kenneth, M49
Meggers, William F., M50
Mell, Patrick Hues, M51
Mendel, Lafayette B., M52
Mills, James Edward, M64
Mirsky, Alfred Ezra, M67
Mitchell, Harold Hanson, M69
Mitchill, Samuel Latham, A55
Moeller, Therald, M70
Moore, Edward Warren, M74
Morfit, Campbell, M75, M76, M77
Morgan, Agnes Fay, M78
Morgan, John, A55
Morley, Edward Williams, M80, M81, M82
Mulliken, Robert S., M87

Nason, Henry B., N1
Nef, John Ulric, N5
Neuberg, Carl, N9
Neurath, Hans, N10
Nickles, Francois Joseph Jerome, N18
Nieuwland, Julius Arthur, N19
Norman, George Miller, N21
Norton, John Pitkin, N25
Noyes, Arthur Amos, M34
Noyes, William Albert, N27

Oenslager, George, O1
Olin, Hubert Leonard, O5
Omansky, Morris, O8
Onsager, Lars, O9

Palmer, Arthur William, P3
Pasteur, Louis, P10
Patterson, Robert Maskell (1787–1854), A55, P11
Patterson, Robert (1743–1824), P11
Pauling, Linus Carl, P12
Peckham, Stephen F., P16

Pemberton, Henry, P18
Perlmann, Gertrude Erika, P26
Peter, Robert, P30
Piccard, Jean Felix, P40
Pickel, J. J., P41
Polanyi, Michael, P50
Potts, William John, P57
Preisler, Paul William, P62
Priestley, Joseph, A55, P64, P65, P66, P67, P68
Proskauer, Bernhard, P71
Pugh, Evan, P72

Rabinowitch, Eugene I., R2
Reese, Charles Lee, Sr., R7, R8
Remsen, Ira, R10
Rheineck, Alfred E., R14
Richards, Ellen Henrietta (Swallow), M35, R18
Richards, Theodore William, R19
Riegelman, Sidney, R23
Rising, Willard Bradley, R24
Robinson, Wirt, R27
Rochow, Eugene George, R29
Rogers, William Barton, R32
Rose, William Cumming, R37
Rossini, Frederick Dominic, R40
Roughton, Francis John Worsley, R42
Ruben, Samuel, K4
Rush, Benjamin, A55

Sanger, Charles Robert, S4
Saunders, Arthur Percy, S5
Schlossberger, Julius Eugen, S11
Schlundt, Herman, S12
Schoch, Eugene Paul, S13
Schuette, Henry August, S16
Schultz, Jack, S17
Schulz, Helmut William, S18
Schweitzer, Paul, S19

Scott, Robert C., S20
Seaborg, Glenn T., S21
Seibert, Florence Barbara, S25
Seidell, Atherton, S26
Severinghaus, Elmer Louis, S28
Shaffer, Philip Anderson, S29
Sheehan, John Clark, S32
Shepard, Charles Upham, S36, S37
Sherwood, Thomas Kilgore, S38
Silliman, Benjamin, Jr. (1816–1885), S44, S45
Silliman, Benjamin, Sr. (1779–1864), A55, S36, S42, S43, S45
Skinner, William Woolford, S48
Small, Lyndon Frederick, S50
Smith, Albert William, S51
Smith, Alexander, S52
Smith, Edgar Fahs, C32, S54, S55
Smith, George Frederick, S56
Smith, George McPhail, S57
Smith, William A., A55
Smithson, James, S59
Snyder, Harry, C70
Stanley, Wendell Meredith, S76
Stearns, Frank Waterman, S77
Stine, Charles M. A., S79
Stinson, Margaret E. Dayton, M35
Storer, Francis H., S80
Strait, Louis A., S84
Stubbs, William Carter, S89, S90, S91
Stumm, Werner, S94
Sumner, James B., S97
Swain, Robert Eckles, S101
Szent-Gyorgyi, Albert, S103, S104

Tatum, Edward Lawrie, T5
Tesoro, Giuliana, W60
Thomas, Roy Zachariah, Sr., T13
Thomson, Elihu, M34
Todd, Clare Chrisman, T18

Torrey, Henry A., T19
Traube, Isidor, T21
Tucker, Willis Gaylord, T29
Twitchell, Ernest, T31

Urey, Harold Clayton, U32

Van Slyke, Donald Dexter, V3
Van Zyl, Gerrit, V5
Vickery, Hubert Bradford, V11
Villars, Donald Statler, V12
Vinograd, Jerome, V13
Voegtlin, Carl, V15
Vogel, Hermann Wilhelm, V16
Voorhees, Louis Augustus, V17

Walton, James Henri, W7
Washburn, Edward Roger, W10
Weber, Harold C., W16
Webster, John White, W17
Weiss, John M., W19
Wetherill, Charles M., W26
Whitby, George S., W31
Widtsoe, John Andreas, W40
Wilber, Francis Augustus, W42
Wiley, Harvey Washington, W44
Willard, Hobert Hurd, W45
Williams, Robert Ramapatnam, W49
Willstatter, Richard Martin, W50
Wilson, Edgar Bright, Jr., W51
Wilson, John Lewis, W52
Wilson, Robert Erastus, W53
Withers, William A., W57
Wood, George B., W62
Woodward, Robert Burns, W66
Wrinch, Dorothy M., W72
Wurtz, Henry, W74

Yoe, John Howe, Y2
Yost, Don M., Y4

SUBJECT INDEX

Agricultural chemistry, A87, B79, C49, C76, F22, F30, G42, H52, H53, J6, K17, K27, L2, M6, N15, N25, O11, P24, P41, P72, S48, S66, S80, S89, V11, V17, W40, W57. See also Dairy chemistry, Food chemistry
  animal feeds, P41
  departments of, U19
Agricultural experiment stations, A87, K70, M3, N16, N17, O11, P22, R50, S90, V11, V17
Alchemy, A20, A88, B64, B66, C47, E16, H75, S86, P9
Alcohol industry, H69
Aluminum, H5
Amino acids, R37. See also Protein chemistry
Analytical chemistry, C34, E15, G32, H93, L5, P3, P57, W42, W45
Anesthetics, J2, J3, M84, W39
Apothecaries. See Pharmacy
Applied chemistry, D50, M77, S94,
Arsenic, S4
Asbestos, B12
Ascorbic acid, L23, S103
Assaying and assayers, A12, A78, C18, C56, C66, C67, G38, K23, M63, N8, U4, U5
Atomic energy policy. See Nuclear energy policy
Atomic weight determination, C38, M80
Aureomycin, G17

Battery additive controversy, L48, W16
Beet sugar. See under Sugar chemistry
Beet sugar refining See under Sugar refining
Benzoate, R10
Biochemistry, G30, H45, N16, S97, W13, W37, W72
  departments of, H33, L32
  history of, C31, E6, O10, S99
Bioluminescence. See Luminiscence
Blasting powder. See under Explosives and explosives industry
Bleaching, J21
Borax, L43, N12
  history of, W65
Brewing industry, A90, E9, F4, G1, G37, H60, K15, K16, M68, M83, N14, P27, P53, P54, R9, R15, S67, S74, T20, V8, V14, W6, Y3
Bromine, B21, S51

Camphoric acid, N27
Carbohydrate chemistry, E15, H45
Carbon-14 dating, J13, K4, L34
Carbon disulfide, T8
Carbon research, M86
Carborundum, A6, A7
Catalytic chemistry, I9
Celluloid, A18, C19, C20
Cellulose, A13
Cement industry, B62, H21, R47, W9, W32, W33
Cement testing, S68
Ceramics and ceramic engineering, A65, O15, R33
Chemical apparatus and instrumentation, P38, P39, V6, W41
Chemical engineering, D49, L40, M28, T7
  colleges, departments, schools of, C16, D34, H37, H38, M29, N24, U24, U25
  computer aided design in, M20
  education, K12
  societies, A51, A52, H30, M36
  students' notes, M1
Chemical genetics, S17

Chemical industry and trade, A7, A67, A84, A85, B11, B17, B19, B21, B23, B33, B34, B35, B41, B91, C12, C15, C30, D29, D32, D33, D37, D40, D60, D61, D62, D63, D65, D66, D67, D69, D70, D71, D73, D75, D76, D78, D79, D81, D83, E8, F18, G5, G24, G39, H63, H88, J12, J17, K10, K35, L1, L24, L43, L61, M13, M27, M71, M73, N8, N12, P21, P35, P53, R4, R7, R13, R45, R46, S20, S27, S35, S50, S53, S66, S71, S79, T2, T8, T15, T17, V4, V10, W27, W48. See also Petroleum refining
  exhibits on, C21
  fine chemicals, R39
  German patents, G4
  in Germany, F28, I1
  heavy chemicals, P59
  history of, D63, D68, H68, N26, W14, W65
  labor unions and, B56, B78, D18, I7, O3, O4
  petrochemicals, H82
  plastics, A18, B13, B97, C19, C20, D49, D58
Chemical information, A61
Chemical physics, J22, M87
  societies, A57
Chemical warfare, F33, G16, G25, H68, M7, M64, S33, U8
  World War I, A9, M26, U3
  World War II, A16, D47, G22, Y2
Chemistry. See also Agricultural chemistry, Applied chemistry, etc.
  colleges, departments, schools of, B38, C16, C17, C22, C37, C51, H34, H85, M31, M54, O2, R11, S30, S102, U11, U13, U14, U17, U22, U24, W3, W25
  essays, lectures, notes in, A2, A56, B3, B27, B39, B54, B55, B67, B100, C79, D52, D53, F1, F23, F24, H12, H15, H66, H86, H98, K13, L21, L54, N23, P49, S57, S79, T26, V1, W26, W54, W62, Y1
  foundation grants for, G4, I6, R30
  in France, N18
  in Germany, E18, H1, S11, W52
  history of, B29, B74, C22, C32, C37, C47, D55, M55, S54, S64, V5, W3, Y1
  journals, A31, B18, F5, H26, J22, N27
  laboratories, B83, B84, D50, H31, H93, M33, M37, U15, U21, W25
  oral history interviews in, M46, N16, N20, U12
  societies, A5, A31, A32, A33, A34, A35, A36, A37, A38, A39, A40, A41, A42, A43, A44, A45, A46, A47, B56, H28, H29, H32, H39, H40, H41, H43, M53, N22
  students' notes in, A14, A22, A25, B51, B53, B63, B68, B86, B90, B98, C24, C36, C72, C25, C84, C87, E4, E17, E19, G44, H76, H95, J16, K11, L16, L33, M16, M30, M48, M56, M62, M65, O7, P56, P57, R7, R31, S14, S36, S72, S88, S92, S93,S97, T28, V9, W36, W56, W68, Y2
  use of statistics in, V12
Chemurgy, C27
Chlorine, S51
Chloroform, S51
Chlorophyll, P8
Clinical pharmacology, G29
Coal chemistry, L55, W74
Coatings chemistry, A85, R14
  in aircraft, A13
Coenzyme A, L39

Coke industry, H48, L1, O6, S27, S73
Colloid chemistry, A23, L44
Combustion research, B14, D16, T7
Copper industry, A60
Cortisone, H59
Cosmetics and cosmetics industry, A19, J4, R13, W18
Crystallography, C34, D35, F5, K37, M15, W72
  societies, A50, M15
Cyclotron technology, L17
Cytochemistry, S17

DNA. See Deoxyribonucleic acid
Dairy chemistry, B6, B7, H65. See also Food chemistry
Deoxyribonucleic acid, R6, W13
Diabetes, L46
Digitalis, G29
Drug industry, A84, B17, C28, C50, D5, D56, E1, G17, G24, G36, H63, H70, J4, J12, K25, L35, L41, M59, P5, P60, R13, S9, S50, T3, W5, W8, W18, W27, W56, W63, Z2. See also Pharmaceutical industry.
  patent medicines, B24, C1, D9, M47, P28, P46, P51, T27
  regulation of, K7, N28
Dyes and dyeing, B27, B36, C20, C53, D63, G7, G20, G21, H36, I2, J10, J21, L53, L61, M19, N2, P51, R41, S8, S47, T2, W47, W70

Electrochemistry, A6, H5, L48, S31, S51, W4, W58
Electron microscopy, A69
Electrotyping, F6
Enzyme chemistry, D39, H57, S97
Ether, N1
Explosives and explosives industry, A85, D59, D62, D64, D65, D67, D75, D76, D80, E8, F37, J2, K18, M41, R7, R12, R24, S86
  blasting powder, C13
  gunpowder, A15, A77, B59, D60, D61, D71, D72, D73, D77, D78, D79, D82, D83, G19, I3, L4, L15, M40, P15, V7
  pyrocellulose, B87
  research, D74, E21, J14, L8, L24, N21
  seismic explosives, J14

Fermentation, N9
Fertilizer chemistry, F30, H23, J76, P22
Fertilizer industry, B33, B34, C78, G18, H54, K10, L57, N8, P1, P14, P42, S19, S87, W28, W35, W55, W69. See also Phosphates and phosphate industry
Filtration, physical chemistry of, G27
Fluoridation of water. See under Water Food and drug legislation, D23, L2, L3. See also Food chemistry and Drug industry
Food chemistry B92, B93, C14, C70, F29, F30, R18, R37, S16, S66, W44, W57. See also Dairy chemistry; names of specific foods
Food chemistry
  additives, D23, P60
  preservation, B95
Fuel chemistry, L7
Furnaces. See under Iron industry

Gas industry, B59, F3, H49, J23, L22
Gas scrubbing, H18
Gas warfare. See Chemical warfare

Gemmology, K38, K39, K40
Geological chemistry, A12, B40, C38, C39, C42, F24, H2, L20, M4, M51, N8, P30, R32, S43, S62, S65
Glass industry, A29, B89, C71, F8, H8, I10, K10, P36, W1
Glue industry, C60
Gold, S51
Gunpowder. *See under* Explosives and explosives industry
Gutta percha, B59

High pressure chemistry, I9
Histamine, G35
Hormones, synthesis of, D85
Hydrocarbon chemistry, S13
Hydrolysis of fats and oils, T31

Illuminating gas. *See* Gas industry
Immunochemistry, H55, S25
Ink manufacture, H47
Inorganic chemistry, A36, A89, G13, H87, M70, S11, Y4
Insecticides. *See* Pesticides
Insulin, P31
Ion exchange, H93
Iron industry, A54, B97, C60, D30, D38, E2, F3, H3, H97, K10, L25, L60, M45, O6, P37, S22, S34, T11, T24, U2. *See also* Steel industry
  furnaces, B50, B101, C46, G47, G48, H11, J19, K6, M22, P25, P32, P45, P55, P58, P69, S83, T16, W23, W71
  history of, A54, B50, B82, M9, P32, S83
  taconite, D10

Labor unions. *See under* Chemical industry and trade; Petroleum refining
Lead industry, A63
Leather chemistry, K13
Leather industry, A21, A28, B69, B102, C11, C53, D30, D54, D60, G7, H58, H94, I3, L9, N29, P61, R17, S10, S40 S67, S82, T12
Lipid chemistry, S16
Lipid metabolism, L23
Lithopone, K35
Low temperature chemistry, G10
Luminescence, L29

Maleic anhydride, W19
Manganese, A8
Manhattan Project, G41, L34
Margarine, B59, S19
Marijuana, A9
Marine biochemistry, F26
Materia medica. *See under* Pharmacy
Metallography, C80, T30
Metallurgy, A26, A63, A75, B104, C8, C26, C30, D19, E10, E22, F12, G2, H5, H19, H50, H79, H84, H90, L11, L50, M9, M21, P48, R21, R25, R44, R48, T25, W58
  engineering, R20
  history of, H9, N6
  nickel alloys, B5, R1
  societies, A53, A71

Metals. *See names of specific metals;* Mines and mining
Meteorites, S62, S63
Mineral industries, B70, D20, M66
Mineral resources, S60
Mineral waters, D20, N29, R43
Mineralogy, B16, B20, B27, B39, B58, B70, B73, C13, C42, C58, F35, H78, J11, K32, K38, K39, K40, M14, N1, P33, R27, S36, S37, S60, S62, S63, S69, U18, V2, W17, W29
Mines and mining, A26, A71, B62, B102, B104, C30, D37, G28, H2, H48, H80, L49, L56, M63, M66, M79, M85, N11, O5, R25, S44, S46, W42, W43
  chrome, T32
  copper, H91, L37
  gold, C18, H91
  iron, I11, J19
  manganese, L47
  mercury, N13
  molybdenum, K28
  tungsten, K28, L47
  zinc, H91
Monosodium glutamate, G51

Naval ordinance, H73
Neural physiology, G35
Nitrate industry, S96
Nitric acid, E8
Nitrogen, A89, F28
Nobel laureates in chemistry. *See individual names of prize winners*
Nuclear chemistry, M17
Nuclear energy policy, A79, A80, A81, A82, A83, A86, B30, B99, E14, F9, F27, H77, L34, R2, S21, W11
Nuclear engineering, B2, M32
  departments of, U26
Nutritional chemsistry, C70, D12, H24, J6, L59, M41, M43, M52, M69, M78, R37, W49. *See also* Food chemistry
Nylon, D58

Oil. *See* Petroleum
Ores. *See also* Assaying and assayers
  processing, B66, B102, C56, G28, G40, H56, I11, P47, R48, S100
  sampling and estimation of, B81, L49, N11, O7, R49, S41, W64
  smelting, A60, A75, H35, H48, H72, L63, M63, M72, M79, P48, S73
Organic chemistry, B60, C52, D25, F13, H7, H98, I4, M10, M30, N23, N27, S32, W66, Y2
Orpet-Lambert poisoning case, E12

Paint and pigments, A11, B75, C33, D69, K35, L30, M42, S39
Paint industry, A84, B17, B57, C50, D18, D27, H22, M23, N3, P53, S15, W1, W27. *See also* Varnish and varnish industry
Paper chemistry, A33, H68
Papermaking, A33, B91, C83, D24, D29, D71, E20, F19, G14, G20, G34, I3, J11, J23, K3, K20, K30, P13, P53, S6, S70, W21, W22, W24, W34, W46
  history of, K20, M2
Patent medicines. *See under* Drug industry
Penicillin, B10, C40, G6, S32
Pesticides, R4, W35
  regulation of, G42
Petrochemical industry. *See under* Chemical industry and trade
Petroleum chemistry, H89, L7, P23
Petroleum pipeline coatings, D17
Petroleum refining, A70, A92, B45, B102, C7, C86, D57, F64, H72, K36, L27, L38, L42, M60, M89, O14, P16, P23, R36, S98, W53
  labor unions and, O3, O4

oil shale, A59, G50, W2
Pharmaceutical chemistry, M75, R23, S84
Pharmaceutical industry, B103, H59, M44, P6, R39, U30, U31. *See also* Drug industry
Pharmacology, A3, C48, C63, D7, G35, H96, M10, M18, T4, U3. *See also* Clinical pharmacology; Veterinary pharmacology
  colleges, departments, schools of, C64, U20, U23
  history of, U23
Pharmacopoeia, A4, U9
Pharmacy, H81, K32, L41, M88, R23, S78
  accreditation in, A30
  associations and societies, A17, A30, A48, G9, H81, M10, O13, P70, R16, W73, Z2
  business of, C25, D36, D46, H81, H92, M25, M61, P4, P51, S75
  colleges, departments, schools of, L62, U16, U27, U28
  education in, B52, H92, L62, P8, W73
  historical materials of, S61, U29
  and materia medica, B27, G33, M48
  in military, B52, L19
  pharmaceutical formulae, A68, A72, B66, B77, D36, F17, G5, G23, H81, K24, P63, T22, T23
  regulation of, L36
Phenol, W19
Phosphate and phosphate industry, C29, C62, G18, H49, M76, P44, S2, W38, W55. *See also* Fertilizer industry
Photochemistry, V16
Photographic chemistry, C40, C57, M49
Photosynthesis, F27, K4
Phthalic anhydride, W19
Physical chemistry, D84, E22, F2, F14, H25, H25, L48, L51, L55, M49, P50, S49, S52, V12
Physiological chemistry, C31, D44, H57, L44, L45, L46, M52, M69
Pigments, biological, D39
Plastics. *See under* Chemical industry and trade
Pollution, B22, C55, G27, G46, H89, K1, L5, M10, O12, P2
  of water, D10, W38, W55
Polymer chemistry, A57, B96, D2, D31, F20, M24, M70, R14
Potash industry, A1, B58, B89, I7, P52
Protein chemistry. *See* Amino acids

Quantum theory in chemistry, D21

Radiochemistry, B65, C82
Radium, H27
Rainmaking, B18
Rare earths, H87, M5, M50, M70
Rayon, S47
Respiratory physiology, R42
Rubber chemistry, O1, W31
Rubber industry, B1, B60, B72, C9, C52, C54, C68, D11, D40, D47, D49, F25, G8, J9, J17, O8, P53, R5, S24, T9, U7, U10
  economics of, B94
Rubber, natural, B59, D9, E5
Rubber, synthetic, B30, D14, D31, L27, M24, R26, T14, U7

Saccharin, R10
Salt, R33
Saltpeter, L4, M69, N26, P15
Sanitary chemistry, M74, U6
Sanitary engineering, B4, B22, H51, T1
Silk, synthetic, B59, L24. *See also* Rayon
Smelting. *See under* Ores

Soap and detergents industry, A73, D13, F7, G7, M76, P35, S87
Soda industry (sodium carbonate), I8
Sodium, B59, S51
Soil chemistry, A8, B32, C70, F30, H23, K21, R22, S81, S91, W15. *See also* Agricultural chemistry
Solvents, L24
    nonaqueous, A89
    pyroxyline, C19
Sorghum, C49
Spectroscopy, C34, C44, H4, L13, L51, M50, M86
Steel industry, A54, C41, G45, H79, L60, O6, P29, P37, P43, T11, W61. *See also* Iron industry
    history of, A54
Stellite, H50
Streptomycin, R50
Sucaryl, A89
Sugar chemistry, B92, H20, N5, N7, S36, S89, S90, S91
    of beet sugar, G51, G53
Sugar refining, A10, A62, B8, B42, B47, B71, C13, C74, D3, D45, D51, G31, G52, H74, H83, K8, K9, K14, L26, L52, L57, M8, M57, M83, N7, P20, P34, R4, R22, R34, R35, S58, S85, T3, T6, W12, W20, W67, Z1
    of beet sugar, A49, B62, C88, D6, G15, G51, G53, L14, M58, N4, P17, S1, S95, W30
    of cane sugar, F31, G43, W30
    history of, C88, G51, L14
Synthetic fibers. *See* Textile chemistry; *names of specific fibers*
Synthetic rubber. *See* Rubber, synthetic

Taconite. *See under* Iron industry
Textile chemistry, A33, D62, D68. *See also* Dyes and dyeing; *names of specific fibers*
    finishing in, J21, S70
    chemicals for, S8. *See also* Dyes and dyeing
Toxicology, T29
Trade unions. *See* Chemical industry and trade, labor unions and; Petroleum refining, labor unions and

United States Atomic Energy Commission, S21, W53
United States Bureau of Chemistry and Soils, W44

Uranium, B64
Uranium isotopes, centrifugation of, S18

Varnish and varnish industry, C6, D9, D27, L18, N3, S15. *See also* Paint industry
Veterinary pharmacology, S3, U20
Vitamins, D12, W49

Water
    analysis of, A8, B28, H29, H51, M28, P3
    desalination of, G27
    fluoridation of, A27, F15, F34, H13, J7, M12, R38, S23
    pollution of, D10, W38, W55
    purification of, B4
Waterproofing, B57, S70
Wine chemistry, G54
Wine industry, C4, C5
Women in chemistry and chemical engineering, I5, M35, M37, M39, M46, W59, W60

X-ray diffraction, C34, C80

Zinc industry, A63

GEOGRAPHICAL INDEX

ALABAMA

Auburn
Alabama Pharmaceutical Association
Basore, Annie Terrell
Crews, J. M.

Maxwell Air Force Base
Air Force units (lower echelon), attached units, and stations overseas

Tuscaloosa
Shelby Iron Company, Shelby, Alabama

ARIZONA

Tucson
Blake, William Phipps
Brinegar, Thomas P.
Douglas, Lewis Williams
Forbes, Robert Humphrey
Miller, Hugo W.
Silliman, Benjamin, Jr.
Whitmore, William Vincent

CALIFORNIA

Berkeley
Bray, William Crowell
Business history, administration, and education: oral history interviews
Butters, Charles
California wine industry: oral history interviews
Fischer, Emil
Fisk, Charles Frederick
Fox, Denis Llewellyn
Giaque, William Francis
Gutleben, Dan
Guymon, James Fuqua
Hassid, William Zev
Hildebrand, Joel Henry
Huffman, Eugene Harvey
Jaffa, Meyer E.
Kamen, Martin David
Lawrence, Ernest Orlando
Lewis, Gilbert Newton
Locke, William Lovering
McMillan, Edwin Mattison
Morgan, Agnes Fay
Nelson, Oscar M. F.
Rising, Willard Bradley Schulz, Helmut William
Stanley, Wendell Meredith
Turner, Francis John
University of California, Berkeley
    College of Chemistry
    oral history interviews
Varian Associates
Villars, Donald Statler

Chico
Sacramento Valley Sugar Company

Los Angeles
Edgeworth, Lovell
Farmer, Moses Gerrish
Johnson, Frederick
Libby, Willard Frank
McCoy, Herbert Newby
Taylor, Frederic William

Palo Alto
Shenk, John Wesley

Pasadena
Goetz, Alexander
Morley family
Vinograd, Jerome
Yost, Don M.

Sacramento
Fay family
Gordon, George
Hilby, Francis Martin
Pioneer Reduction Works, Nevada County, California

San Diego
Urey, Harold Clayton

San Francisco
American Sugar Refinery Company, San Francisco, California
California Wine Association
Dill, Marshall
Riegelman, Sidney
Strait, Louis A.
Varney, Thomas
W. P. Fuller and Company

San Marino
Catlin, Amos Parmalee
Hague, James Duncan
Kunz, George Frederick
Nevada Mining Companies
Oliver-Gowen Collection
Sill, Harley A.
Sutro, Adolph Heinrich
Wood, Henry Ellsworth

Stanford
Babbitt, James Aloysius
Black, Joseph
Brasch, Frederick E.
Findlay, Alexander
I. G. Farben Trust, Propaganda Division
Luck, James Murray
New Almaden Mine, California
Peckham, Stephen Farnum
Shekerjian, Haig
Swain, Robert Eckles

COLORADO

Boulder
Darley family

Denver
American Shale Refining Company
Boettcher, Charles
Gold and Silver Extraction Company
Hill, Crawford
Kirkland, George W.
Knight, John
Lyon, James E.
Morland and Warwick, Villaldama, Mexico
National Sugar Manufacturing Company
Ruter, Charles
Wadleigh, Frank A.

CONNECTICUT

Hartford
Bulkeley, Gershom
Connecticut: State Library, Hartford
Hartford residents' correspondence, 1807-1947
Trinity College: Student essays

Middletown
Atwater, Wilber Olin

New Haven
Arnold, Benedict
Boltwood, Bertram Borden
Brewer, William Henry
Brown, William Robinson
Candee, Leverett
Century of Progress International Exposition New York World's Fair collection
Davis, John William
Foskett & Bishop Company, New Haven, Connecticut
Mendel, Lafayette Benedict
Norton, John Pitkin
Onsager, Lars
Peters, John Punnett
Plattner, Karl Friedrich
Rockwell, Alfred Perkins
Russell Process Company
School records, 1715, 1770-1963
Silliman family
Underhill, Frank Pell
Vanderbilt, Cornelius
Vickery, Hubert Bradford

Salisbury
Pettee family

DELAWARE

Greenville
Aeronautics: The Lammot du Pont, Jr., collection of aeronautics
Amalgamated Leather Companies, Wilmington, Delaware
Atlas Powder Company collection
Baugh & Sons Company
Brinton, Clement Starr
Brown, Harry Fletcher
Cardon (A.) and Company
Carpenter, Walter Samuel
Carter & Scattergood, Philadelphia Pennsylvania
Curtis & Brother Company, Newark, Delaware
Delaware industrial miscellany
Du Pont de Nemours (E. I.) and Company
Du Pont, Alfred Victor
Du Pont de Nemours, Eleuthère Irénée
Du Pont de Nemours, Pierre Samuel
Du Pont, Ernest
Du Pont Eugene, 1840-1902, and Eugene du Pont Jr., 1873-195
Du Pont family
Du Pont, Francis Gurney
Du Pont, Francis Victor
Du Pont, Henry
Du Pont, Lammot
Explosives
Fulton, Gardiner
Gilpin, Joshua
Haldeman family
Harrison Brothers & Company
Henry Bower Chemical Manufacturing Company
Industrial workers of Brandywine Valley and Wilmington, Delaware, oral history
Johnson, Norman G.
Joseph Bancroft & Sons Company, Wilmington, Delaware
Kimball, Philip J.

Krebs, August Sonnin
Krebs, Henrik Johannes
La Motte, Arthur
Lukens Iron and Steel Company, Coatesville, Pennsylvania
Maxim, Hudson
Morris family
National Paint, Varnish, & Lacquer Association, Inc., Washington, D.C.
Norman, George Miller
Pheonix Steel Corporation
Raskob, John Jakob
Reese, Charles Lee
Repauno Chemical Company, Repauno, New Jersey
Rhoads (J. E.) & Sons
Savery, Thomas H.
Schrack (C.) and Company, Philadelphia, Pennsylvania
Sun Oil Company
Taylor-Wharton Iron & Steel Co., High Bridge, New Jersey
Warner, Irving
Whann family
Wood, Frederick William

Newark
Drake, Quaesita Cromwell

Wilmington
American Iron and Steel Institute collection
Du Pont, Eugene, Jr.
Du Pont de Nemours (E. I.) and Company
   Employee Relations Department
   Experimental Station, Chemical Department
   Office of the President
   Textile Fibers Department
   William P. Allen, Executive
   Willis F. Harrington, Executive.
Greenewalt, Crawford
LeMaistre, Frederic J.
Philadelphia Quartz Company
Principio Company
Reese, Charles Lee, Sr.
Stine, Charles M. A.
Wetherill, George D. (and Co.), Philadelphia
Wright, Samuel G.

DISTRICT OF COLUMBIA

Acheson, Edward Goodrich
Ainsworth, Frederick Crayton
Albany Billiard Ball Company
Alchemy collection
American Chemical Society
Amory family
Arrhenius, Svante August
Baekeland, Leo H.
Becker, George Ferdinand
Black, Joseph
Bolton, Henry Carrington
Browne, Charles Albert Celluloid Company
Celluloid Corporation
Clarke, Frank Wigglesworth
Coblentz, William Weber
Collected scientific manuscripts
Cook, George Smith
Cooper, Hewitt & Company. Ringwood, New Jersey
Cottrell, Frederick Gardner
Draper family
Draper, John William
DuBois, J. Harry: History of plastics collection
Du Pont Company: Nylon collection
Dushman, Saul
Foreman, Edward R.
Funk, Casimir
Gibbes, Lewis Reeves
Gilchrist, Huntington
Hare, Robert
Hitchcock, Ethan Allen
Holley, Alexander Lyman
Houdry, Eugene Jules
Jackson, Charles Thomas
Jenckes, Thomas Allen
Kunz, George Frederick
Langmuir, Irving
Lipmann, Fritz Albert
Little, Arthur Dehon
Loeb, Jacques
Mayer, Ralph
Medical history collection
Mineral industries
Morfit, Campbell
Morley, Edward Williams
Morton, William Thomas Green
Murphree, Eger Vaughan
Physical sciences history collection
Physical sciences information files
Piccard family
Priestley, Joseph
Shepard, Charles Upham
Smithson, James
Smithsonian Institution
   Department of Mineral Sciences
   Division of Medical Sciences
   Division of Meteorites
   Division of Mineralogy and Petrology
   Division of Physical Sciences
   Office of the Smithsonian Chemist
   Section of Agriculture, Division of Agriculture and Mining
Szent-Gyorgyi, Albert
Whitehorn, John Clare
Wiley, Harvey Washington
Williams, Robert Ramapatnam
Wilson, Robert Erastus
Wurtz, Henry

GEORGIA

Athens
Mell, Patrick Hues

Atlanta
Bellinger, Frederick
Herty, Charles Holmes
Jacobs, Joseph, pharmacist, of Atlanta, Georgia

Savannah
LeConte-Furman family papers

ILLINOIS

Chicago
Association of Cambridge Scientists
Association of Los Alamos Scientists
Association of Oak Ridge Engineers and Scientists
Association of Pasadena Scientists
Association of Scientists for Atomic Education
Atomic Scientists of Chicago
Brown, Rachel Fuller Brown
Bulletin of Atomic Scientists
Emergency Committee of Atomic Scientists
Federation of American Scientists
Franck, James
Hogness, Thorfin
Mulliken, Robert S.
Nef, John Ulric, 1862–1915
Nef, John Ulric, 1899–
Northwestern University
Polanyi, Michael
Rabinowitch, Eugene I.
Slotin, Louis A., Memorial Fund collection
Swift, Harold H.
University of Chicago: Physical Sciences Division
Washington Association of Scientists

Evanston
Ipatieff, Vladimir N.
Lamb, George G.
Mah, Richard S. H.
Northwestern University
   Technological Institute
   Department of Chemical Engineering

Springfield
C. Wakefield and Company. (Bloomington, Illinois)
Dryden Georg Bascom
Eisenschiml, Otto
Gaskill, James R. M.
Reisch Brewing Company (Springfield)

Urbana
Adams, Roger
Andrews, Andrew Irving
Audrieth, Ludwig Frederick
Babbitt, Harold Eaton
Bartow, Edward
Bartow, Virginia
Bauer, Frederick Charles
Carter, Herbert Edmund
Chedsey, William R.
Clark, George Lindenberg
Derick, Clarence G.
Englis, Duane T.
Fuson, Reynold Clayton
Hopkins, B. Smith
Laitinen, Herbert August
McChesney, Joseph Henry
Mitchell, Harold Hanson
Moeller, Therald
Noyes, William Albert
Palmer, Arthur William
Ridenour, Louis Nicot
Rolfe family
Rose, William Cumming
Sampson, Jesse
Shamel, Charles Harmonas
Smith, George Frederick
Talbot, Arthur Newell
Thompson, Charles Manfred
University of Illinois: Department of Chemistry and Chemical Engineering
Waldo, Edward Hardenbergh
Wardall, William J.

INDIANA

Bloomington
Media-Related Perceptions of Contemporary Problems

Evansville
Mead Johnson and Company, Evansville, Indiana

Kokomo
Haynes Edwoo

Notre Dame
Callahan, Patrick Henry
Cullity, Bernard D.
D'Alelio, G. Frank
Nieuwland, Julius Arthur

Notre Dame, Indiana (*continued*)
Rossini, Frederick Dominic
University of Notre Dame Radiation Laboratory

Vincennes
Badollet, M. S.

West Lafayette
Brown, Herbert C.

IOWA

Des Moines
Olin, Hubert Leonard

Iowa City
Weeks, Carl

West Branch
Hoover, Herbert Clark

KANSAS

Lawrence
Spencer, Kenneth Aldred
Watkins, Jabez Bunting

KENTUCKY

Bowling Green
McReynolds family

Lexington
Means family
Peter, Robert
Seaton family

Louisville
Falls City Brewing Company, Louisville, Kentucky
Kelly, Wayne Clinton
Kennedy, James Arthur
Yandell family

LOUISIANA

Baton Rouge
Arceneaux, Claude Joseph
Boyd, Overton F.
Dugas and LeBlanc, Paincourtville, Louisiana
Freeland, Emile C.
Harris, James Courtland
Keller, Anatole J.
Keller, Arthur G.
Kerr, W. W.
Krotz Springs Cycling Plant
Link, Louis
Louisiana Sugar Planters Association
McCutcheon, Samuel
Pharr family
Rolston, William A., Sr.
Smithfield plantation records
Stubbs, William C.
Whipple, William

Lafayette
Griffin, Lucille Mouton

New Orleans
De Milt, Clara
Marine Paint and Varnish Company, Inc.

Thibodaux
Lepine, J. Wilson

MAINE

Brunswick
Cleaveland, Parker

Orono
Gilbert family

Portland
Albis Company, Portland, Maine
Barrows, Horace Aurelius
Robison family

MARYLAND

Baltimore
Abel, John Jacob
Adams White Lead Company, Baltimore, Maryland
Alexander, F. W.
Alexander, John H.
Amelung family
American Smelting and Refining Company
Atkinson and Smith families
Baltimore Chemical Works
Baugh Chemical Company
Druggists' records, 1784 1895
Gittings family
Glyol Chemical Company, Baltimore
Hellerman, Leslie
Herter, Christian A.
Jessop, Joshua
Kelley, William J.
Kirkwood, Frank Coates
Lehninger, Albert L.
Lucas-White collection
Maryland Chemical Works account books
McKim, William Duncan
Morfit family
Porter, G. Harvey
Remsen, Ira
Resinol Chemical Company, Baltimore, Maryland
Shoemaker family
Shriver (A. K.) & Sons Tannery, Union Mills, Maryland
Thompson family
Thompson, James
Tyson family

Beltsville
Knight, Harry Granger
Skinner, William Woolford

Bethesda
Black, Joseph
Book of Secrets
Cullen, William
Fordyce, George
Griffin, Edward Lawrence
Harte, Robert A.
Hastings, Albert Baird
Heidelberger, Michael
Maver, Mary Eugenie
Medical lecture notes, 1746 1864
Oral history collection, 1926, 1951 68
Pasteur, Louis
Playfair, Lyon, Baron
Seidell, Atherton
Small, Lyndon Frederick
Szent-Gyorgyi, Albert
United States. Hygienic Laboratory
Van Slyke, Donald Dexter
Voegtlin, Carl
Wood, George Bacon
Wormald, William

MASSACHUSETTS

Amherst
Hopkins, Arthur John
Shepard, Charles Upham

Andover
Pearson, Eliphalet

Boston
Adams Sugar Refinery, Boston
Dane, Dana, and Company, Boston
Davis, T. M., firm, Cambridge, Massachusetts
Fall River Iron Works, Fall River, Massachusetts
Goostray, Stella
Harvard University. Graduate School of Business Administration
Higginson, Henry Lee
Hooper (Samuel) and Company, Boston
Indian Head Mills, Inc.
Jackson, Charles Thomas
Monsanto Chemical Company
Pacific Guano Company
Phoenix Glass Company, Boston
Pierson (Isaac G.) and Brothers
Powers & Weightman
Pratt, Zadock
Stearns, Frank Waterman
United States Rubber Company
White (J & J) Paper Company, Canton, Massachusetts

Cambridge
Ashdown, Avery Allen
Bartlett, Paul D.
Baxter, Gregory Paul
Blatchford, Seward, & Griswold
Conant, James Bryant
Cooke, Josiah Parsons
Coolidge, Albert Sprague
Deutch, John M.
Edsall, John Tileston
Eliot, Charles William
Ferry, Ronald M.
Fieser, Louis P.
Forbes, George Shannon
Frondel, Clifford
Gorham, John
Harvard University
  Association of Harvard Chemists
  Boylston Chemical Club
  Chemical Engineering Society
  Chemical Laboratories
  Davy Club
  Department of Biochemistry and Molecular Biology
  Department of Chemistry
  Graduate School of Business Administration: Baker Library
  Graduate School of Engineering
  Harvard Chemical Club
  Harvard-Radcliffe Chemical Society
  Harvard-Tech Chemical Club
  Office of the President: James B. Conant
  Rumford Chemical Society
Hazen, Allen
Henderson, Lawrence Joseph
Institute of Women's Professional Relations
Jackson, Charles Loring
Jones, Grinnell
Kohler, Elmer Peter
Lamb, Arthur Becket
Lord, Richard Collins
Lovering, Mary Campbell
Massachusetts Institute of Technology
  Office of the President
  Office of the President: Richard, Ellen Swallow
  Department of Chemical Engineering
  Department of Chemistry

    Department of Nuclear Engineering
    Department of Physics
    Ten Club
    Women's Laboratory
McAdams, William Henry
Moore, Edward Warren
Omansky, Morris
Pinkham (Lydia E.) Medicine Company
Potts, Joseph D.
Recombinant DNA Controversy
Richards, Theodore William
Rochow, Eugene George
Rogers, William Barton
Sanger, Charles Robert
Sheehan, John Clark
Sherwood, Thomas Kilgore
Sondericker, Jerome
Storer, Francis H.
Stumm, Werner
Taylor, Charles Fayette
Torrey, Henry A.
United States Chemical Warfare Association, Boston Section
Watson, James Dewey
Weber, Harold C.
Webster, John White
Wilson, Edgar Bright, Jr.
Women in Science
Women in Science and Engineering Oral History Collection
Woodward, Robert Burns

North Andover
Barron, Springall and Company, Dexter, Maine
Louisville, Textiles, Inc.
Rothermel, Johannes
Talbot Mills, North Billerica, Massachusetts
William G. Johnson Company, Uncasville, Connecticut
Wright, Alexander, John, and Peter

Northhampton
Richards, Ellen Henrietta (Swallow)
Wrinch, Dorothy Maud

Plymouth
LeBaron, Francis

Salem
Abbot, Stephen
Choate family
Cutler, Manasseh

Springfield
Skinner (William) & Son, Holyoke, Mass.

Sturbridge
Hull, Alfred

Taunton
Leonard Iron Works, Taunton, Mass.

Waltham
Seaborg, Glenn Theodore

Woods Hole
Crowell family

Worcester
Apothecary's notebook, ca. 1800–864
Clark University: Department of Chemistry
Goddard, Robert Hutchings

MICHIGAN

Ann Arbor
American Chemical Society: Michigan Section
Babst, Earl D.
Bachmann, Werner Emmanuel
Barkman family
Bartell, Floyd Earl
Brownell, Lloyd Earl
Christman, Adam Arthur
Clauss, Julius A.
Crawford, Fred Lewis
Doty, Wirt P.
Du Bois, Alfred
Edwards, John Haldane
Fajans, Kasimir
Fletcher, Frank Ward
Gilchrist family
Harris, William P.
Holden, Edward Fuller
Jay, Philip
Kraus, Edward Henry
Langley family
Langley, John Williams
Laporte, Otto
Lawrence, Howard Cyrus
Leslie, Eugene Hendricks
Lewis, Howard Bishop
Michigan Sugar Company, Bay City, Michigan
Mummery, Arthur E.
Parsons, Henry Betts
Pettee, William Henry
Pontiac, Michigan. Weed's Drug Store
Prenderghast
Reason, Walter M.
Russell, Israel Cook
Tower family
Traverse City, Michigan: C. A. Bugbee Drug Company
Ulster Iron Works
University of Michigan
    Chemical Laboratory
    College of Pharmacy
    Department of Chemistry
    Department of Mineralogy and Petrology
Upjohn Company, Kalamazoo, Michigan
Upjohn family
West family
Willard, Hobert Hurd
Winchell, Alexander
Worcester, Dean Conant

Detroit
American Chemical Society: Detroit Section
Blackburn, Sam
Day, Walter
Detroit Society for Coatings Technology
Michigan College Chemistry Teachers Association
Parke, Davis & Company
Tregaskis, Richard

East Lansing
Donnell, John W.
Gutleben, Dan
Kedzie, Frank Stewart
Michigan State University
    Department of Chemistry
    Miscellaneous papers
    Miscellaneous records
Michigan Sugar Company, Bay City

Holland
Godfrey, Almon T.
Hope College, Holland, Michigan: Department of Chemistry
Kleinheksel, J. Harvey
Van Zoeren, Gerrit John
Van Zyl, Gerrit

Kalamazoo
Kalamazoo Paper Company

MINNESOTA

Minneapolis
Wulling, Frederick John

St. Paul
American Chemical Society: Minnesota Section
American Crystal Sugar Company
Chester-Kent, Inc., St. Paul
Craig family
Davis, Edward Wilson
Glenn family
Grain Belt Breweries, Inc., Minneapolis
Jones, Richard Uriah
Longyear (E. J.) Company, Minneapolis
Noyes, Winthrop Gilman
O'Shaughnessy, Ignatius Aloysius
Schaefer family
Whited, Oric Ogilvie
Wilson, John Lewis (1898–1983) and family
Yoerg Brewing Company (St. Paul, Minnesota)

MISSOURI

Columbia
American Chemical Society: University of Missouri Section
Babb, H. B.
Brown, H. Clifford
Hamilton, T. M. ("Ted")
Luckey, Thomas D.
Spillman, William Jasper
University of Missouri
    Department of Agricultural Chemistry
    Department of Veterinary Physiology and Pharmacology

Independence
Dean, Reginald S.

St. Louis
Academy of Science of St. Louis
American Chemical Society: St. Louis Section
American Zinc Company
Ashley, William Henry
Aufrichtig, Alois
Belcher family
Cori, Carl Ferdinand
Graham, Helen (Tredway)
Preisler, Paul William
Rosen, Ralph
Schlundt, Herman
Schweitzer, Paul
Shaffer, Paul Anderson

MONTANA

Helena
Granite-BiMetallic Consolidated Mining Company
Hauser, Samuel T.
Helena Mining and Reduction Company
Holter, Anton M.
Kessler, Charles Nicholas
Kessler, Nickolas
Linforth, F. A.
Lucas, John R.
Montana Ore Purchasing Company
Parchen, Henry Martin
Sizer, Frank L.
Spogen, Dominic
United States: Assay Office, Helena, Montana
Virginia Brewery

Montana (*continued*)

Missoula
Group Against Smog and Pollution, Missoula, Montana
Missoula Brewing Company

NEBRASKA

Lincoln
American Potash Company
Blackman, Elmer Ellsworth
Chase, Warren Tinker
Chemurgy project
Day, Stephen Delevan
Gray, William
Hamilton, Cliff Struthers
Hummel, Ray Orvin
Mikkelsen, Niels
Roberts, Paul Henley
Washburn, Edward Roger
Weakly, Harry Elmer

NEVADA

Reno
Consolidated Virginia Mining Company
Inyo Development Company
Nevada Salt and Borax Company

NEW HAMPSHIRE

Concord
Lane, Samuel

Hanover
Gilder, William Henry King, Victor L.

NEW JERSEY

Freehold
Dover Forge Iron Works, Berkeley Township, Ocean County, New Jersey
Schanck family

New Brunswick
Beck, Lewis C.
Beck, Lewis Caleb
Bogert, Nicholas I. Marsellus
Brakeley, John H.
Condit family
Cox, Joseph Warren
Darcy, Timothy Johnes
Hendrickson family
Hommell, Philemon E.
Johnston, James A.
Maclean family
Rumford Chemical Works
Rutgers University: College of Agriculture and Experiment Station
Vermeule, Adriana, Jr.
Voorhees, Louis Augustus
Wilbur, Francis Augustus

Newark
Academy of Medicine of New Jersey
Bradley, Joseph P.

Princeton
Baruch, Bernard Mannes

NEW MEXICO

Albuquerque
Clark, John Dustin

NEW YORK

Albany
Burden Iron Company, Troy, N.Y.
Griswold, John Augustus
John L. Thompson, Sons & Company (Troy, N.Y.)
New York State Agricultural Society
Williams family

Buffalo
Mrozowski, Stanislaw W.

Clinton
Saunders family

Coxsackie
Austin family

Hyde Park
Roosevelt family
Taussig, Charles William
United States: Special Committee to Study the Rubber Situation

Ithaca
Babcock, Stephen Moulton
Bancroft, George
Bancroft, Wilder Dwight (B101)
Blomquist, Alfred Theodore
Bowman, Henry
Brown, Matthew
Burt, Le Van Merchant
Caldwell, George Chapman
Carded Woolens Manufacturers Association
Chamot, Emil Monnin
Curtis, Charles Elbert
Debye, Peter Joseph Wilhelm
Dexter, Simon Newton
Hammond family
Herrington, Barbour Lawson
Hitchins, Clayton S.
Ithaca Glass Works, Ithaca, New York
LaForte, Benoist
Lavoisier, Antoine Laurent
Maynard, Leonard Amby
New York State College of Agriculture agricultural leaders project: oral history interviews
New York: Agricultural Experiment Station, Geneva
Peduto, Michael L.
Penn Chemical Company
Semet-Solvay Company
Shaw, Albert Duane
Student notebooks
Sugar beet industry, New York State
Sumner, James Batcheller
Taylor, Edward R.
West Virginia Pulp and Paper Company
White, Canvass
White, Hugh

New York
Adelberg & Raymond
Alexander, Jerome
American Crystallographic Association
American Physical Society: Division of High Polymer Physics
Benedum and the oil industry: oral history collection
Berent family
Brookhaven National Laboratory
Brookhaven National Laboratory: Office of the Director
Browne, Charles Albert
Campbell, William
Chandler, Charles Frederick
Columbia University. Department of Chemistry
Cornell University Medical Center: Department of Pharmacology
Curie, Marie (Sklodowska)
Debye, Peter Joseph Wilhelm
Du Vigneaud, Vincent
Egleston, Thomas
Ehrlich, Paul
Fankuchen, Isidor
Frank, Adolf
Gold, Harry
Gutleben, Dan, collector
Haber, Fritz
Halford, Ralph Stanley
Halsband, Ruth Alice
Hamilton Manufacturing Company, Hamilton, New York
Hammett, Louis Plack
Hanford, Franklin
Howe, Henry Marion
Jones, Alfred Goldsborough
Journal of Chemical Physics
Kunz, George Frederick
Levison, Wallace Goold
Mackintosh, James Buckton
McLachlan, Dan, Jr.
Mallmann, Paul
Maxim, Hudson
Mees, Charles Edward Kenneth
Meggers, William Frederick
Nobel laureates on scientific research: oral history collection
Pauling, Linus Carl
Proskauer family
Randall (W & R), Cortland, New York
Rhinelander family
Smith, Alexander
Stabler-Leadbeater Apothecary Shop, Alexandria, Va.
Sullivan, John F.
Tucker, Willis Gaylord
Vogel, Hermann Wilhelm
Willstatter, Richard Martin
Wolcott, Edson Ray

North Tarrytown
Bergmann, Max
Craig, Lyman Creighton
Dunham, Lawrence Boardman
Gasser, Herbert Spencer
International Education Board
Jacobs, Walter Abraham
Kunitz, Moses
Loeb, Jacques
Longsworth, Lewis Gibson
MacInnes, Duncan Arthur
Mirsky, Alfred Ezra
Perlmann, Gertrude Erika
Rockefeller Foundation
Tatum, Edward Lawrie

Poughkeepsie
Vassar, Matthew

Schenectady
Ruder, William Ernst

Syracuse
Archbold, John Dustin
Delaney, James J.
Dun & Bradstreet
Fairfield, N.Y.: Seminary

Troy
Cloke, John B.

Harteck, Paul
Horsford, Eben Norton
Mason, William Pitt
Nason, Henry B.
Rensselaer Polytechnic Institute School of Science

West Point
Robinson, Wirt

NORTH CAROLINA

Chapel Hill
Cowles, Calvin Josiah
Culver, Stephen Berry
Dabney family
Davis, Aaron
Gilchrist, Peter Spence
Hawkins family
Hedrick, Benjamin Sherwood
Howell, Edward Vernon
Judge, John
Kimberly, John
LeConte, Joseph
Long, William Lunsford
MacNider, William DeBerniere
Panknin, Charles F.
Person, Alice (Morgan)
Stubbs, William Carter
Winfield, John Augustus
Withers, Robert W.
Zoeller, Edward V.

Durham
Baugh & Sons Company, Philadelphia, Pennsylvania
Brotherton, William H.
Bunsen, Robert Wilhelm Eberhard
Chicora Mining and Manufacturing Company, Charleston, South Carolina
Cox, Talton L.
Craig, Locke
Duke University, Durham, North Carolina: Library
Hedrick, Benjamin Sherwood
Jones, Meriwether
National Dye Works
Nycum, John and John Q.
Sawyer, Francis A. and Jonathan
Sawyer Woolen Mills, Dover, N.H.
Terrell, James Wharey
Walker, Elbridge Gerry
Wightman, Joseph Milner
Worth family

Greenville
American Chemical Society: Eastern North Carolina Section
Crisp, Lucy Cherry
Evans, William Ernest
Grimes, J. Bryan
North Carolina Academy of Science
Waynick, Capus Miller
Whitfield, James Vivian

Raleigh
Cowles, Calvin Josiah
Fessenden, Reginald A.
Kilgore, Benjamin Wesley
Pickel, J. J.
Withers, William A.

NORTH DAKOTA

Fargo
Ladd, Edwin Freemont
Rheineck, Alfred E.

Grand Forks
Abbott, George Alonzo
Gustafson, Ben G.
Ladd, Edwin Freemont
McCrae, James Archibald

OHIO

Akron
B. F. Goodrich Company, Akron, Ohio
D'Ianni, James D.
General Tire and Rubber Company, Akron, Ohio
Oenslager, George
Whitby, George S.

Cincinnati
Lloyd, John Uri
Twitchell, Ernest

Cleveland
Barnes, George E.
Beeman Chemical Company, Cleveland, Ohio
Case Institute of Technology: Department of Chemistry and Chemical Engineering
Case Western Reserve University: Department of Chemistry
Darrow, Fritz Sage
Ebaugh, William Clarence
Gordon, Louis
Gilchrist, Harry L.
Grasselli family
Hall, Charles Martin
Jones, Day, Cockely & Reavis, Cleveland, Ohio
Levene, Phoebus Aaron Theodor
Licking County papers
Maron, Samuel Herbert
Morley, Edward Williams
Smith, Albert William
Ullman family
Van Horn, Frank Robertson
Waite, Frederick Clayton
Western Reserve University: Department of Chemistry

Columbus
Buckeye Steel Castings Company, Columbus, Ohio
Day, Jesse Erwin
Federal Glass Company, Columbus, Ohio
Garford, Arthur Lovett
Garrett, Alfred Benjamin
Hegler family
Lord, Nathaniel Wright
Martin, Sumner Leroy
Ohio State University: Department of Chemistry
Orton, Edward
Pence, Edward H.
Seiberling, Frank A.
Society of Separatists of Zoar
Stout, Wilbur Elihu

Fremont
Trommer Extract of Malt Company, Fremont, Ohio

Milan
Edison, Thomas Alva

OKLAHOMA

Norman
Bienfang, Ralph David
Cushing Refining Company, Cushing, Oklahoma
De Barr, Edwin
Lillie, Foress B.

OREGON

Corvallis
Oregon: Agricultural Experiment Station, Corvallis

Eugene
Andrus, Leonard A.
Dodson, William Daniel Boone
Drake, Lee D.
Fries, Amos Alfred
McBain, Bertram Telfar
Woodman, Ruth Cornwall

Portland
Fisher family
Oregon Environmental Council
Oregon State Pharmaceutical Association

PENNSYLVANIA

Bethlehem
General Economy, Northampton Co., Pennsylvania

Carlisle
Alexander, Samuel
Hastings, John
Priestley, Joseph

Erie
American Chemical Society: Erie Section

Harrisburg
Barclay family
Ephrata collection
Pine Grove Furnace collection

Haverford
Parrish, Maxfield

Lebanon
Coleman family

Philadelphia
Alden (L. H.) and Company, Aldenville, Pa.
American Chemical Society
    Division of Cellulose, Paper and Textile Chemistry
    Division of Chemical Education
American Institute of Chemical Engineers
American Philosophical Society
Bache, Franklin
Barton, Benjamin Smith
Bergmann, Max
Berry family
Booth, James Curtis
Bruson, Herman A.
Chymia
Clark, William Mansfield
Clark, Hans Thacher
Cohn, Mildred
Collins family
Cooper, Thomas
Darrach, William
Davy, Sir Humphry
Donohue, Jerry
Drabkin, David L.
Drinker, Cecil K.
Drinker, Henry
Duhamel du Monceau, Henri Louis
Dumas, Jean Baptiste A.
Edsall, John Tileston
Erdmann, Otto
Flory, Paul J.
Gershenfeld, Louis
Gibbs, Wolcott
Grahame, Israel J.

Philadelphia, Pennsylvania (*continued*)
Grubb family
Grubb, Peter
Hare, Robert
Hare-Willing family
Hartman, Frank A.
Harvey, Edmund Newton
Hunt, Reid
Jessup, Augustus Edward
Keefer, Horace A.
Lee, George F.
Lewis family
Lloyd, Malcolm
Long, Cyril Norman Hugh
Loury, Homer Hiram
Lutz family
McCulloch, Warren Sturgis
Marshall, Christopher
Mills, Frederick Ira
Morfit, Campbell
Neuberg, Carl
Nickles, Francois Joseph Jerome
Paschall family
Patterson family
Peale family
Pemberton family
Penington family
Pennsylvania Historical Society
Perot Malting Company, Philadelphia
Potts family
Potts, William John
Potts, William McCleery
Priestley, Joseph
Professional Circle, Philadelphia, Pa.
Rosengarten and Denis, Philadelphia
Roughton, Francis John Worsley
Roussel family
Schlossberger, Julius Eugen
Schultz, Jack
Seibert, Florence Barbara
Severinghaus, Elmer Louis
Silliman, Benjamin, Sr.
Smith, Edgar Fahs
Smith, E. F., Collection
Sorby, Henry
Straits Sugar Company
Survey of Sources for the History of Biochemistry and Molecular Biology
Thompson family
Traube, Isidor
University of Pennsylvania: Department of Chemistry
Weiss, John M.
Wetherill, Charles M.
Wheatley, Charles Moore
Willcox family
Zeller, Edwin Adrian

Pittsburgh
Craig, Isaac
Huntingdon Co., Pa., iron and forge industry, 1820
Oliver Iron and Steel Company, Pittsburgh
Strunz Soap Company, Pittsburgh
Vernon-Benshoff Company, Pittsburgh

Swarthmore
Stout, Jacob

University Park
Agricultural chemistry lecture notebooks
Frear, William
Herrick, John Peirce
Labor education
McFarland, David Ford
Pennsylvania State University
   Agricultural Experiment Station
   Petroleum Refining Laboratory
   School of Agriculture
Priestley, Joseph
Pugh, Evan

York
Margaretta Furnace, Margaretta, Pennsylvania

RHODE ISLAND
Providence
Bacon, Nathaniel Terry
Dexter, Henry Bowers
Hill, Walter Nickerson
Rumford Chemical Works
Special Fabrics Company, Saylesville, Rhode Island

SOUTH CAROLINA
Charleston
Carson family
Coosaw family
Gourdin-Young family papers
Piedmont Guano Company (Charleston, S.C.)
St. Amand, Clarence W.

Columbia
Bratton family
Klein, J. J.
Law, Thomas Cassels
McBryde, John McLaren
Mills family
Pinckney family

Rock Hill
Thomas family

TENNESSEE
Johnson City
East Tennessee and Western North Carolina Railroad Company
East Tennessee Medicine Company (Johnson City)

Knoxville
Kefauver, Estes
Middle and West Tennessee papers

Memphis
American Chemical Society: Memphis Section
Bailey, Alton Edward

Nashville
American Society for Testing Materials
Dudley, William Lofland
Ingersoll, Arthur William
McGill, John Thomas
Stritch family

TEXAS
Arlington
Brazel, David
International Union of Operating Engineers, Big Spring, Texas
Oil, Chemical and Atomic Workers International Union
Oil, Chemical and Atomic Workers International Union: Local 4-228, Port Neches

Austin
Harper, Henry Winston
Herschel family
Schoch, Eugene Paul
White, Reuben G.

Galveston
Bodansky, Meyer
Collection of scrapbooks relating to medical subjects
University of Texas: Department of Pharmacology and Toxicology
Wood, Henry A.

Houston
Autry, James Lockhard

UTAH
Salt Lake City
American Institute of Chemical Engineers: Great Salt Lake Section
Cutler, Thomas Robinson
Eyring, Henry
Guthrie, Boyd
Hinckley, Robert H.
Larsen, Gustive O.
Moss, Frank E.
Taylor, John
Widtsoe family
Wilder, Edward B.

VIRGINIA
Blacksburg
Hussey, Robert E.
Ivanhoe Mining and Smelting Corporation, Ivanhoe, Virginia

Charlottesville
Bahn, Gilbert Schuyler
Conservation Council of Virginia
Harmon, Robert Rogers
Hench, Atcheson Laughlin
Hench, Philip Showalter
Kerr, George Alexander
Miller, Alamby M.
Montgomery Court House ledgers
Nelson, Wilbur Armistead
Student chemistry notebooks
Yoe, John Howe

Richmond
Beverley family
Bouldin, Thomas Tyler
Claiborne, Herbert A.
Eppes, Richard
Gwathmey, Allan Talbott
Jeffress, Elizabeth Talbott (Gwathmey)
Lewis, Richmond Addison
Stuart, Alexander Hugh Holmes
Tredegar Company, Richmond
Tucker, Henry St. George

Williamsburg
Stubbs, William Carter

WASHINGTON
Chewelah
Travis, Eugene Porter

Du Pont
Du Pont (E. I.) Company: Du Pont, Washington

Granite Falls
Assay Office, Granite Falls, Washington

Lopez Island
Port Stanley Kelp Plant, Washington

Pullman
Adams, Mark F.
American Chemical Society: Northern Intermountain Section
Armstrong, Lyndon King
Brewster, Carl Milton
Chapman, M. J.
French, Charles C.
Fulmer, Elton
Galland-Burke Brewing and Malting Company, Spokane, Washington
Harston, C. B.
Henco Brewery, Spokane, Washington
New York Brewery, Spokane, Washington
Pacific Northwest Pollution Control Association
Ricketts, Bernard G.
Spokane Brewing and Malting Company
Todd, Clare Chrisman

Seattle
American Chemical Society
   Northwest Regional Meeting
   Puget Sound Section
American Institute of Mining and Metallurgical Engineers: North Pacific Section
Babb, Albert L.
Barkdull, Calvin H.
Benson, Henry K.
Cross, Paul C.
Howe Sound Mining Company, Holden Village, Washington
Neurath, Hans
Roberts, Milnor Oakes
Scott, Robert C.
Seattle-King County Department of Public Health
Smith, George McPhail
United States: Bureau of the Mint, Branch Assay Office, Seattle
University of Washington
   Chemical Engineering Department
   Nuclear Engineering Department
   Pharmacy College
   Pharmacy School

Tacoma
Asarco Smelter, Tacoma, Washington
Hooker, Albert H., Jr.

WEST VIRGINIA
Morgantown
Camden, Johnson Newlon
Noyes, Bradford
Smith, Donald E.
West Virginia University Library collection of historical materials
West Virginia University

WISCONSIN
Appleton
Koops, Matthias

Green Bay
Walter (George) Brewing Company, Appleton, Wisconsin

Madison
Aluminum Company of America
American Association of Colleges of Pharmacy
American Chemical Society
   Division of History of Chemistry
   Division of Inorganic Chemistry
   Wisconsin Section
American College of Apothecaries
Arnold, Frederick
Babcock, Stephen Moulton
Baumgartner, Jacob
Brays, William H.
Collier, Peter
Daniels, Farrington
Davis, William Hammatt
Effinger Brewing Company, Baraboo, Wisconsin
Everest, David Clark
Ferry, John D.
Finke, Almore H.
Frank, Joseph Otto
Frisch, John G.
Glenz, Adolph H.
Hardgrove, Timothy A.
Kaftan, Arthur
Kahlenberg, Louis
Kimberly-Clark Corporation
King, Franklin Hiram
Kotch, Alex
Krauskopf, Francis Craig
Lyman, Rufus Ashley
McKay, Frederick Sumner
Mathews, Joseph Howard
Pearce, Charles Chester
Rho Chi Society
Schuette, Henry August
Stewart, Francis Edward
Tanner, Herbert Battles
Tatum, Arthur Lawrie
United States Pharmacopoeial Convention
University of Wisconsin: School of Pharmacy Library
Walton, James Henri

Milwaukee
Prange, Louis H.

Platteville
Potosi Brewing Company, Potosi, Wisconsin

Racine
Davis, Marguerite

WYOMING
Laramie
Altshuler, Henry Irving
Browne, E. Wayles, Jr.
Garvan, Francis P.
Midwest Oil Corporation
Ridley, Grahame Brooke
Rose, Paul C.
Storie, Raymond Earle

Center for History of Chemistry

CHOC is a joint endeavor of the American Chemical Society, the American Institute of Chemical Engineers, and the University of Pennsylvania, established in January 1982 to discover and disseminate information about historical resources and to encourage research, scholarship, and popular writing in the history of chemistry, chemical engineering, and the chemical process industries.

The aims of the Center are to develop a program of interviews and undertake oral histories of major developments in modern chemistry; to locate historical manuscripts and archival records in the hands of individuals, societies, trade associations, and companies important in the history of chemistry, chemical engineering, and the chemical process industries; to encourage the preservation of these records in appropriate repositories; to offer aid in the appraisal, arrangement, and description of such records; to develop a comprehensive data base describing such collections, with finding aids and other reference tools, at the Edgar Fahs Smith Memorial Collection; and to make known the achievements of chemists, chemical engineers, and the chemical industry.

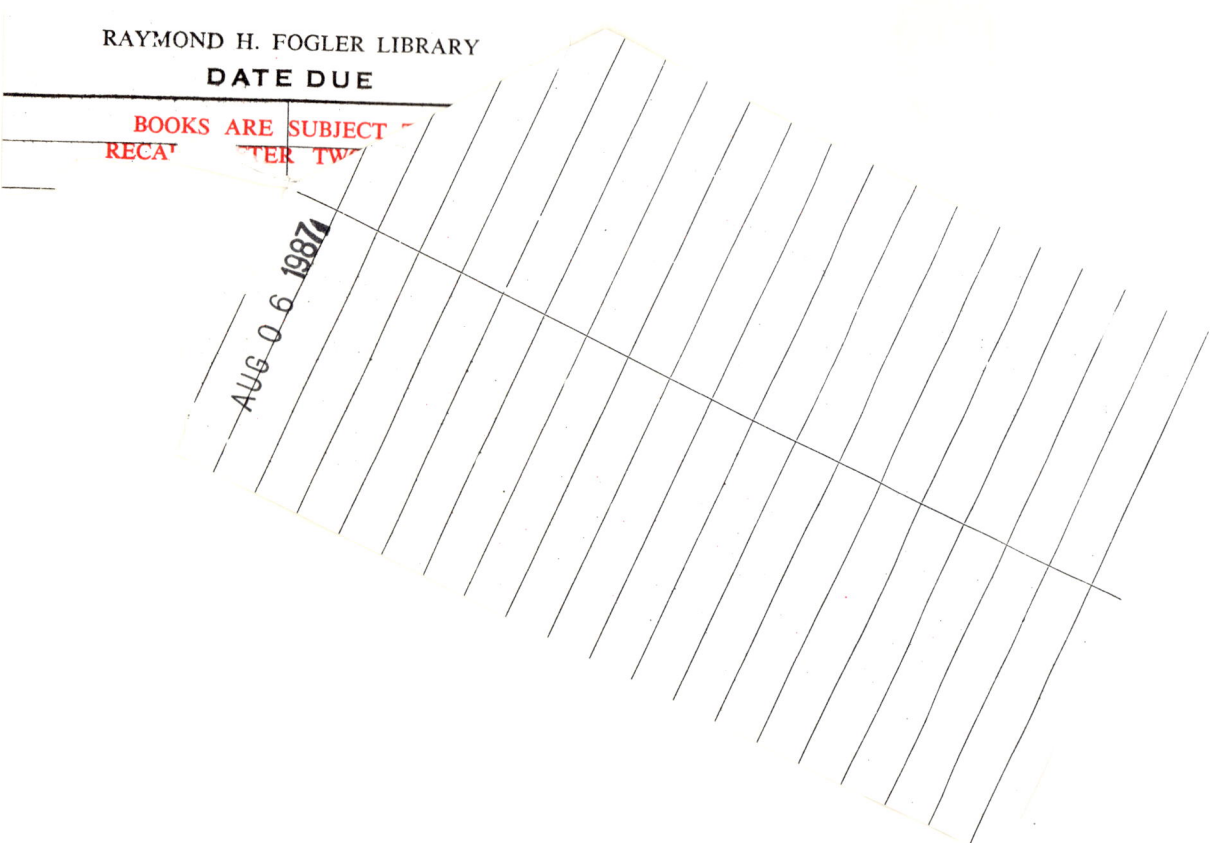